T0309327

The Dynamical Mordell–Lang Conjecture

**Mathematical
Surveys
and
Monographs**

Volume 210

The Dynamical Mordell–Lang Conjecture

**Jason P. Bell
Dragos Ghioca
Thomas J. Tucker**

American Mathematical Society
Providence, Rhode Island

EDITORIAL COMMITTEE

Robert Guralnick   Benjamin Sudakov
Michael A. Singer, Chair   Constantin Teleman
Michael I. Weinstein

2010 *Mathematics Subject Classification*. Primary 11-02, 11G25, 11G35, 14A10, 37-02, 37F10, 37P55, 37P20.

For additional information and updates on this book, visit
www.ams.org/bookpages/surv-210

Library of Congress Cataloging-in-Publication Data
Bell, Jason P., 1974–
    The Dynamical Mordell-Lang Conjecture / Jason P. Bell, Dragos Ghioca, Thomas J. Tucker.
      pages cm. — (Mathematical surveys and monographs ; volume 210)
    Includes bibliographical references and index.
    ISBN 978-1-4704-2408-4 (alk. paper)
    1. Mordell conjecture. 2. Curves, Algebraic. 3. Arithmetical algebraic geometry. 4. Geometry, Algebraic. I. Ghioca, Dragos, 1978– II. Tucker, Thomas J., 1969– III. Title.
QA565.B45 2016
516.3′52—dc23
                               2015036689

To Jason's wife, Jessica, and kids, Chris and Caitlin

To Dragos' mother, Lidia

To Tom's wife, Amanda

Contents

Preface

This book originated from the authors' desire to give an explanation of several recent applications of p-adic analysis to number theory and especially to arithmetic geometry. Central to this end has been the work done by several people (including the authors) to prove the *Dynamical Mordell-Lang conjecture*, which gives predictions about how the orbits of points in a variety under self-maps should intersect subvarieties. As the name suggests, this can be interpreted as a dynamical analogue of the classical Mordell-Lang Conjecture (proved by Faltings and Vojta) concerning intersections between finitely generated subgroups and subvarieties in a semiabelian variety.

Many results working towards this conjecture have used p-adic analysis, and we describe all known (to us) partial results up to this point in time—both those using p-adic analysis and those using alternative approaches—towards the Dynamical Mordell-Lang Conjecture. In some cases, we present entire proofs of results, while in other cases only a sketch is given, and in certain cases only a brief overview of the idea of the proof is provided. Our choice should not be interpreted as our opinion about the relative importance of the included results, but is instead an editorial choice regarding which material we thought best fits the overarching theme of this book.

We also give other applications of p-adic analysis to number theory and arithmetic geometry. In these cases, our list of applications is not meant to be exhaustive, but rather our goal is to show the wide reach of applications and potential applications of p-adic analysis to arithmetic geometry. While the uses of p-adic analytic methods we give do not always explicitly relate to the Dynamical Mordell-Lang Conjecture, we have generally favored applications of p-adic analysis to problems with some relation to the Dynamical Mordell-Lang Conjecture.

We thank all our colleagues with whom we wrote many of the papers whose results are detailed in this book; obviously, without the joint efforts we put towards solving the Dynamical Mordell-Lang Conjecture we would not have had a topic for this book. So, we thank Rob Benedetto, Ben Hutz, Par Kurlberg, Jeff Lagarias, Tom Scanlon, Yu Yasufuku, Umberto Zannier, and Mike Zieve. We are also grateful to the referees for their careful reading of a previous version of this book, and for suggesting many improvements for our work. Last, but definitely not least, we thank our families for their love and support while writing this book.

Notation

We let $\mathbb{Z}$, $\mathbb{Q}$, $\mathbb{R}$ and $\mathbb{C}$ be the sets of integer, rational, real, respectively, complex numbers. $\mathbb{N}_0$ is the set of all nonnegative integers, while $\mathbb{N}$ is the set of all positive integers.

An *arithmetic progression* is a set of the form $\{a+rn\}_{n\in\mathbb{N}_0}$, where the *common difference* r may be equal to 0 (in which case the set consists of a single element). If the common difference r is nonzero, then the arithmetic progression is infinite. Note that in the literature, sometimes one calls such a sequence a *one-sided* arithmetic progression in order to distinguish it from a *two-sided arithmetic progression*, which is a set of the form $\{a + rn\}_{n\in\mathbb{Z}}$. However, since in this book we mainly encounter one-sided arithmetic progressions and only occasionally encounter two-sided arithmetic progressions, our convention is to call arithmetic progression a sequence $\{a + rn\}_{n\in\mathbb{N}_0}$, while a sequence $\{a + rn\}_{n\in\mathbb{Z}}$ is called a two-sided arithmetic progression.

For a matrix A, we denote by A^t its transpose.

For a set U, we denote by id_U the identity function on U.

For any field K, we denote by $\mathrm{char}(K)$ its characteristic. By $\overline{K}$ we denote a fixed algebraic closure of K.

For any subfield $K \subseteq \overline{\mathbb{Q}}$, we denote by $\mathfrak{o}_K$ the ring of algebraic integers contained in K. If K is a number field, and $\mathfrak{p}$ is a prime ideal of K, then $k_\mathfrak{p}$ is the residue field corresponding to $\mathfrak{p}$, i.e., $k_\mathfrak{p} \xrightarrow{\sim} \mathfrak{o}_K/\mathfrak{p}$.

The usual affine space of dimension m is denoted by $\mathbb{A}^m$; for any field K, we have that $\mathbb{A}^m(K)$ consists of all m-tuples of points with coordinates in K. Similarly, we denote by $\mathbb{P}^m$ the projective space of dimension m; for any field K, we have that $\mathbb{P}^m(K)$ consists of all equivalence classes of $(m+1)$-tuples of points with coordinates in K not all equal to 0, under the equivalence relation

$$[x_0 : x_1 \cdots : x_m] \sim [y_0 : y_1 : \cdots : y_m]$$

if and only if there exists a nonzero scalar $c \in K$ such that

$$y_i = cx_i \text{ for all } i = 1, \ldots, m.$$

By *affine variety* we mean a subset of an affine space defined by a set of algebraic equations. Note that we do not ask a priori the variety be irreducible. Similarly, by *projective variety* we mean a subset of a projective space defined by a set of algebraic equations. We endow both the affine space and the projective space with the Zariski topology where the closed sets are precisely the (affine, respectively projective) varieties. We say that X is a quasiprojective variety if it is the open subset of a projective subvariety of some projective space. We say that a variety X is defined over a field K if it may be defined by a set of equations with coefficients in K. For a variety X defined over a field K, we denote by $X(K)$ the set of K-rational points of X.

We denote by $\mathbb{G}_a$ the affine line $\mathbb{A}^1$ endowed with the additive group law; we extend this law coordinatewise to $\mathbb{G}_a^n$. We denote by $\mathbb{G}_m$ the (Zariski open subset of the affine line) $\mathbb{A}^1 \setminus \{0\}$, i.e., the affine line without the origin, endowed with the multiplicative group law. Similarly to $\mathbb{G}_a^n$, we extend the multiplicative group law to $\mathbb{G}_m^n$.

An *abelian variety* is an irreducible projective variety which has the structure of an algebraic group.

For a set X, a map $\Phi : X \longrightarrow X$ is called a *self-map*. In general, for a self-map $\Phi : X \longrightarrow X$ and for any integer $n \geq 0$, we denote by Φ^n the n-th compositional iterate of Φ, i.e. $\Phi^n = \Phi \circ \cdots \circ \Phi$ (n times), with the convention that Φ^0 is the identity map. The *orbit* of a point $x \in X$ is denoted as $\mathcal{O}_\Phi(x)$ and it is the set of all $\Phi^n(x)$ for $n \in \mathbb{N}_0$.

A *dynamical system* consists of a topological space X endowed with a continuous self-map Φ.

For two real-valued functions f and g, we write $f(x) = o(g(x))$ if $\lim_{x \to \infty} f(x)/g(x) = 0$. Similarly, we write $f(x) = O(g(x))$ if the function $x \mapsto f(x)/g(x)$ is bounded as $x \to \infty$.

In a metric space $(X, d(\cdot, \cdot))$, for $x \in X$ and $r \in \mathbb{R}_{>0}$ we denote by $D(x, r)$ the open disk

$$D(x, r) = \{y \in X : d(x, y) < r\}.$$

We denote by $\overline{D}(x, r)$ the closure of $D(x, r)$.

CHAPTER 1

Introduction

In this chapter we describe various instances of the Dynamical Mordell-Lang Conjecture which appear in seemingly different areas. We conclude our Introduction by giving a brief overview of the rest of the book.

1.1. Overview of the problem

We start by presenting several arithmetic questions which are all connected, though this may not be so obvious *a priori*. All these questions have in common the following theme: we have a dynamical system Φ on a topological space X, and then for a point $\alpha \in X$ and a closed subset V of X, we ask for what values of $n \in \mathbb{N}_0$ we have $\Phi^n(\alpha) \in V$? The underlying theme of this book is that all the questions we consider have, or are conjectured to have, the same answer to the above question: *finitely many arithmetic progressions*. We also recall our convention that an arithmetic progression of common difference equal to 0 is simply a *singleton*.

The cases we consider are the following ones:

(1) Find all $n \in \mathbb{N}_0$ such that $a_n = 0$ where $\{a_n\}_{n \in \mathbb{N}_0}$ is a linear recurrence sequence. Say that the recurrence relation verified by the sequence is given for all $n \geq 0$ by

$$a_{n+m} = c_1 a_{n+m-1} + \cdots + c_m a_n,$$

for some given complex numbers $c_1, \ldots, c_m$. Then the ambient space is the affine space $\mathbb{A}^m$ with the Zariski topology, while the dynamical system is the one given by

$$\Phi\left((x_1, \ldots, x_m)\right) = (x_2, \ldots, x_m, c_1 x_m + \cdots + c_m x_1),$$

the starting point of the iteration is

$$x := (a_0, \ldots, a_{m-1}),$$

and $V \subset \mathbb{A}^m$ is the hyperplane given by the equation $x_1 = 0$. In Section 1.2 and Subsection 2.5.1, we explain this example in greater detail. In Section 2.5 we prove that the answer to this question is always a *finite union of arithmetic progressions*. A related, but more general problem involving (multi-dimensional) polynomial-exponential equations is discussed in Section 1.3.

(2) Find all $n \in \mathbb{N}_0$ such that given a matrix $A \in M_n(\mathbb{C})$ acting on the complex affine space $\mathbb{A}^n(\mathbb{C})$, a point $\alpha \in \mathbb{A}^n(\mathbb{C})$, and a subvariety $V \subset \mathbb{A}^n$, then $A^n \alpha \in V(\mathbb{C})$. This case is discussed in Section 1.4 and it turns out to be equivalent with the problem (1) discussed above (see the equivalence proven in Proposition 2.5.1.4).

(3) Find all $n \in \mathbb{N}_0$ such that given an endomorphism Φ of a quasiprojective variety X defined over $\mathbb{C}$, a point $\alpha \in X(\mathbb{C})$, and a subvariety V of X, then $\Phi^n(x) \in V(\mathbb{C})$. This problem, called the *Dynamical Mordell-Lang Conjecture* generalizes both of the above problems described above (see Section 1.5 for a first discussion of this conjecture). It is expected the answer to this question is again *finitely many arithmetic progressions*.

(4) Given a power series

$$f(z) := \sum_{n=0}^{\infty} a_n z^n$$

which satisfies a linear differential equation with polynomial coefficients, describe the set

$$S_f := \{n \in \mathbb{N}_0 \colon a_n = 0\}.$$

Rubel [**Rub83**, Problem 16] conjectured that S_f is a finite union of arithmetic progressions. We discuss this problem in Subsection 3.2.1, and show that a positive answer for an extension of the above Dynamical Mordell-Lang Conjecture to rational maps would solve Rubel's question.

1.2. Linear recurrence sequences

Let $\{F_n\}_{n \geq 0}$ be the Fibonacci sequence defined by

$$F_0 = 0, \; F_1 = 1 \text{ and } F_{n+2} = F_{n+1} + F_n \text{ for all } n \geq 0.$$

Also, let $\{a_n\}_{n \geq 0}$ be the sequence defined recursively by

$$a_{n+2} = 5a_{n+1} - 6a_n,$$

where $a_0 = \frac{7}{12}$ and $a_1 = \frac{3}{2}$.

QUESTION 1.2.0.1. *What are the numbers which appear in both of the sequences* $\{F_m\}_{m \in \mathbb{N}_0}$ *and* $\{a_n\}_{n \in \mathbb{N}_0}$?

We can compute easily the first elements in both sequences:

$$F_0 = 0, \; F_1 = 1, \; F_2 = 1, \; F_3 = 2, \; F_4 = 3, \; F_5 = 5, \; F_6 = 8, \; F_7 = 13, \; \ldots$$

and

$$a_0 = \frac{7}{12}, \; a_1 = \frac{3}{2}, \; a_2 = 4, \; a_3 = 11, \; a_4 = 31, \; a_5 = 89, \; a_6 = 259 \ldots.$$

One observes that $F_{11} = 89 = a_5$, and it is a reasonable question to ask whether this is the only answer to Question 1.2.0.1. This is a hard question since one would have to solve the equation $F_m = a_n$ in nonnegative integers m and n (for more details, see [**Eve95**]). Moreover, since it is easy to find a formula for the general term of both of these sequences (see Proposition 2.5.1.4), Question 1.2.0.1 reduces to finding $m, n \in \mathbb{N}$ such that

$$\frac{1}{\sqrt{5}} \cdot \left(\left(\frac{1 + \sqrt{5}}{2} \right)^m - \left(\frac{1 - \sqrt{5}}{2} \right)^m \right) = 2^{n-2} + 3^{n-1}.$$

On the other hand, if we were to ask the easier question of when the above equality holds when $m = n$, the answer would be *never* since (by a simple inductive argument) one can show that $a_k > F_k$ for all $k \in \mathbb{N}$.

In general, given two linear recurrence sequences $\{a_m\}_{m\in\mathbb{N}_0}$ and $\{b_n\}_{n\in\mathbb{N}_0}$, one would like to understand whether there exists an underlying *structure* for the solutions $(m, n) \in \mathbb{N} \times \mathbb{N}$ for which $a_m = b_n$. Or, at least, for the easier case, one would like to understand the structure of the set of all $n \in \mathbb{N}$ such that $a_n = b_n$. It is immediate to see that this last case reduces to understanding when a given linear recurrence sequence $\{c_n\}_{n\in\mathbb{N}_0}$ (in this case, $c_n = a_n - b_n$) takes the value 0. Then the answer is that if there exist infinitely many $n \in \mathbb{N}$ such that $c_n = 0$, then there exists an *infinite* arithmetic progression $\{\ell + nk\}_{n\in\mathbb{N}_0}$ such that $c_{\ell+nk} = 0$. This will be proven in Section 2.5. As described in Section 1.1, the proper dynamical setting for this example is as follows: given a linear recurrence sequence

$$\{a_m\}_{m\in\mathbb{N}_0} \subset \mathbb{C}$$

which satisfies the relation

$$a_{n+m} = c_1 a_{n+m-1} + \cdots + c_m a_n,$$

for some given complex numbers $c_1, \ldots, c_m$, then the dynamical system is the one given by the map

$$\Phi((x_1, \ldots, x_m)) = (x_2, \ldots, x_m, c_1 x_m + \cdots + c_m x_1)$$

acting on the m-dimensional affine complex space $\mathbb{A}^m$. Then finding all $n \in \mathbb{N}_0$ such that $a_n = 0$ is equivalent with finding all $n \in \mathbb{N}_0$ such that

$$\Phi^n((a_0, \ldots, a_{m-1})) \in V(\mathbb{C}),$$

where $V \subset \mathbb{A}^m$ is the hyperplane given by the equation $x_1 = 0$.

1.3. Polynomial-exponential Diophantine equations

Let $m, k \in \mathbb{N}$, let $F \in \mathbb{Z}[x_1, \ldots, x_m, y_1, \ldots, y_k]$, and let $r_1, \ldots, r_k \in \mathbb{Z}$. A polynomial-exponential equation has the form

$$F(j_1, \ldots, j_m; r_1^{n_1}, \ldots, r_k^{n_k}) = 0,$$

where the variables $j_1, \ldots, j_m \in \mathbb{Z}$, respectively $r_1, \ldots, r_k \in \mathbb{N}_0$. In general, there might be *many* solutions to the above equation, especially if the degree of f in x_i is 1 for at least one variable x_i. But, even if $\deg_{x_i} f = 1$, there might be no solutions due to some local constraints such as in the following case:

$$(1.3.0.1) \qquad 21x_1^3 x_3 - 7 \cdot 3^{n_1} x_2 + 14 \cdot 5^{n_2} x_3^2 - 49 x_1 x_3 + 2 = 0,$$

when there are no solutions $x_1, x_2, x_3 \in \mathbb{Z}$ and $n_1, n_2 \in \mathbb{N}_0$ by considering the congruence modulo 7 for the equation (1.3.0.1). Now, even if one assumes $j_1 = j_2 = \cdots = j_m = j$, and that the polynomial f has the variables x_i and y_j separated, the problem is not easier. Even also assuming that $r_1 = r_2 \cdots = r_k$ does not simplify the problem much. For example, we discuss in Chapter 13 the following special case:

$$g(x) = \sum_{i=1}^{k} c_i p^{n_i},$$

where $g \in \mathbb{Z}[x]$, $c_1, \ldots, c_k \in \mathbb{Z}$ and p is a prime number. Essentially, one *expects* that if $g(x)$ has *few* nonzero p-adic digits, then x (or a linear function evaluated at x) would also have *few* p-adic digits. However, this is far from being proven even in simple cases such as $g(x) = x^2$ and $k \geq 5$ (for more details, see [**BBM13, CZ00, CZ13**] and the references therein).

On the other hand, if one assumes that

$$j_1 = \cdots = j_m = n_1 = n_2 = \cdots = n_k,$$

then the problem reduces essentially to the one discussed in Section 1.2 (see also Section 2.5). Thus one obtains that if there exist infinitely many $n \in \mathbb{Z}$ such that

$$(1.3.0.2) \hspace{3cm} H\left(n, r_1^n, r_2^n \ldots, r_k^n\right) = 0,$$

where $H \in \mathbb{Z}[z_0, z_1, z_2, \ldots, z_k]$, then there exists an infinite arithmetic progression $\{\ell + nk\}_{n \in \mathbb{N}_0}$ such that each element of it is a solution to (1.3.0.2).

1.4. Linear algebra

Let A be an invertible matrix in $\mathrm{GL}_r(\mathbb{C})$, let V be a linear subspace of $\mathbb{C}^r$, and let $z \in \mathbb{C}^r$.

QUESTION 1.4.0.1. *Is there a simple description of the set of positive integers n such that $A^n z \in V$?*

We note that the problem discussed in this section could easily be asked for an arbitrary subvariety defined over $\mathbb{C}$ of the affine space $\mathbb{A}^r$; however this more general question reduces to the case V is a linear subvariety.

If V is a line passing through the origin of $\mathbb{C}^r$, then once there exist two distinct nonnegative integers $m < n$ such that

$$(1.4.0.2) \hspace{3cm} A^m z \in V \text{ and } A^n z \in V,$$

then we immediately conclude that V is fixed by A^{n-m} and therefore

$$A^{m+\ell(n-m)} z \in V \text{ for all } \ell \in \mathbb{N}_0.$$

In particular, if k_0 is the smallest positive integer k such that A^k fixes V, and if m_0 is the smallest nonnegative integer m such that $A^m z \in V$, then $A^n z \in V$ if and only if $n = m_0 + \ell k_0$ for some nonnegative integer ℓ.

Things are not so simple in general. For example, when V is a line that does not pass through the origin, it is easy to see that you can have distinct m and n such that (1.4.0.2) holds without getting an entire arithmetic progression of such integers, just by choosing a line V which passes through two arbitrary points $A^m z$ and $A^n z$. But in the case of lines not passing through the origin, once you have a large finite number of integers n such that

$$(1.4.0.3) \hspace{3cm} A^n z \in V,$$

you must have an infinite arithmetic progression of such n. There is even an explicit bound on that number due to Beukers-Schlickewei [**BS96**], which is likely nowhere near sharp. In fact, under the assumption that each eigenvalue of A is either equal to 1 or is not a root of unity, and furthermore for each two distinct eigenvalues λ_i and λ_j of A we have that λ_i/λ_j is not a root of unity, Beukers-Schlickewei [**BS96**] show that there are at most 61 integers $n \in \mathbb{N}_0$ such that (1.4.0.3) holds. The general case of an arbitrary matrix A follows easily from this special case.

1.5. Arithmetic geometry

The subject of this book is a geometric generalization (see Conjecture 1.5.0.1) of all of the above problems, and it is also connected to the classical Mordell-Lang Conjecture (see Chapter 3). In each of the three problems discussed in Sections 1.2 to 1.4, we deal with a geometric object: the line V in Section 1.4, or the hypersurface $F = 0$ in Section 1.3, or the hyperplane $x_1 = 0$ in the affine space $\mathbb{A}^m$ as in Section 1.2. And we want to understand when an arithmetic dynamical system intersects the geometric object. The arithmetic dynamical system is the iteration of the matrix A in Section 1.4, or the input of an integer number into the equation $F = 0$ (which is a discrete dynamical system simply because all integers are obtained from 0 by repeated operations of either $z \mapsto z + 1$ or $z \mapsto z - 1$), or a linear recurrence sequence as in Section 1.2. And in each case one obtains that once there exist infinitely many instances of the intersection between the geometric object and the arithmetic dynamical object, then there is a *structure* for the intersection which is given by *finitely many arithmetic progressions*. This principle is formally stated in the Dynamical Mordell-Lang Conjecture (for more details, see Chapter 3).

CONJECTURE 1.5.0.1 (Dynamical Mordell-Lang Conjecture). *Let X be a quasi-projective variety defined over $\mathbb{C}$, let Φ be any endomorphism of X, let $\alpha \in X(\mathbb{C})$, and let $V \subseteq X$ be any subvariety. Then the set of all $n \in \mathbb{N}_0$ such that $\Phi^n(\alpha) \in V(\mathbb{C})$ is a union of finitely many arithmetic progressions.*

We note that the Dynamical Mordell-Lang Conjecture can be formulated over any field K of characteristic 0 (see Conjecture 3.1.1.1); however such a formulation reduces to proving the case when $K = \mathbb{C}$ (see Proposition 3.1.2.1).

A special case of Conjecture 1.5.0.1 that is known is when X is an abelian variety, and Φ is the translation-by-P endomorphism of X for some point $P \in X(\mathbb{C})$. In this latter case we encounter the cyclic case of the classical Mordell-Lang Conjecture (for more details, see Section 3.4).

We present below a few cases of Conjecture 1.5.0.1; all our examples are set in the ambient space $X = \mathbb{A}^3$ in which case there is at this time no general proof of the Dynamical Mordell-Lang Conjecture.

EXAMPLE 1.5.0.2. Consider the endomorphism

$$\Phi : \mathbb{A}^3 \longrightarrow \mathbb{A}^3$$

given by

$$\Phi(x, y, z) = (x^2 + x, y^2 + y, z^2 + z).$$

Let $V \subset \mathbb{A}^3$ be the plane given by the equation

$$x + y + z = 1.$$

Then for *most* points $\alpha \in \mathbb{A}^3(\overline{\mathbb{Q}})$, the set

$$S := \{n \in \mathbb{N}_0 \colon \Phi^n(\alpha) \in V(\overline{\mathbb{Q}})\}$$

is finite. For example, this can be seen immediately if all three coordinates of α are integers (in which case, at the very most, S has 1 element). However, if α is an arbitrary point in $\mathbb{A}^3(\overline{\mathbb{Q}})$, then it is much harder to prove that S is always a finite union of arithmetic progressions (possibly with common difference equal to 0). However, we will see later (see Corollary 7.0.0.1) that for *any* subvariety

$V \subseteq \mathbb{A}^3$, the set S is a finite union of artithmetic progressions. Furthermore, using the classification of periodic curves under the coordinatewise action of a polynomial done by Medvedev-Scanlon [**MS14**], one can show that in the case of the above plane V, the set S is finite assuming α is not preperiodic. Now, if α is preperiodic, the question of whether S contains an infinite arithmetic progression is equivalent with finding three preperiodic points a, b and c for the action of the polynomial

$$f(z) := z^2 + z,$$

such that

$$a + b + c = 1.$$

This last question is a deep question related to the problem of unlikely intersections in dynamics which we discuss in Subsection 14.2.2.

EXAMPLE 1.5.0.3. Consider the endomorphism

$$\Phi : \mathbb{A}^3 \longrightarrow \mathbb{A}^3$$

given by

$$\Phi(x, y, z) = (x^5, y^3, z^5).$$

Let $\alpha = (0, i, 0)$ and let $V \subset \mathbb{A}^3$ be the surface given by the equation

$$x^3 + y + z^3 = i.$$

We easily see that α is periodic under the action of Φ and moreover, $\Phi^n(\alpha) \in V$ if and only if n is an even nonnegative integer. Actually, using Theorem 9.3.0.1 one can show that for any $\alpha \in \mathbb{A}^3(\mathbb{C})$ and for any complex subvariety $V \subseteq \mathbb{A}^3$, the set S of all $n \in \mathbb{N}_0$ such that $\Phi^n(\alpha) \in V(\mathbb{C})$ is a finite union of arithmetic progressions. Furthermore, according to the classical Mordell-Lang conjecture for an algebraic torus (proven by Laurent [**Lau84**]; see also Section 3.4), one obtains that the above set S is finite unless V contains a translate of a positive dimensional algebraic torus.

EXAMPLE 1.5.0.4. Consider the endomorphism

$$\Phi : \mathbb{A}^3 \longrightarrow \mathbb{A}^3$$

given by

$$\Phi(x, y, z) = (x^2 + y, y^2 + z, z^2 + x).$$

Let $\alpha = (1, 1, 1)$ and $S \subset \mathbb{A}^3$ be the surface given by the equation

$$x + y^2 + z^3 = x^2 + y^3 + z.$$

It is immediate to see that the entire orbit $\mathcal{O}_\Phi(\alpha)$ is contained in the surface S, and the reason for this is that V contains the line L given by the equation

$$x = y = z,$$

which is fixed by the action of Φ. However, if V is an arbitrary subvariety of $\mathbb{A}^3$, and also α is an arbitrary point in $\mathbb{A}^3(\mathbb{C})$, then it is not known whether Conjecture 1.5.0.1 holds. In some sense, the endomorphism Φ from this Example lies outside all the presently known cases of the Dynamical Mordell-Lang Conjecture (see Chapter 3 for more details).

Examples 1.5.0.2 to 1.5.0.4 give us a glimpse into the philosophy behind the Dynamical Mordell-Lang Conjecture. With the notation as in Conjecture 1.5.0.1, the set

$$S := S(\Phi, V, \alpha) := \{n \in \mathbb{N}_0 \colon \Phi^n(\alpha) \in V\}$$

is finite unless one of the following two conditions holds:

(1) α is preperiodic under the action of Φ and V contains a point from the periodic cycle of $\mathcal{O}_\Phi(\alpha)$; or
(2) V contains a positive dimensional subvariety W which is periodic under the action of Φ, and moreover, W intersects $\mathcal{O}_\Phi(\alpha)$ (for the definition of periodic subvarieties, see Section 2.2).

It is easy to see that either (1) or (2) above yield a corresponding infinite set $S = S(\Phi, V, \alpha)$. Moreover, it is also clear that if either case (1) holds, or V itself is a periodic variety, then the set S consists of finitely many arithmetic progressions. So, the content of the Dynamical Mordell-Lang Conjecture is to prove that when α is not preperiodic, then the *only* possibility for the set S to be infinite is when condition (2) holds.

1.6. Plan of the book

We sketch here the contents of the remaining chapters of our book. Also, at the end of this section, we suggest several plans for studying from this book.

1.6.1. Description of each chapter. In Chapter 2 we present the necessary background material for the rest of the book, focusing on the notions from algebraic and arithmetic geometry, valuation theory, p-adic analysis and their applications to the problems studied in our book. Of special importance for our study is Theorem 2.5.4.1, which also constituted the starting point for our p-adic approach to the Dynamical Mordell-Lang Conjecture. Theorem 2.5.4.1 is the classical result of Skolem [**Sko34**] (later generalized by Mahler [**Mah35**] and Lech [**Lec53**]) that solves the problem discussed in Section 1.2: given a linear recurrence sequence $\{a_n\} \subset \mathbb{C}$, the set of nonnegative integers n such that $a_n = 0$ is a finite union of arithmetic progressions.

In Chapter 3 we discuss Conjecture 1.5.0.1 and its connection to the classical Mordell-Lang conjecture (proven by Faltings [**Fal83**]) and to the Denis-Mordell-Lang Conjecture (see [**Den92a**]). We also discuss a multi-dimensional problem stemming from the Dynamical Mordell-Lang Conjecture, which turns out to be false in general, but sometimes, in outstanding cases, such as the classical Mordell-Lang conjecture itself, it has a positive answer. We explore in more depth this multi-dimensional analogue of the Dynamical Mordell-Lang Conjecture in Chapter 5. Also, in Chapter 5 we prove an interesting instance of Conjecture 1.5.0.1 when $X = \mathbb{A}^2$, V is the diagonal line, and

$$\Phi(x, y) := (f(x), g(y))$$

for arbitrary polynomials $f, g \in \mathbb{C}[z]$ (see [**GTZ08, GTZ12**]). The proof of the main result from Chapter 5 is one of the very few instances when a special case of the Dynamical Mordell-Lang Conjecture is proven without using a p-adic approach; other special cases of Conjecture 1.5.0.1 proven without the explicit use of p-adic analysis are the works of Ng and Wang [**NW13**] and Xie [**Xie14, Xieb**].

In Chapter 4 we expand on the discussion from Section 2.5 by giving a geometric twist of the classical result of Skolem-Mahler-Lech regarding the occurrence of zeros in an arithmetic progression; essentially, Theorem 2.5.4.1 can be easily reformulated in terms of automorphisms of $\mathbb{P}^n$ (see Denis [**Den94**]). In particular, we show that given an étale endomorphism Φ of a quasiprojective variety X defined over $\mathbb{C}$, and given a point $\alpha \in X(\mathbb{C})$ there exists a prime number p, a suitable embedding into $\mathbb{Q}_p$, and a positive integer k such that the map $n \mapsto \Phi(kn)$ is p-adic analytic (see [**BGT10**] whose main result builds on previous work of Bell [**Bel06**]). This method of finding a p-adic analytic parametrization of the orbit of a point is called the *p-adic arc lemma*.

In Chapter 6 we present the main results from p-adic dynamics which allow us to parametrize the orbit of a point under a non-étale endomorphism Φ of a quasiprojective variety X. The main results are for rational maps Φ acting on $X = \mathbb{P}^1$, and they are due to Rivera-Letelier [**RL03**]. As an application of these p-adic analytic parametrizations we obtain several interesting results in Chapters 7 and 11 (see [**BGKT10, BGKT12, BGHKST13**]). In Chapter 8 we present heuristics regarding the general case of the Dynamical Mordell-Lang Conjecture (for more details, see [**BGHKST13**]). In particular, these heuristics suggest that the p-adic approach *might* not work to prove the general case of Conjecture 1.5.0.1.

There are fewer instances of p-adic analytic parametrizations of orbits under endomorphisms Φ of higher dimensional varieties X; the main result in this area is an older theorem of Herman and Yoccoz [**HY83**]. However, this last result is sufficient for us to prove certain special cases of the Dynamical Mordell-Lang Conjecture in Chapter 9.

In Chapter 10 we present two results of Scanlon [**Sca, Sca11**] towards the Dynamical Mordell-Lang Conjecture which both use analytic parametrizations of the orbit – one of the parametrizations using p-adic analysis, and the other one using real analytic functions. In Chapter 10, we also discuss briefly Xie's proofs [**Xie14, Xieb**] of the special cases of the Dynamical Mordell-Lang Conjecture for endomorphisms of $\mathbb{A}^2$. We point out right from the beginning that Xie proved one of the most outstanding open case of the dynamical Mordell-Lang Conjecture – the case of endomorphisms of $\mathbb{A}^2$ – however, due to our emphasis in this book for the p-adic analytic approach to the Dynamical Mordell-Lang Conjecture and also due to the fact that Xie's results are very recent (actually, [**Xieb**] was not even released when we submitted our first draft of our book), we do not include a thorough description of Xie's theorems. But we encourage the reader interested in the problem we study in this book to consult Xie's almost 100 pages preprint [**Xieb**] (which uses a previously released almost equally long preprint [**Xiea**]); in a way, together [**Xiea**] and [**Xieb**] contain enough material for another book on the topic of the Dynamical Mordell-Lang Conjecture!

In Chapter 11 we prove a weaker version of the Dynamical Mordell-Lang Conjecture. In the highest possible generality (even surpassing the world of algebraic geometry and dealing with continuous self-maps Φ on Noetherian spaces X), we prove in Theorem 11.4.2.2 that for any closed subset $Y \subseteq X$ and for any $\alpha \in X$, the set

$$S := \{n \in \mathbb{N}_0 \colon \Phi^n(\alpha) \in Y\}$$

is a union of finitely many arithmetic progressions along with a set of Banach density 0. In other words, we prove that if X contains no closed subset which is both periodic under Φ and also intersects the orbit of α, then the set S has Banach density 0. The Dynamical Mordell-Lang Conjecture (when Φ is an endomorphism of a quasiprojective variety X) asks that S must be finite in this case. Nevertheless, Theorem 11.4.2.2 (see also [**BGT15b**] where this result was first published) presents very strong evidence towards Conjecture 1.5.0.1. Furthermore, also in Chapter 11 we present various strengthenings of Theorem 11.4.2.2 for various special cases of the Dynamical Mordell-Lang Conjecture (for more details, we refer the reader to [**BGKT10**] and [**BGT15b**]).

In Chapter 12 we discuss the Denis-Mordell-Lang Conjecture which may be viewed as a hybrid between the classical Mordell-Lang problem and Conjecture 1.5.0.1 over a field of positive characteristic. We continue the exploration of the Dynamical Mordell-Lang Conjecture in characteristic p in Chapter 13; very little is known for the characteristic p analogue of Conjecture 1.5.0.1 (see Conjecture 13.2.0.1) even for the case of endomorphisms of $\mathbb{G}_m^n$.

In Chapter 14 we discuss various other questions in arithmetic geometry whose solution was obtained (or it *might* be obtained) using a p-adic analytic approach. Among these questions we mention the Dynamical Manin-Mumford Conjecture and the unlikely intersection problem in dynamics (see also [**Zan12**] for more details on related problems in arithmetic geometry). In Chapter 15 we conclude by speculating on the future of the Dynamical Mordell-Lang Conjecture.

1.6.2. Suggested plans for studying. Of course, we hope the interested reader will find time to read our entire book, but in case one wants to read only a subset of our book in order to gain understanding to some of the more important results and methods discussed in the book, we present in each of the following Subsections a possible reading plan. We leave out from our suggested reading plans most of the background Chapter 2 and also the Chapters 14 and 15 which talk about related questions to the Dynamical Mordell-Lang Conjecture and also speculate about its future. Obviously, we encourage all readers to read Chapter 2 in its entirety to familiarize with the notions from algebraic geometry and number theory that we use in our book. Also, we hope Chapters 14 and 15 will motivate the reader to study in the future various cases of the Dynamical Mordell-Lang Conjecture or related questions from arithmetic dynamics. On the other hand, we encourage all readers to include Chapter 3 in their study of our book since it provides a comprehensive introduction into the Dynamical Mordell-Lang Conjecture and its connection to other important questions from arithmetic geometry.

The plans listed in the subsequent Subsections are given a name that informally describes the goal of each such reading plan.

1.6.3. The p-adic arc lemma. The central method used in this book for attacking the Dynamical Mordell-Lang Conjecture is the p-adic arc lemma, which is formally introduced in Chapter 4. So, if a reader wants to understand this important tool, we recommend one to read first Section 2.5 (where a very special case of the p-adic arc lemma, known as the Skolem's method, is introduced) and then to proceed to Chapter 4. Then, the reader could study Chapters 6 and 7, and also Sections 11.5 and 11.11 for more examples of use of the p-adic arc lemma and of similar p-adic uniformization techniques for an orbit of a point. Finally, the reader should read Chapter 8 to understand the limitations one encounters in the use of the p-adic arc lemma for the Dynamical Mordell-Lang Conjecture.

1.6.4. The Dynamical Mordell-Lang for algebraic groups. The motivation for Conjecture 1.5.0.1 comes from the classical Mordell-Lang Conjecture; for more details, see Chapter 3. So, a reader could focus on the connection between these two conjectures, in particular on the special case of Conjecture 1.5.0.1 when the ambient variety is a semiabelian variety. Hence, one can read in Chapter 9 the proof of the Dynamical Mordell-Lang Conjecture for endomorphisms of semiabelian varieties. This proof from chapter 9, which avoids the use of the p-adic arc lemma (after all, an alternative proof of the Dynamical Mordell-Lang Conjecture for all semiabelian varieties is found in Chapter 4; see Corollary 4.4.1.2) has the advantage of extending to proving some special cases of Question 3.6.0.1, which is a question generalizing both the classical Mordell-Lang conjecture and the Dynamical Mordell-Lang Conjecture.

Then one can read about a hybrid version of the classical Mordell-Lang conjecture and the Dynamical Mordell-Lang Conjecture, which is the Denis-Mordell-Lang conjecture in the context of Drinfeld modules. Actually, this conjecture of Denis [**Den92a**] was the starting point for formulating Conjecture 1.5.0.1 in [**GT09**]. Finally, one can read in Chapter 13 about a characteristic p version of the Dynamical Mordell-Lang Conjecture, and also read about partial results on this conjecture in the context of semiabelian varieties using the same approach as in Chapter 9 but also using an alternative approach coming from the theory of automata.

1.6.5. The intersection of orbits. While trying to prove a very special case of Conjecture 1.5.0.1 for lines in the plane under the coordinatewise action of two one-variable polynomials, the authors of [**GTZ08**] discovered a more general result regarding the intersection of two orbits under the action of two polynomials. Briefly, the results of [**GTZ08**] (and their extension from [**GTZ12**]) say that if two polynomials $f, g \in \mathbb{C}[z]$ of degrees larger than 1 have the property that there exist $\alpha, \beta \in \mathbb{C}$ such that

$$\mathcal{O}_f(\alpha) \cap \mathcal{O}_g(\beta) \text{ is infinite,}$$

then there exist linear polynomials $\mu, \nu \in \mathbb{C}[z]$, some polynomial $h \in \mathbb{C}[z]$ of degree larger than 1, and positive integers m and n such that

$$f = \mu \circ h^m \text{ and } g = \nu \circ h^n.$$

These results can be viewed as a possible bridge towards Question 3.6.0.1, which is presented in Chapter 3. The interested reader will find all about these questions and results in Chapter 5, which is mainly self-contained.

CHAPTER 2

Background material

In this chapter we present the basic definitions, notation, and conventions that we use from algebraic geometry and number theory.

2.1. Algebraic geometry

In this section we recall briefly the basic notions of algebraic geometry which we will use in the book; for more details see the classical textbooks of Hartshorne [**Har77**], of Hindry and Silverman [**HS00**, Part A] and of Shafarevich [**Sha74**].

2.1.1. Varieties. Let K be a field, let $\overline{K}$ be a fixed algebraic closure of it, and let N be a positive integer.

We denote by $\mathbb{A}^N = \mathbb{A}_K^N$ the N-dimensional affine space (over K) which is the set of all tuples $(x_1, \ldots, x_N)$ with $x_i \in K$. We let $\mathbb{P}^N = \mathbb{P}_K^N$ be the set of equivalence classes of $\left(\mathbb{A}^{N+1} \setminus \{(0, \ldots, 0)\} \right) / \sim$, where we have

$$[x_0 : x_1 : \cdots : x_N] \sim [y_0 : y_1 \cdots : y_N]$$

if there exists a nonzero scalar $c \in K$ such that $y_i = cx_i$ for each $i = 0, \ldots, N$.

An *affine subvariety* V of $\mathbb{A}^N$ defined over K is the set of all points

$$(x_1, \ldots, x_N) \in \mathbb{A}^N$$

which is the common zero set of all polynomials in a given ideal I of $K[z_1, \ldots, z_N]$, i.e., for each polynomial $f \in I$, we have

$$f(x_1, \ldots, x_N) = 0.$$

For any variety $V \subset \mathbb{A}^n$, we let the *vanishing ideal* $I(V)$ of V be the set of all polynomials $f \in \overline{K}[x_1, \ldots, x_N]$ such that $f(\alpha) = 0$ for each $\alpha \in V(\overline{K})$. We define its *affine coordinate ring* $\overline{K}[V]$ be $\overline{K}[x_1, \ldots, x_N]/I(V)$. We define the *Zariski topology* on $\mathbb{A}^N$ as the topology for which the closed sets are the affine subvarieties of $\mathbb{A}^N$.

A *projective subvariety* V of $\mathbb{P}^N$ defined over K is the set of all

$$[x_0 : x_1 : \cdots : x_N] \in \mathbb{P}^N,$$

which is the common zero set of all homogeneous polynomials in a given ideal I of $K[z_0, z_1, \ldots, z_N]$. Note that if f is a homogeneous polynomial in I, then for any other representation $[y_0 : y_1 : \cdots : y_N]$ in $\mathbb{P}^N$ of the point $[x_0 : x_1 : \cdots : x_N]$ we have

$$f(y_0, \ldots, y_N) = 0.$$

We define the Zariski topology on $\mathbb{P}^N$ for which the closed sets are the projective subvarieties of $\mathbb{P}^N$.

Either in $\mathbb{P}^n$ or in $\mathbb{A}^n$, the zero locus of a single polynomial equation $F = 0$ is called a hypersurface; if F is a linear polynomial, then it is called a hyperplane.

A *(quasiprojective) variety* is a Zariski open subset of a projective variety. If $X \subset \mathbb{P}^m$ is a quasiprojective variety, for any subfield $L \subset \overline{K}$, we denote by

$$X(L) \subset \mathbb{P}^m(L) \cap X$$

the set of L-valued (or L-rational) points of X. Sometimes we identify X with the set of its $\overline{K}$-points. We endow X with the inherited Zariski topology from $\mathbb{P}^m$. A *subvariety* Y of X is a quasiprojective variety contained in X. In particular, we allow for the possibility that Y is either an open or a closed subset of X with respect to the Zariski topology; however, unless otherwise stated, all subvarieties of a (quasiprojective) variety X are closed. If $Y \subset X$ is a possibly non-closed subset, its *Zariski closure* (in X) is denoted by $\overline{Y}$.

Inside a variety X endowed with the Zariski topology, we say that a subvariety Y defined over K is *irreducible* (over K) if it cannot be expressed as a union of two proper closed subsets, i.e., if

$$Y = Y_1 \cup Y_2,$$

with Y_1 and Y_2 closed subvarieties defined over K, then either

$$Y = Y_1 \text{ or } Y = Y_2.$$

If K is algebraically closed, then we say that Y is *geometrically irreducible*. Each quasiprojective variety is a finite union of irreducible quasiprojective varieties. Since $\mathbb{P}^m$ can be covered by $(m+1)$ copies of $\mathbb{A}^m$, then each (quasiprojective) variety can be covered by finitely many (possibly non-closed) affine subvarieties. This allows us to define various local properties for varieties by restricting ourselves to irreducible affine varieties. In Subsection 2.1.2, in order to define maps between varieties, we use that each variety can be covered by finitely many affine subsets.

2.1.2. Rational and regular functions. Let $X \subset \mathbb{P}^n$ be an irreducible quasiprojective variety defined over K, and let $L \subset \overline{K}$ be a subfield. We say that a function

$$f : X \longrightarrow \mathbb{A}^1$$

is *regular* (defined over L) at the point $\alpha \in X$ if there exists an affine open set $U \subset X$ containing α and homogeneous polynomials

$$A, B \in \overline{K}[x_0, \ldots, x_n]$$

(respectively in $L[x_0, \ldots, x_n]$) of the same degree such that

$$B(x) \neq 0 \text{ for all } x \in U$$

and

$$f(x) = \frac{A(x)}{B(x)} \text{ for all } x \in U.$$

It is immediate to see that once a function is regular at a point, then it is regular on a Zariski open subset of X, and moreover, if X is irreducible, then it is regular on a dense Zariski open subset of X. If

$$f : X \longrightarrow \mathbb{A}^1$$

is regular at each point of X, then it is called a *regular map*, or *regular function* (or a morphism to $\mathbb{A}^1$); the set of all regular functions on X is denoted by $\mathcal{O}_X$. If $X \subset \mathbb{A}^n$ is an affine variety, then

$$\mathcal{O}_X \stackrel{\sim}{\to} \overline{K}[X].$$

We call

$$f : X \longrightarrow \mathbb{A}^m$$

regular at the point $\alpha \in X$, if

$$f = (f_1, \ldots, f_m)$$

and each f_i is regular at α. We say that

$$f : X \longrightarrow \mathbb{A}^m$$

is regular if it is regular at each point of X.

If $Y \subseteq X$ is an irreducible subvariety of the irreducible quasiprojective variety X defined over K, then the *local ring* of X along Y, denoted $\mathcal{O}_{X,Y}$ is the set of pairs (U, f) where U is an open subset of X such that

$$Y \cap U \neq \emptyset$$

and f is a regular function on U modulo the equivalence relation where we identify

$$(f_1, U_1) \sim (f_2, U_2)$$

if

$$f_1 = f_2 \text{ on } U_1 \cap U_2.$$

If $Y = X$ we obtain the *(rational) function field* of X, also denoted

$$\overline{K}(X) := \mathcal{O}_{X,X};$$

if we restrict to functions defined over a subfield $K \subset L \subset \overline{K}$, then the corresponding function field is $L(X)$. If $Y = \{\alpha\}$ is a point, then we obtain the local ring $\mathcal{O}_{X,\alpha}$ of X at α.

2.1.3. Morphisms and rational maps between varieties. Let $X \subset \mathbb{P}^n$ and $Y \subset \mathbb{P}^m$ be quasiprojective varieties. We call a map $\Phi : X \longrightarrow Y$ regular (or a *morphism*) if for each point $\alpha \in X$, there exists an affine open subset V which contains $f(x)$, and there exists an affine open subset U of X containing α such that Φ restricts on U to a map

$$\Phi : U \longrightarrow V$$

which is regular. We say that Φ is an *isomorphism* if there exists another morphism of varieties

$$\Psi : Y \longrightarrow X$$

such that

$$\Phi \circ \Psi \text{ and } \Psi \circ \Phi$$

are the corresponding identity maps. If

$$\Phi : X \longrightarrow X$$

is a morphism, then it is called an *endomorphism*.

A *rational map* Φ between two quasiprojective varieties X and Y is a map which restricts to a morphism on an open subset U of X, i.e.,

$$\Phi : U \longrightarrow Y$$

is a morphism. We say that Φ is *birational* if it has a rational inverse, i.e. there exist open subsets $U \subset X$ and $V \subset Y$ and a rational map Ψ from Y to X such that $\Phi : U \longrightarrow V$ and $\Psi : V \longrightarrow U$ are morphisms and Φ is the inverse of Φ, i.e.,

$$\Phi \circ \Psi = \mathrm{id}_V \text{ and } \Psi \circ \Phi = \mathrm{id}_U .$$

Two varieties X and Y are *birationally equivalent* if there exists a birational map between them. Two irreducible varieties X and Y defined over the same field K are birationally equivalent if and only their function fields are isomorphic as K-algebras.

Let $\Phi : X \longrightarrow Y$ be a morphism of varieties, let $x \in X$, and let $y = \Phi(x)$. Then we have an induced morphism on the rings of regular functions, i.e.

$$\Phi^* : \mathcal{O}_{Y,y} \longrightarrow \mathcal{O}_{X,x} \text{ given by } \Phi^*(f) = f \circ \Phi.$$

Similarly, we get an induced morphism

$$\Phi^* : \mathcal{O}_Y \longrightarrow \mathcal{O}_X$$

and a morphism between the corresponding function fields

$$\Phi^* : \overline{K}(Y) \longrightarrow \overline{K}(X)$$

in the case that the image of Φ is dense.

In the special case of a rational map

$$\Phi : \mathbb{P}^1 \longrightarrow \mathbb{P}^1$$

defined over a field K, we have two equivalent representations of it, either as

(1) $\Phi([X : Y]) = [F(X,Y) : G(X,Y)]$ for two coprime homogeneous polynomials F and G with coefficients in K of degree d; or

(2) $\Phi(t) = f(t)/g(t)$ for coprime polynomials $f(t), g(t) \in K[t]$ where the polynomials F and G from (1) are defined by

$$F(X,Y) = Y^d f(X/Y) \text{ and } G(X,Y) = Y^d g(X/Y).$$

Then the *degree* of Φ is d.

2.1.4. Properties of morphisms and of rational maps. Let Φ be a rational map between two quasiprojective irreducible varieties X and Y. We define the *indeterminacy locus* $I(\Phi)$ of Φ be the intersection of all closed subsets $Z \subset X$ with the property that Φ is defined on $X \setminus Z$. Clearly, $I(\Phi)$ is a closed subset of X. Now, let $U \subset X$ be an open (dense) subset such that $\Phi : U \longrightarrow Y$ is a morphism. We say that Φ is *dominant* if $\Phi(U)$ (or equivalently, $\Phi(U')$ for any other open subset of X on which Φ restricts to a morphism) is Zariski dense in Y.

Let X and Y be affine varieties defined over K and $\Phi : X \longrightarrow Y$ be a morphism. Let

$$\Phi^* : \overline{K}[Y] \longrightarrow \overline{K}[X]$$

be the induced morphism of $\overline{K}$-algebra, which endows $\overline{K}[X]$ with a structure of $\overline{K}[Y]$-module. We call the morphism Φ *finite*, if $\overline{K}[X]$ is a finitely generated $\overline{K}[Y]$-module. More generally, if X and Y are quasiprojective varieties and

$$\Phi : X \longrightarrow Y$$

is a morphism, then we say that Φ is finite if for each open affine subset $V \subset Y$, the set $U := \Phi^{-1}(V)$ is also affine and the restriction map

$$\Phi : U \longrightarrow V$$

is finite (as defined above).

If $\Phi : X \longrightarrow Y$ is a finite and dominant morphism, then there exists a dense open subset U of Y such that for each $y \in U$, the fiber $\Phi^{-1}(y)$ has precisely

$d := [\overline{K}(X) : \overline{K}(Y)]$ points (where we view $\overline{K}(X)$ embedded into $\overline{K}(Y)$ through the map

$$\Phi^* : \overline{K}(Y) \longrightarrow \overline{K}(X),$$

which is injective since Φ is dominant). A finite map is *closed*, i.e. the image of any closed subvariety is also closed.

2.1.5. Dimension. Let X be an irreducible quasiprojective variety defined over K. The *dimension* of X (denoted $\dim(X)$) is the transcendence degree over $\overline{K}$ of the function field $\overline{K}(X)$. If $Y \subset X$ is a subvariety, we say that its *codimension* is d, where

$$d := \dim(X) - \dim(Y).$$

The *degree* of a subvariety $V \subseteq \mathbb{P}^n$ (denoted $\deg(V)$) is the number of points (over an algebraically closed field K) contained in the intersection of V with $\dim(V)$ generic hyperplanes in $\mathbb{P}^n$. In particular, if V is a hypersurface given by the equation $f = 0$, where $f \in K[X_0, \dots, X_n]$ is a homogeneous polynomial of degree d, then

$$\deg(V) = d$$

and if V is a finite collection of points, then $\deg(V)$ is the cardinality of V.

2.1.6. Tangent subspace, nonsingular points and morphisms between varieties. Let X be an irreducible quasiprojective variety defined over K, and let $\alpha \in X(K)$. We denote by $\mathcal{O}_\alpha := \mathcal{O}_{X,\alpha}$ the ring of all functions on X regular at α. We let $\mathfrak{m}_\alpha := \mathfrak{m}_{X,\alpha}$ be the maximal ideal of $\mathcal{O}_\alpha$ containing all $f \in \mathcal{O}_\alpha$ such that $f(\alpha) = 0$. Then the *tangent space* of X at α (denoted $T_{X,\alpha}$) is isomorphic to the dual of $\mathfrak{m}_\alpha / \mathfrak{m}_\alpha^2$ (i.e., the space of $\overline{K}$-linear functions defined on $\mathfrak{m}_\alpha / \mathfrak{m}_\alpha^2$). We say that the point $\alpha \in X$ is *smooth* (or *nonsingular*) if

$$\dim_{\overline{K}} \mathfrak{m}_\alpha / \mathfrak{m}_\alpha^2 = \dim(X).$$

Since each point belongs to an open affine subset V, the notion of smoothness agrees with the classical notion of smoothness from differential geometry, i.e. if V is defined by the (independent) equations

$$f_1(x_1, \dots, x_n) = \cdots = f_m(x_1, \dots, x_n) = 0,$$

then the point $x \in V$ is smooth if and only if the rank of the *Jacobian*

$$\left(\frac{df_j}{dx_i}(x) \right)$$

is $n - m$, which is also the dimension of V (and therefore of X since X is irreducible and V is dense in X). To see that

$$\dim(V) = n - m,$$

note that V is an affine variety defined by m independent equations in n variables. The set of all smooth points on X is denoted by X^{smooth}, and it is a dense Zariski open subset of X. A variety X is *smooth* if

$$X^{\mathrm{smooth}} = X.$$

Let α be a nonsingular point on the variety X of dimension n. Functions $u_1, \dots, u_n \in \mathcal{O}_\alpha$ are called *local parameters* for α if they form a $\overline{K}$-basis for the vector space $\mathfrak{m}_\alpha / \mathfrak{m}_\alpha^2$. Then each $f \in \mathcal{O}_\alpha$ (defined over K) is uniquely represented as a power series in variables $u_1, \ldots, u_n$ with coefficients in K, i.e., we have an embedding of

$$\mathcal{O}_\alpha \hookrightarrow K[[u_1, \ldots, u_n]].$$

For a morphism Φ between two quasiprojective varieties X and Y defined over K, $\alpha \in X$ and $\beta = \Phi(\alpha)$, we have an induced morphism on the tangent spaces

$$D\Phi : T_{X,\alpha} \longrightarrow T_{Y,\beta},$$

which is called the *differential* of Φ. The differential of Φ is induced by

$$\Phi^* : \mathcal{O}_{Y,\beta} \longrightarrow \mathcal{O}_{X,\alpha};$$

note that Φ^* also maps $\mathfrak{m}_{Y,\beta}$ into $\mathfrak{m}_{X,\alpha}$, and therefore we have an induced map

$$\Phi^* : \mathcal{O}_{Y,\beta}/\mathfrak{m}_{Y,\beta} \longrightarrow \mathcal{O}_{X,\alpha}/\mathfrak{m}_{X,\alpha},$$

which is the dual of $D\Phi$. An isomorphism

$$\Phi : X \longrightarrow Y$$

induces an isomorphism of tangent spaces, and therefore,

$$\Phi\left(X^{\text{smooth}}\right) \subseteq Y^{\text{smooth}}.$$

2.1.7. Differentials. Let X be a variety defined over K, and let $f \in K(X)$. Then for each x in the domain of f, we have a tangent map

$$df(x) : T_{X,x} \longrightarrow T_{\mathbb{A}^1, f(x)} = K.$$

This differential satisfies the usual Leibniz rules

$$d(f + g) = df + dg \text{ and } d(f \cdot g) = f \cdot dg + g \cdot df.$$

A *regular differential* 1-*form* ω on X has the property that for all $x \in X$, there exists an open set $x \in U \subset X$ and there exist (finitely many) functions f_i, g_i regular on U such that

$$\omega = \sum_i f_i \cdot dg_i$$

on U. We denote by $\Omega^1[X]$ the set of all regular 1-forms on X; this is both a $\overline{K}$-vector space and an $\mathcal{O}(X)$-module. When X is a smooth projective curve, $\Omega^1[X]$ is finite-dimensional as a $\overline{K}$-vector space and the *(geometric) genus* of X is defined to be the dimension of this space.

2.1.8. Some properties of varieties. A variety X is said to be *complete* (over the field K) if every projection

$$\pi : X \times Y \longrightarrow Y$$

is a closed map; projective varieties are always complete.

A variety X defined over K is called *normal* at the point x if the local ring $\mathcal{O}_{X,x}$ is integrally closed; then we call X normal if it is normal at each point. A smooth variety is always normal. If X is affine, then X is normal if and only if $K[X]$ is integrally closed. For any variety X there exists a *normalization* X' of it which is a normal variety equipped with a morphism

$$\Psi : X' \longrightarrow X$$

having the universal property that for any normal variety Z and any dominant morphism

$$\Phi : Z \longrightarrow X$$

there exists a unique morphism

$$\Phi' : Z \longrightarrow X'$$

such that

$$\Phi = \Psi \circ \Phi'.$$

Intuitively, X' is the smallest normal variety mapping onto X.

2.1.9. Algebraic groups. An *algebraic group* G defined over a field K is a variety with a distinguished point $e \in G(K)$, and morphisms $m : G \times G \longrightarrow G$ and $i : G \longrightarrow G$ (corresponding respectively to multiplication and inversion in G) satisfying the axioms of a group law, i.e.

- $m(e, x) = m(x, e) = x$;
- $m(i(x), x) = m(x, i(x)) = e$;
- $m(m(x, y), z) = m(x, m(y, z))$.

In the case when G is an affine variety, the main examples we will use are the *additive group* $\mathbb{G}_a$ and the multiplicative group $\mathbb{G}_m$. We call *(algebraic) torus* any algebraic group (not necessarily connected) which is isomorphic with an algebraic subgroup of $\mathbb{G}_m^N$ for some $N \in \mathbb{N}$.

In the case when G is projective (and connected), then G is an *abelian variety*; in this case the group law on G is indeed abelian and we usually denote by 0 its identity element. We will also work with *semiabelian varieties* S, which have the property that over $\overline{K}$ they fit into the following short exact sequence of algebraic groups:

$$1 \longrightarrow \mathbb{G}_m^n \longrightarrow S \longrightarrow A \longrightarrow 0,$$

where A is an abelian variety, and $\mathbb{G}_m^n$ is a power of the multiplicative group $\mathbb{G}_m$. Hence a semiabelian variety is a generalization of both powers of the multiplicative group and also of abelian varieties.

Any endomorphism of a semiabelian variety is the composition of an algebraic group endomorphism and a translation.

2.1.10. Divisors. A (Weil) *divisor* of the variety X is a formal sum

$$D := \sum_{i=1}^{s} k_i \cdot Y_i,$$

where each $k_i \in \mathbb{Z}$ and each Y_i is an irreducible codimension-one subvariety of X. If X is defined over K, then we say that the divisor D is also defined over K if it is invariant under the action of $\mathrm{Gal}(\overline{K}/K)$. We call the divisor D *effective* if $k_i \geq 0$ for each i.

Assume now that the set of singular points on X has codimension at least 2, then for each irreducible codimension-one subvariety $Y \subset X$, the ring $\mathfrak{o}_{X,Y}$ of rational functions on X which are regular on Y is a discrete valuation ring. Therefore, for each $f \in K(X)$ we can associate its order of vanishing along Y, which is a well-defined integer number. So, for each rational function f on X we associate its *support*, which is a divisor

$$\sum_{i} k_i \cdot Y_i,$$

where $k_i \neq 0$ if either f vanishes or $1/f$ vanishes along Y_i; we call k_i the order of vanishing (respectively the order of the pole of f) along Y_i. We call *principal divisor* (and denote it (f)) the divisor

$$(f) = \sum_i k_i \cdot Y_i$$

associated to a rational function f, where each k_i is the order of f along Y_i. We say that two divisors D_1 and D_2 are *linearly equivalent* if there exists a rational map f such that

$$D_1 - D_2 = (f).$$

If X is smooth, we let $\mathrm{Pic}(X)$ be the *Picard group* for the variety X, which is the group of equivalence classes of divisors modulo linear equivalence.

Let $\Phi : X \longrightarrow Y$ be a morphism of smooth varieties. Then we have a well-defined induced morphism

$$\Phi^* : \mathrm{Pic}(Y) \longrightarrow \mathrm{Pic}(X);$$

the definition of Φ^* is easier to be given using Cartier divisors and then one uses the equivalence between Cartier divisors and Weil divisors in the case of smooth varieties.

2.1.11. Ramification. Let $\Phi : X \longrightarrow Y$ be a finite map of smooth projective varieties, let $x \in X$, and let

$$f^{\#} : \mathcal{O}_{Y,f(x)} \longrightarrow \mathcal{O}_{X,x}$$

be the induced maps of local rings. Let $\mathfrak{m}$ be the maximal ideal of $\mathcal{O}_{Y,f(x)}$ and let $\mathfrak{m}' := f^{\#}(\mathfrak{m})$. We say that f is *unramified* at x if $\mathfrak{m}'$ is the maximal ideal of $\mathcal{O}_{X,x}$ and the induced map

$$f^{\#} : \mathcal{O}_{Y,f(x)}/\mathfrak{m} \longrightarrow \mathcal{O}_{X,x}/\mathfrak{m}'$$

is a finite separable field extension. We say that f is unramified, if it is unramified at each point of X.

With the above notation, we say that f is *flat* at x, if

$$f^{\#} : \mathcal{O}_{Y,f(x)} \longrightarrow \mathcal{O}_{X,x}$$

is a flat map of rings; if f is flat at each point of X, then we say that f is flat. We recall that for two rings A and B,

$$f : A \longrightarrow B$$

is a *flat map* if it is a ring homomorphism which makes B a flat A-module. Also, in general, an A-module B is called *flat* if tensoring with B over A preserves any exact sequence of A-modules.

We say that $f : X \longrightarrow Y$ is *étale* if it is both flat and unramified.

2.1.12. Sheaves.

DEFINITION 2.1.12.1. Let X be a topological space. A *presheaf* $\mathcal{F}$ on X consists of the following data:

- for every open subset U in X, we have a set $\mathcal{F}(U)$; and

- for all open subsets $V \subset U \subset X$, a restriction map

$$r_{U,V} : \mathcal{F}(U) \longrightarrow \mathcal{F}(V)$$

such that

$$r_{U,U} = \mathrm{id}_{\mathcal{F}(U)} \text{ and } r_{U,W} = r_{V,W} \circ r_{U,V}.$$

DEFINITION 2.1.12.2. A morphism of presheaves

$$f : \mathcal{F}_1 \longrightarrow \mathcal{F}_2$$

is a collection of maps

$$f(U) : \mathcal{F}_1(U) \longrightarrow \mathcal{F}_2(U)$$

such that for every $V \subset U$, the maps $f(U)$ and $f(V)$ are compatible with the restriction maps

$$r_{2;U,V} = f(V) \circ r_{1;U,V}.$$

If the presheaves $\mathcal{F}_1$ and $\mathcal{F}_2$ are presheaves of groups (or rings, or modules), then we also ask that each $f(U)$ is a group (or ring, or module) homomorphism.

DEFINITION 2.1.12.3. Let X be a topological space. A sheaf $\mathcal{F}$ on X is a presheaf with the property that for every open subset $U \subset X$ and every open covering

$$U = \cup_i U_i,$$

the following properties hold:

- for each $x, y \in \mathcal{F}(U)$ such that $r_{U,U_i}(x) = r_{U,U_i}(y)$ for each i, then $x = y$; and
- given a collection of elements $x_i \in \mathcal{F}(U_i)$ such that for each i and j we have $r_{U_i, U_i \cap U_j}(x_i) = r_{U_j, U_i \cap U_j}(x_j)$, then there exists $x \in \mathcal{F}(U)$ such that $r_{U,U_i}(x) = x_i$ for each i.

A classical example of sheaves is the sheaf $\mathcal{O}_X$ of regular functions on a variety X equipped with the Zariski topology. So, for each open subset $U \subset X$ we define $\mathcal{O}_X(U)$ be the set of all regular functions on U. We say that

$$f : \mathcal{F} \longrightarrow \mathcal{G}$$

is a morphism of sheaves if $\mathcal{F}$ and $\mathcal{G}$ are both sheaves, while f is a morphism of presheaves.

DEFINITION 2.1.12.4. The *stalk* $\mathcal{F}_x$ of a sheaf $\mathcal{F}$ at a point $x \in X$ is the direct limit of the $\mathcal{F}(U)$ over all open subsets U containing x (where the limit is taken with respect to the restriction maps $r_{U,V}$).

If $\mathcal{F}$ is the sheaf of regular functions, then $\mathcal{F}_x = \mathcal{O}_{X,x}$.

DEFINITION 2.1.12.5. Let X be a variety. An $\mathcal{O}_X$-*module* is a sheaf $\mathcal{F}$ on X such that

- for every open $U \subset X$, $\mathcal{F}(U)$ is a module over the ring $\mathcal{O}_X(U)$; and
- for every open $V \subset U \subset X$, the map $r_{U,V} : \mathcal{F}(U) \longrightarrow \mathcal{F}(V)$ is a homomorphism of $\mathcal{O}_X(U)$-modules.

Let $\mathcal{F}$ be an $\mathcal{O}_X$-module on the variety X. We say that $\mathcal{F}$ is *locally free* if each point in X has a neighborhood over which $\mathcal{F}$ is free. The *rank* of a locally free sheaf $\mathcal{F}$ is the integer r such that

$$\mathcal{F}(U) \xrightarrow{\sim} \mathcal{O}_X(U)^r$$

for all sufficiently small open subsets U. A locally free sheaf of rank 1 is called an *invertible sheaf*, or a *line bundle*. One can use line bundles to define $\mathrm{Pic}(X)$ even when X is not smooth.

Given a sheaf $\mathcal{F}$ on a topological space X, and an open subset $Y \subset X$, we define the *restricted sheaf* $\mathcal{F}|_Y$ on Y (endowed with the inherited topology from X) as follows:

- for each open $U \subset Y$, we define $\mathcal{F}|_Y(U) := \mathcal{F}(U)$; and
- for all open subsets $V \subset U \subset Y$, we define $r_{Y,U,V} := r_{X,U,V}$.

Given a sheaf $\mathcal{F}$ on a topological space X, and a continuous function $f : X \longrightarrow Y$, we can construct the *push-forward* $f_*\mathcal{F}$ of the sheaf $\mathcal{F}$ on the topological space Y as follows:

- for every open set $U \subset Y$, we define $f_*\mathcal{F}(U) := \mathcal{F}(f^{-1}(U))$; and
- for all open subsets $V \subset U \subset Y$ we define the map

$$r_{Y,U,V} : f_*\mathcal{F}(U) \longrightarrow f_*\mathcal{F}(V) \text{ by } r_{Y,U,V} := r_{X,f^{-1}(U),f^{-1}(V)}.$$

For two sheaves $\mathcal{F}_1$ and $\mathcal{F}_2$ on X, we define their tensor product $\mathcal{F}_1 \otimes \mathcal{F}_2$ be the sheaf $\mathcal{F}$ for which

$$\mathcal{F}(U) := \mathcal{F}_1(U) \otimes_{\mathcal{O}_X(U)} \mathcal{F}_2(U)$$

for each open subset $U \subseteq X$. For a sheaf $\mathcal{F}$ on X and a positive integer n, we define the tensor power $\mathcal{F}^{\otimes n}$ be the tensor product of $\mathcal{F}$ with itself n times.

2.1.13. Schemes.

DEFINITION 2.1.13.1. Let R be an integral domain. The *spectrum* $\mathrm{Spec}(R)$ is a pair consisting of a topological space (also denoted $\mathrm{Spec}(R)$) and a sheaf $\mathcal{O}_R$. The topological space $\mathrm{Spec}(R)$ is the set of prime ideals of R endowed with a topology whose closed sets are the sets

$$V(I) := \{\mathfrak{p} \in \mathrm{Spec}(R) : I \subset \mathfrak{p}\}$$

for any ideal $I \subset R$. The sheaf $\mathcal{O}_R$ is characterized by

$$\mathcal{O}_R\left(\mathrm{Spec}(R) \setminus V((f))\right) := R_f$$

for any nonzero $f \in R$, where

$$R_f := S^{-1} \cdot R$$

for the multiplicative set

$$S := \{1, f, f^2, \cdots\},$$

DEFINITION 2.1.13.2. A *ringed space* is a pair $(X, \mathcal{O}_X)$ consisting of a topological space X and a sheaf of rings $\mathcal{O}_X$ on X. It is a *locally ringed space* if for all $x \in X$, the stalk $\mathcal{O}_x$ is a local ring. The sheaf $\mathcal{O}_X$ is called the structure sheaf of the ringed space.

DEFINITION 2.1.13.3. A *morphism* of ringed spaces is a pair

$$(f, f^\#) : (X, \mathcal{O}_X) \longrightarrow (Y, \mathcal{O}_Y)$$

where $f : X \longrightarrow Y$ is continuous and

$$f^\# : \mathcal{O}_Y \longrightarrow f_* \mathcal{O}_X$$

is a morphism of sheaves over Y. It is a morphism of locally ringed spaces if further for all $x \in X$, the map $f^\#$ induces a local ring homomorphism

$$f_x^\# : \mathcal{O}_{Y,f(x)} \longrightarrow \mathcal{O}_{X,x},$$

i.e., $(f_x^\#)^{-1}(\mathfrak{m}_x) = \mathfrak{m}_{f(x)}$, where $\mathfrak{m}_x$ and $\mathfrak{m}_{f(x)}$ are the corresponding maximal ideal of the local rings $\mathcal{O}_{X,x}$ and $\mathcal{O}_{Y,f(x)}$.

A typical example of a locally ringed space is $(\mathrm{Spec}(R), \mathcal{O}_R)$ for any integral domain R, or an algebraic variety with its sheaf of regular functions. A locally ringed space of the form $(\mathrm{Spec}(R), \mathcal{O}_R)$ is called an *affine scheme*. Any morphism of affine schemes corresponding to integral domains R and S is induced by a ring homomorphism between R and S.

DEFINITION 2.1.13.4. A *scheme* is a locally ringed space $(X, \mathcal{O}_X)$ that can be covered by open affine subsets U such that $(U, \mathcal{O}_X|_U)$ is isomorphic to some affine scheme $(\mathrm{Spec}(R), \mathcal{O}_R)$.

A morphism of schemes is a morphism of locally ringed spaces that are schemes.

A morphism $\varphi : X \longrightarrow Z$ of schemes is an *immersion* if it gives an isomorphism between X and an open subset of a closed subset of Z.

If S is a scheme and X is another scheme endowed with a morphism of schemes

$$f : X \longrightarrow S,$$

then we call X an *S-scheme*. If Y is another S-scheme (with respect to a morphism $g : Y \longrightarrow S$), then a *morphism of S-schemes* between X and Y is a morphism of schemes

$$\Phi : X \longrightarrow Y$$

satisfying

$$f = g \circ \Phi.$$

This notion extends the definition of morphisms of varieties since each variety over a field K is a $\mathrm{Spec}(K)$-morphism.

For an S-scheme X, we define $X(S)$ be the set of all scheme morphisms

$$\alpha : X \longrightarrow S;$$

this notion replaces the notion of points of varieties. If $S = \mathrm{Spec}(R)$ for some ring R, for each $\mathfrak{p} \in S$, we let $k(\mathfrak{p})$ be the fraction field of $R/\mathfrak{p}$. Then we let $X_\mathfrak{p}$ be the *fiber* of X above $\mathfrak{p}$ be defined as the fiber product of X and $\mathrm{Spec}(k(\mathfrak{p}))$ over S. Furthermore, if $\mathfrak{p} = (0)$, then $k(\mathfrak{p}) = \mathrm{Frac}(R)$ and $X_\mathfrak{p}$ is called the generic fiber of X. Conversely, let X be a variety defined over K (i.e., a K-variety). Let R be a subring of K whose field of fractions is K. We let $S := \mathrm{Spec}(R)$, and we say that $\mathcal{X}$ is a *model* of X over S if $\mathcal{X}$ is an S-scheme whose generic fiber is isomorphic to X.

As a matter of notation, if C is a curve defined over an algebraically closed field K, then we denote by $\mathbb{P}^1_C$ the model of $\mathbb{P}^1$ over C, i.e, there exists a morphism

$$\pi : \mathbb{P}^1_C \longrightarrow C$$

such that for each $x \in C(K)$, $\pi^{-1}(x) \xrightarrow{\sim} \mathbb{P}^1$ over K.

2.1.14. Ample line bundles. Assume now that X is an S-scheme, where $S = \operatorname{Spec}(R)$ for some ring R.

The set of *global sections* is the set $\mathcal{F}(X)$. We say that $\mathcal{F}$ is *very ample* if there exist global sections $s_0, \ldots, s_N \in \mathcal{F}(X)$ such that the map

$$\iota_{\mathcal{F}} : X \longrightarrow \mathbb{P}^N \text{ given by } \iota_{\mathcal{F}}(x) := [s_0(x) : \cdots : s_N(x)]$$

is an immersion. We say that $\mathcal{F}$ is *ample* if $\mathcal{F}^{\otimes n}$ is very ample for some positive integer n.

2.2. Dynamics of endomorphisms

In this short section we recall the basic notions for the dynamics of an endomorphism of a quasiprojective variety.

2.2.1. Orbit of a point. For a quasiprojective variety X, for an endomorphism Φ of X, and for a point α of X, we denote by $\mathcal{O}_{\Phi}(\alpha)$ the orbit of α under Φ, i.e., the set of all iterates $\Phi^n(\alpha)$ for $n \in \mathbb{N}_0$.

2.2.2. Periodic and preperiodic points. We say that the point $\alpha \in X$ is *(Φ-)preperiodic* if $\mathcal{O}_{\Phi}(\alpha)$ is finite; the set of all preperiodic points of Φ is denoted by $\operatorname{Prep}_{\Phi}(X)$. If $\Phi^n(\alpha) = \alpha$ for some $n \in \mathbb{N}$, then we say that α is *(Φ-)periodic*. Sometimes we call n *a period* of α; when n is the smallest such positive integer, then it is called *the (minimal) period* of α.

For a periodic point α under a map $\Phi : X \longrightarrow X$, we define the *periodic cycle* of α be the finite set

$$\{\alpha, \Phi(\alpha), \ldots, \Phi^{\ell-1}(\alpha)\},$$

if ℓ is the smallest positive integer n such that $\Phi^n(\alpha) = \alpha$.

2.2.3. Periodic and preperiodic subvarieties. We extend the notion of (pre)periodicity to arbitrary (i.e., not necessarily closed) subvarieties V of X. Hence we say that V is *Φ-preperiodic* if there exist distinct nonnegative integers m and n such that

$$\overline{\Phi^m(V)} \subseteq \overline{\Phi^n(V)}.$$

If there exists $m \in \mathbb{N}$ such that

$$\overline{\Phi^m(V)} \subseteq \overline{V},$$

then we say that V is *Φ-periodic*. In particular, if

$$\Phi^k(V) \subseteq V,$$

then we say that V is periodic, and we call k a *period* of V. Intuitively, one could consider an alternative definition for periodic subvarieties by asking that V is periodic under the action of Φ if

$$(2.2.3.1) \qquad\qquad \Phi^k(V) = V,$$

for some $k \in \mathbb{N}$. However, there are many examples of subvarieties V which are mapped by a power Φ^k of Φ into themselves without having that the equality from (2.2.3.1) holds. On the other hand, in the Dynamical Mordell-Lang Conjecture (see Conjecture 1.5.0.1), each subvariety V which is periodic according to our definition, i.e.

$$\Phi^k(V) \subseteq V \text{ for some } k \in \mathbb{N}$$

yields an infinite set

$$(2.2.3.2) \qquad\qquad S := \{n \in \mathbb{N}_0 \colon \Phi^n(\alpha) \in V\}$$

for any $\alpha \in V$; moreover, the set S is a finite union of arithmetic progressions of common difference k. So, in this case, the common difference of the arithmetic progressions from the conclusion of Conjecture 1.5.0.1 is the *period* of the subvariety V. The whole content of the Dynamical Mordell-Lang Conjecture (see also its reformulation from Conjectures 3.1.3.1 and 3.1.3.2) is to prove that only when V contains a positive dimensional periodic subvariety we can have that

$$V \cap \mathcal{O}_\Phi(\alpha) \text{ is infinite for some non-preperiodic point } \alpha.$$

For more about this interpretation of the Dynamical Mordell-Lang Conjecture, we refer the reader to Chapter 3.

Now, if V is a closed subvariety, then it is easy to see that the property that V is periodic can be reformulated as follows:

$$V \text{ is periodic if and only if } \Phi^{-n}(V) \supseteq V \text{ if and only if } \Phi^n(V) \subseteq V.$$

First of all, it is immediate to see the equivalence of the last two statements:

$$\Phi^{-n}(V) \supseteq V \text{ if and only if } \Phi^n(V) \subseteq V.$$

Indeed, applying Φ^n to the left inclusion yields the right inclusion, and applying Φ^{-n} to the right inclusion and also using that

$$V \subseteq \Phi^{-n}(\Phi^n(V))$$

yields the left inclusion. So, we are left to prove that

$$V \text{ is periodic if and only if } \Phi^{-n}(V) \supseteq V.$$

Now, if $V \subseteq \Phi^{-n}(V)$ then $\Phi^n(V) \subseteq V$ and thus

$$\overline{\Phi^n(V)} \subseteq \overline{V} = V.$$

Conversely, if

$$\overline{\Phi^n(V)} \subseteq \overline{V} = V,$$

then

$$V \subseteq \Phi^{-n}\left(\overline{\Phi^n(V)}\right) \subseteq \Phi^{-n}(V),$$

as desired. We note that if V is not necessarily a closed subvariety, then we still have that

$$V \subseteq \Phi^{-n}(V) \text{ yields } \overline{\Phi^n(V)} \subseteq \overline{V},$$

but the converse is not necessarily true.

Furthermore, if Φ is a finite morphism, and thus all its iterates are finite and hence closed, then a closed irreducible subvariety V is Φ-periodic if and only if there exists $n \in \mathbb{N}$ such that

$$\Phi^n(V) = V.$$

Similarly, in this case, V is preperiodic if and only if there exist $m, n \in \mathbb{N}$ with $0 \le m < n$ such that

$$\Phi^m(V) = \Phi^n(V).$$

So, for closed subvarieties and finite morphisms, the intuitive notion of periodicity for subvarieties matches also our more general definition of periodic subvarieties. While our subvarieties will almost always be closed, it is not necessarily true that the endomorphisms we consider in this book, and more generally the ones appearing in the Dynamical Mordell-Lang Conjecture are finite morphisms; this is the reason why we opted for our notion of periodicity for subvarieties.

Finally, sometimes it is useful to define the *orbit of a subvariety* $V \subseteq X$ under the action of an endomorphism Φ of X. We define thus $\mathcal{O}_\Phi(V)$ be the orbit of V under Φ, which is a set consisting of all closed subvarieties $\overline{\Phi^n(V)}$ for $n \in \mathbb{N}_0$. Then, according to our definition for preperiodic subvarieties, V is preperiodic under the action of Φ if and only if $\mathcal{O}_\Phi$ is a finite collection of closed subvarieties.

2.2.4. Polarizable endomorphisms. Let X be a variety and $\Phi : X \longrightarrow X$ an endomorphism. We say that Φ is *polarizable* if there exists an ample line bundle $\mathcal{L}$ on X such that $\Phi^*(\mathcal{L}) \overset{\sim}{\to} \mathcal{L}^d$ for some integer $d > 1$.

2.3. Valuations

In this section we introduce the basic concepts from p-adic analysis required in this book; for more details on p-adic numbers, and p-adic analysis we refer the reader to [**Rob00**].

2.3.1. p-adic numbers. Let p be a prime number. For each nonzero integer n we define $\mathrm{ord}_p(n)$ be the exponent of p in n. Then we define

$$|n|_p := p^{-\mathrm{ord}_p(n)}.$$

We extend this definition for all nonzero rational numbers:

$$\left| \frac{m}{n} \right|_p = \frac{|m|_p}{|n|_p}.$$

By convention, we let $|0|_p = 0$. We call $| \cdot |_p$ be the *p-adic absolute value* (or norm); this induces a metric on $\mathbb{Q}$. We let $\mathbb{Q}_p$ be the completion of $\mathbb{Q}$ with respect to the p-adic norm; we denote by $| \cdot |_p$ the extension of the p-adic absolute value to $\mathbb{Q}_p$. The set of all $x \in \mathbb{Q}_p$ such that $|x|_p \leq 1$ is the ring of *p-adic integers* $\mathbb{Z}_p$. There is an explicit description of the p-adic integers, as follows. Each $x \in \mathbb{Z}_p$ corresponds to a sequence

$$\{x_n\}_{n \in \mathbb{N}_0} \subseteq \mathbb{Z}$$

with the property that for each $n \in \mathbb{N}$ we have

$$x_n \equiv x_{n-1} \pmod{p^n}.$$

Essentially, x is the limit of the sequence x_n with respect to the p-adic norm. Two such sequences $\{x_n\}_{n \in \mathbb{N}_0}$ and $\{y_n\}_{n \in \mathbb{N}_0}$ correspond to the same number in $\mathbb{Z}_p$ if for each $n \in \mathbb{N}_0$ we have

$$x_n \equiv y_n \pmod{p^{n+1}}.$$

Alternatively, an element $x \in \mathbb{Z}_p$ can be uniquely represented as an infinite sum of powers of p, i.e.,

$$x = a_0 + a_1 p + \cdots + a_n p^n + \cdots,$$

where for each $n \in \mathbb{N}_0$ we have $a_n \in \{0, \ldots, p-1\}$. The connection with the previous representation of x as a limit of a sequence $\{x_n\}_{n \in \mathbb{N}_0}$ is made by taking

$$x_n = \sum_{i=0}^{n} a_i p^i$$

for each $n \in \mathbb{N}_0$. Conversely, given a sequence $\{x_n\}_{n \in \mathbb{N}_0}$, we let

$$b_n \in \{0, \ldots, p^n - 1\}$$

for each $n \in \mathbb{N}$ such that

$$x_{n-1} \equiv b_{n-1} \pmod{p^n}.$$

Then a_m is the m-th digit of b_n (for any $n \geq m$) in the p-adic basis. The fact that

$$b_k \equiv x_k \equiv x_\ell \equiv b_\ell \pmod{p^{1+\min\{k,\ell\}}}$$

guarantees that a_m is well-defined (independent of the choice of $n \geq m$).

An element $x \in \mathbb{Z}_p$ is a *p-adic unit* if $|x|_p = 1$. Each nonzero element of $\mathbb{Q}_p$ is written uniquely as $p^\alpha \cdot u$, where $\alpha \in \mathbb{Z}$ and u is a p-adic unit.

For any finite extension K of $\mathbb{Q}_p$ there exists a unique extension of $|\cdot|_p$ to a p-adic absolute value on K. By taking direct limits, we can therefore extend the p-adic absolute value of $\mathbb{Q}_p$ to $\overline{\mathbb{Q}}_p$. Also, we can embed $\overline{\mathbb{Q}}$ into a fixed algebraic closure of $\mathbb{Q}_p$ and therefore on each number field we have an extension of the p-adic absolute value.

2.3.2. Hensel's Lemma. The following result (Hensel's Lemma) is essential for solving polynomial congruence equations.

LEMMA 2.3.2.1 (Hensel). *Let p be a prime number, let f be a polynomial with coefficients in $\mathbb{Z}_p$, and let $x_0 \in \mathbb{Z}_p$ be such that*

$$|f(x_0)|_p < |f'(x_0)|_p^2.$$

Then there exists $\alpha \in \mathbb{Z}_p$ such that $f(\alpha) = 0$.

PROOF. If $f(x_0) = 0$, then we are done. So, from now on assume that $f(x_0) \neq 0$. Furthermore, since $\mathbb{N}_0$ is dense in $\mathbb{Z}_p$, we may assume without loss of generality that $x_0 \in \mathbb{N}_0$. By hypothesis we know that $f'(x_0) \neq 0$, and thus there is some $\delta \in \mathbb{N}_0$ such that

$$|f'(x_0)|_p = p^{-\delta}.$$

So we know that

$$|f(x_0)|_p \leq p^{-2\delta - 1}.$$

We construct α through approximation, i.e., we construct a sequence

$$\{x_n\}_{n \in \mathbb{N}_0} \subset \mathbb{Z}$$

such that for each $n \in \mathbb{N}_0$ we have

$$|f(x_n)|_p \leq p^{-2\delta - 1 - n}$$

and also

$$|x_{n+1} - x_n|_p \leq p^{-\delta - 1 - n}.$$

Thus if we let $\alpha \in \mathbb{Z}_p$ be the limit of the sequence with respect to the p-adic limit, then we have

$$|f(\alpha)|_p < p^{-n}$$

for each $n \in \mathbb{N}_0$ and so $f(\alpha) = 0$.

We construct the sequence $\{x_n\}_{n \in \mathbb{N}_0}$ as follows. We already have x_0 for which

$$|f(x_0)|_p < |f'(x_0)|_p^2 = p^{-2\delta}$$

and thus

$$|f(x_0)|_p \leq p^{-1-2\delta}.$$

We assume that we constructed $x_n \in \mathbb{Z}$ with the above properties. We let

$$x_{n+1} = x_n + p^{n+\delta+1} \ell$$

for some $\ell \in \mathbb{Z}$ for which we solve next. Clearly,

$$|x_{n+1} - x_n|_p \leq p^{-\delta-1-n}.$$

Then expanding $f(x_{n+1})$ around $z = x_n$ gives

(2.3.2.2) $\qquad f(x_{n+1}) \equiv f(x_n) + p^{n+\delta+1}\ell f'(x_n) \pmod{p^{2n+2\delta+2}}.$

Since

$$x_n \equiv x_0 \pmod{p^{\delta+1}}$$

and f' has coefficients in $\mathbb{Z}_p$, we have

$$|f'(x_n) - f'(x_0)|_p \leq p^{-\delta-1}$$

and because

$$|f'(x_0)|_p = p^{-\delta},$$

we conclude that

$$|f'(x_n)|_p = p^{-\delta}.$$

Thus

$$f'(x_n) = p^\delta \cdot u_n,$$

where $|u_n|_p = 1$ with $u_n \in \mathbb{Z}_p$. On the other hand, by the inductive hypothesis, there exists $v_n \in \mathbb{Z}_p$ such that

$$f(x_n) = p^{n+2\delta+1} \cdot v_n.$$

So, using (2.3.2.2) we conclude that

$$|f(x_{n+1})|_p \leq p^{-2\delta-2-n}$$

if and only if

$$v_n + \ell u_n \equiv 0 \pmod{p}.$$

The above last congruence has a solution ℓ because $|u_n|_p = 1$. This concludes the proof of Hensel's Lemma. $\qquad \square$

2.3.3. A complete, algebraically closed field. The algebraic closure $\overline{\mathbb{Q}}_p$ of $\mathbb{Q}_p$ is not complete; however its completion $\mathbb{C}_p$ is both algebraically closed and complete. We include the proof of this classical result, for the sake of completeness.

PROPOSITION 2.3.3.1. *The completion $\mathbb{C}_p$ of $\overline{\mathbb{Q}}_p$ is also algebraically closed.*

PROOF. Let $x \in \overline{\mathbb{C}_p}$; then there exists $d \in \mathbb{N}$ and there exist $c_i \in \mathbb{C}_p$ for $i = 0, \ldots, d-1$ such that

$$x^d + c_{d-1}x^{d-1} + \cdots + c_1 x + c_0 = 0.$$

Since $\overline{\mathbb{Q}}_p$ is dense in its completion $\mathbb{C}_p$, then for each $i = 0, \ldots, d-1$ there exists a sequence

$$\{\lambda_{i,n}\}_{n \geq 1} \subset \overline{\mathbb{Q}}_p \text{ such that } \lambda_{i,n} \to c_i.$$

Because $\overline{\mathbb{Q}}_p$ is algebraically closed we have that all solutions to the polynomial equation

(2.3.3.2) $\qquad t^d + \lambda_{d-1,n}t^{d-1} + \cdots + \lambda_{1,n}t + \lambda_{0,n} = 0$

lie in $\overline{Q}_p$. Thus we may factor the polynomial $P_n(t)$ appearing in (2.3.3.2) as follows:

$$P_n(t) := (t - \beta_{1,n}) \cdots (t - \beta_{d,n}),$$

with $\beta_{1,n}, \ldots, \beta_{d,n} \in \overline{\mathbb{Q}}_p$. By construction,

$$P_n(x) = P_n(x) - x^d - c_{d-1}x^{d-1} - \cdots - c_1 x - c_0 = \sum_{i=0}^{d-1}(\lambda_{i,n} - c_i)x^i.$$

In particular, we see that

$$|P_n(x)|_p \leq \max_{0 \leq i \leq d-1}\left(|\lambda_{i,n} - c_i|_p \cdot |x|_p^i\right) \to 0$$

as $n \to \infty$ (because $\lambda_{i,n} \to c_i$ in $\overline{\mathbb{Q}}_p$). On the other hand,

$$|P_n(x)|_p = \prod_{i=1}^{d}|x - \beta_{i,n}|_p,$$

and so we see that for each n we can find some solution $t = x_n$ to Equation (2.3.3.2) such that $|x - x_n|_p \to 0$ as $n \to \infty$, thus proving that x lies in the completion of $\overline{\mathbb{Q}}_p$. So, $\mathbb{C}_p$ is both complete and algebraically closed, as claimed. □

It is elementary to see that for each $\lambda \in \mathbb{Q}_p$ such that

$$|\lambda|_p = 1,$$

there exists a positive integer d such that

$$|\lambda^d - 1|_p \leq \frac{1}{p},$$

which yields easily that

$$|\lambda^{dp} - 1|_p < \frac{1}{p}.$$

Actually, one can prove the following more general result, which we will use later in our book.

PROPOSITION 2.3.3.3. *Let $\lambda \in \mathbb{C}_p$ such that $|\lambda|_p = 1$. Then there exists a positive integer d such that $|\lambda^d - 1|_p < \frac{1}{p}$.*

PROOF. The only real difficulty in the proof of Proposition 2.3.3.3 lies in the fact that the closed unit ball in $\mathbb{C}_p$ is not compact. We can get around this by reducing to a compact set since $\overline{\mathbb{Q}}_p$ is dense in $\mathbb{C}_p$. Thus we may assume that $\lambda \in \overline{\mathbb{Q}}_p$. Now let $K = \mathbb{Q}_p(\lambda)$. Then K is a finite extension of $\mathbb{Q}_p$. We let $\mathcal{O}_K$ denote the subring of K containing all elements of norm at most 1. Then $\mathcal{O}_K$ is a finite free $\mathbb{Z}_p$-module of rank $[K : \mathbb{Q}_p]$. In particular, $\mathcal{O}_K$ is compact, as it is the continuous image of $\mathbb{Z}_p^{[K:\mathbb{Q}_p]}$. By construction, $\lambda \in \mathcal{O}_K$. We now consider the sequence of powers of λ. By the Bolzano-Weierstrass theorem, there exists a convergent subsequence and hence there exist positive integers m and n with $m < n$ such that

$$|\lambda^m - \lambda^n|_p < \frac{1}{p}.$$

Since $|\lambda|_p = 1$ we then see

$$|\lambda^d - 1|_p < \frac{1}{p},$$

where $d = n - m$. This completes the proof. □

2.3.4. p-adic analytic functions. When we talk about p-adic analytic functions, we generally consider functions defined on open subsets of $\mathbb{C}_p$. For $w \in \mathbb{C}_p$ and $r > 0$, we denote by $D(w, r)$ the open disk centered at w and of radius r, i.e.

$$D(w, r) := \{z \in \mathbb{C}_p : |z - w|_p < r\}.$$

If $w \in \mathbb{Q}_p$, sometimes we consider the above open disk inside $\mathbb{Q}_p$, and still denote it by $D(w, r)$.

DEFINITION 2.3.4.1. Let B be an open subset of $\mathbb{C}_p$, and let $f : B \longrightarrow \mathbb{C}_p$. We say that f is *(p-adic) analytic* at $w \in B$ if there exists $r > 0$ such that f restricted on $D(w, r)$ is given by a convergent power series:

$$f(z) = \sum_{i=0}^{\infty} c_i (z - w)^i.$$

If f is analytic at each point of B, then we says that f is *(p-adic) analytic* on B.

The following result is proven in [**Rob00**, Section 5.4].

LEMMA 2.3.4.2. *The p-adic logarithmic function*

$$\log_p(1 + z) := \sum_{i=0}^{\infty} \frac{(-1)^i z^{i+1}}{i + 1}$$

is analytic on $D(0, 1)$, and the p-adic exponential map

$$\exp_p(z) = \sum_{i=0}^{\infty} \frac{z^i}{i!}$$

is analytic on $D\left(0, p^{-1/(p-1)}\right)$. Therefore, for each $u_0 \in \mathbb{C}_p$ such that

$$|u_0 - 1|_p < p^{-1/(p-1)}$$

we can define the function

$$f(z) = u_0^z = \exp_p(z \cdot \log_p(1 + (u_0 - 1))),$$

which is analytic for all $z \in \overline{D}(0, 1)$. Moreover, since

$$|u_0 - 1|_p < p^{-1/(p-1)},$$

then

$$|\log_p(u_0)|_p \leq |u_0 - 1|_p < p^{-1/(p-1)}.$$

2.3.5. Mahler series. For any integer $k \geq 0$ we define the polynomial function $z \mapsto \binom{z}{k}$ given by

$$\binom{z}{k} := \frac{z(z - 1) \cdots (z - k + 1)}{k!},$$

where by convention, $\binom{z}{0} = 1$.

When studying p-adic analytic maps, it is often more useful to work with Mahler series rather than conventional power series. A *Mahler series* is a function

$$f : \mathbb{Z}_p \to \mathbb{C}_p \text{ given by } f(z) = \sum_{j=0}^{\infty} c_k \binom{z}{k},$$

where $c_0, c_1, \ldots$ are elements of $\mathbb{C}_p$ with $|c_n|_p \to 0$ as $n \to \infty$. We note that since a Mahler series is a uniform limit of polynomials, then it is continuous on $\mathbb{Z}_p$.

The reason for which it is often advantageous to use Mahler series expansions rather than power series expansions when studying continuous and p-adic analytic maps on $\mathbb{Z}_p$ is that the maps

$$z \mapsto \binom{z}{i}, \qquad i = 0, 1, \dots$$

form a $\mathbb{Z}_p$-module basis for the polynomials that map $\mathbb{Z}_p$ into itself and because of this fact, the coefficients of the Mahler expansion of a function are often more readily obtained than the coefficients of its power series expansion. More generally, one can show that every continuous map on $\mathbb{Z}_p$ is given by Mahler series. To show this, we let $\mathcal{C}_p$ denote the collection of continuous maps from $\mathbb{Z}_p$ into $\mathbb{C}_p$. The *forward difference operator* is

$$\Delta : \mathcal{C}_p \to \mathcal{C}_p \text{ given by } \Delta(f)(z) = f(z+1) - f(z).$$

Iteration of Δ then gives

$$(2.3.5.1) \qquad \Delta^n(f)(z) = \sum_{j=0}^{n} \binom{n}{j}(-1)^{n-j} f(z+j).$$

THEOREM 2.3.5.2. *If $f : \mathbb{Z}_p \to \mathbb{C}_p$ is continuous then $|\Delta^j(f)(0)|_p \to 0$ and*

$$f(z) = \sum_{j} \Delta^j(f)(0) \binom{z}{j}.$$

In particular, $f(z)$ has a Mahler series expansion.

PROOF. Let M denote the maximum of $|f(z)|_p$ on $\mathbb{Z}_p$. We claim that for each $d \geq 0$, there is some natural number $N = N(d)$ such that the maximum of $\Delta^k(f)(z)$ on $\mathbb{Z}_p$ is less than M/p^d whenever $k \geq N(d)$. We prove this by induction on d. The case when $d = 0$ is immediate. Suppose next that the claim holds whenever $d < m$. Then by assumption, there is some N such that whenever $k \geq N$, we have

$$|\Delta^k(f)(z)|_p < M/p^{m-1}$$

for all $z \in \mathbb{Z}_p$. Let $g(z) = \Delta^N(f)(z)$. Since $\mathbb{Z}_p$ is compact, $g(z)$ is uniformly continuous on $\mathbb{Z}_p$, and so there is some j such that

$$|g(z + p^j) - g(z)|_p < M/p^m$$

for every $z \in \mathbb{Z}_p$. We have

$$\Delta^{p^j + N}(f)(z) = \Delta^{p^j}(g)(z) = \sum_{i=0}^{p^j} \binom{p^j}{i}(-1)^{p^j - i} g(z+i),$$

and since $|\binom{p^j}{i}|_p \leq p^{-1}$ for $i = 1, \dots, p^j - 1$, we see that $\Delta^{p^j}(g)(z) - (g(z+p^j) - g(z))$ can be written as a sum of terms whose absolute value is less than M/p^m and so the maximum of $\Delta^{p^j}(g)(z)$ on $\mathbb{Z}_p$ is at most M/p^m. It follows immediately that

$$|\Delta^k(f)(z)|_p < M/p^m$$

for all $z \in \mathbb{Z}_p$ whenever $k \geq p^j + N$. The claim follows by induction. Therefore

$$h(z) := \sum_{j=0}^{\infty} \Delta^j(f)(0) \binom{z}{j}$$

is a Mahler series and hence is continuous on $\mathbb{Z}_p$. We claim that $h(z) = f(z)$. Since $h(z)$ and $f(z)$ are continuous and $\mathbb{N}_0$ is dense, it is sufficient to show that $h(n) = f(n)$ for every nonnegative integer n. By uniform convergence, we have

$$\Delta^i(h)(z) = \sum_{j=0}^{\infty} \Delta^{j+i}(0) \binom{z}{j}$$

for every nonnegative integer i. This then gives that

$$\Delta^i(h)(0) = \Delta^i(f)(0) \text{ for every } i \geq 0.$$

A straightforward induction argument now gives that $h(n) = f(n)$ for every nonnegative integer n and the result follows. $\qquad\square$

Since every continuous function on $\mathbb{Z}_p$ has a Mahler series expansion, it is clear that Mahler series are not in general analytic. The following result gives a criterion, which when met, ensures that a Mahler series is analytic on $\mathbb{Z}_p$.

THEOREM 2.3.5.3. *Let*

$$f(z) = \sum_{j=0}^{\infty} c_j \binom{z}{j}$$

be a Mahler series and suppose that

$$|c_n/n!|_p \to 0 \text{ as } n \to \infty.$$

Then $f(z)$ is the Mahler series expansion of a map that is analytic on $\mathbb{Z}_p$.

PROOF. Let $f_n(z) = \sum_{j=0}^{n} c_j \binom{z}{j}$. We may write

$$f_n(z) = \sum_{i=0}^{n} b_{i,n} z^i \text{ with } b_{i,n} \in \mathbb{C}_p.$$

Since $j!\binom{z}{j} \in \mathbb{Z}_p[z]$, we see that

$$|b_{i,n+1} - b_{i,n}|_p \leq |c_n/n!|_p \to 0 \text{ as } n \to \infty.$$

Thus for each i there is some $b_i \in \mathbb{C}_p$ such that $b_{i,n} \to b_i$ as $n \to \infty$; moreover,

$$|b_i - b_{i,n}| \leq \max_{j \geq n} |c_j/j!|_p \to 0 \text{ as } n \to \infty.$$

In particular, $|b_n|_p \to 0$ as $n \to \infty$, since $b_{n,n+1} = 0$. Let

$$h(z) = \sum_{i=0}^{\infty} b_i z^i.$$

Then $h(z)$ is analytic on $\mathbb{Z}_p$ and by construction we have that $f_n(z)$ converges uniformly to both $h(z)$ and $f(z)$ on $\mathbb{Z}_p$. It follows that $h(z) = f(z)$ and so $f(z)$ is analytic on $\mathbb{Z}_p$. $\qquad\square$

In order to use this result, one must have an understanding of the behavior of $|n!|_p$ for a prime p. A simple argument shows that the exponent of the highest power of p dividing $n!$ is exactly

$$\lfloor n/p \rfloor + \lfloor n/p^2 \rfloor + \lfloor n/p^3 \rfloor + \cdots.$$

Bounding this sum with a geometric series, we immediately obtain the bound

(2.3.5.4) $|n!|_p \geq p^{-n/(p-1)}.$

2.3.6. Arbitrary absolute values. Let K be a field of arbitrary characteristic. We call $|\cdot|_v$ an *absolute value* or a *norm* on K if

(1) $|ab|_v = |a|_v \cdot |b|_v$;
(2) $|a+b|_v \leq |a|_v + |b|_v$ (this is called the *triangle inequality* property);
(3) $|a|_v = 0$ if and only if $a = 0$.

We say that two absolute values $|\cdot|_1 := |\cdot|_{v_1}$ and $|\cdot|_2 := |\cdot|_{v_2}$ on K are *equivalent* if there exists a positive real number c such that

$$|x|_1 = |x|_2^c \text{ for all } x \in K.$$

As proven in [**Lan02**, Prop. 1.1, Ch. XII], two absolute values are equivalent if and only they define the same topology on K. An equivalence class of absolute values is called a *place*. Often when we refer to a place of a field K we fix an absolute value from the equivalence class of absolute values represented by the given place.

If we have the triangle inequality (which is hypothesis (2) in the definition of an absolute value) replaced by the stronger inequality, also called the *ultrametric inequality*:

$$|a+b|_v \leq \max\{|a|_v, |b|_v\},$$

then $|\cdot|_v$ is called a *non-archimedean* absolute value; otherwise, $|\cdot|_v$ is called an *archimedean* absolute value. Sometimes, for a non-archimedean absolute value $|\cdot|_v$ we consider the corresponding *valuation*

$$v : K \longrightarrow \mathbb{R} \cup \{\infty\}$$

given by $v(0) = \infty$, while for each nonzero $x \in K$ we have

$$v(x) = -\log |x|_v.$$

Let $K \subset L$ be two fields and let v be a place of K while w is a place of L. If the restriction of $|\cdot|_w$ on K is equal to $|\cdot|_v$, then we say that w is an *extension* and we denote this by $w|v$. We also say that the place w *lies above* v (or equivalently that v *lies below* w).

We denote by K_v the completion of K with respect to the norm $|\cdot|_v$. If $K = \mathbb{Q}$ and $|\cdot|_v$ is the usual absolute value, then the completion is $\mathbb{R}$, while if $|\cdot|_v = |\cdot|_p$ we obtain the p-adic numbers.

Let v be a non-archimedean absolute value, and let $\mathbb{C}_v$ be an algebraically closed, complete field with respect to the v-adic norm $|\cdot|_v$. The following simple result on zeros of analytic functions can be found in [**Gos96**, Proposition 2.1, p. 42] (see also [**GT08b**, Lemma 3.4] and [**Rob00**, Subsection 6.2.1]). We include a short proof for the sake of completeness.

LEMMA 2.3.6.1. *Let*

$$F(z) = \sum_{i=0}^{\infty} a_i z^i$$

be a power series with coefficients in $\mathbb{C}_v$ that is convergent in an open disk B of positive radius around the point $z = 0$. Suppose that F is not the zero function. Then the zeros of F in B are isolated.

PROOF. Let w be a zero of F in B. We may rewrite F in terms of $(z - w)$ as a power series

$$F(z) = \sum_{i=1}^{\infty} b_i (z-w)^i$$

that converges in a disk B_w of positive radius around w. Let m be the smallest index n such that $b_n \neq 0$. Since F is convergent in B_w, there exists a positive real number r such that for all $n > m$, we have

$$\left| \frac{b_n}{b_m} \right|_v < r^{n-m}.$$

Then, for any $u \in B_w$ such that

$$0 < |u - w|_v < \frac{1}{r},$$

we have

$$|b_m(u-w)^m|_v > |b_n(u-w)^n|_v \text{ for all } n > m.$$

Hence

$$|F(u)|_v = |b_m(u-w)^m|_v \neq 0.$$

Thus $F(u) \neq 0$, and so, F has no zeros other than w in a non-empty open disk around w. □

We will use the above lemma several times in the book, mainly in the case when $\mathbb{C}_v = \mathbb{C}_p$.

2.3.7. S-integers. Let K be a field, and let S be a set of places of K. We call $x \in K$ an S-*integer* if for each absolute value $|\cdot|_v$ which is not from S, we have $|x|_v \leq 1$. The classical example is that when K is a number field and S is a finite set containing all the archimedean places of K; then the celebrated result of Siegel [**Sie29**] says the following.

THEOREM 2.3.7.1 (Siegel). *Let K be a number field, and let S be a set containing all the archimedean places of K and also at most finitely many non-archimedean places of K. Let $C \subset \mathbb{A}^2$ be a plane curve containing infinitely many points with both coordinates S-integral. Then C contains a* Siegel curve, *i.e., an irreducible curve of genus 0 with at most two points at infinity.*

The above result is valid also when K is replaced by a finitely generated extension of $\mathbb{Q}$ (for more details, see Chapter 5). For a polynomial $F(x, y)$, we say that an irreducible polynomial $G(x, y)$ dividing $F(x, y)$ is a *Siegel factor* of $F(x, y)$ if the plane curve given by the equation $G(x, y) = 0$ is a Siegel curve.

2.4. Chebotarev Density Theorem

In this short section we state the *Chebotarev Density Theorem* which is used later in the book; for more details, see [**SL96**].

THEOREM 2.4.0.1. *Let L/K be a Galois extension of number fields, and let $G := \mathrm{Gal}(L/K)$. Let $C \subseteq G$ be closed under conjugation, and define*

$$\Pi_C(x, L/K) := \#\{\mathfrak{p} : N(\mathfrak{p}) \leq x, \mathfrak{p} \text{ is unramified in } L/K, \text{ and } \sigma_\mathfrak{p} \subseteq C\},$$

where $N(\mathfrak{p})$ is the $(K/\mathbb{Q})$-norm of the prime ideal $\mathfrak{p}$ of K, and $\sigma_\mathfrak{p}$ is the Frobenius conjugacy class corresponding to $\mathfrak{p}$ in $\mathrm{Gal}(L/K)$. Then

$$\lim_{x \to \infty} \frac{\Pi_C(x, L/K)}{\Pi_G(x, L/K)} = \frac{|C|}{|G|}.$$

2.5. The Skolem-Mahler-Lech Theorem

In this section we present an important result regarding linear recurrence sequences: given a linear recurrence sequence $\{a_n\}_{n\in\mathbb{N}_0}$, the set of $n \in \mathbb{N}$ such that $a_n = 0$ is a union of at most finitely many arithmetic progressions (when $a_n \in \mathbb{Z}$ for each n, the result was proven by Skolem [**Sko34**], and then extended to arbitrary algebraic numbers by Mahler [**Mah35**], and finally to complex numbers by Lech [**Lec53**]).

2.5.1. Background on linear recurrences. In this subsection, we recall some of the basic facts about linear recurrence sequences.

DEFINITION 2.5.1.1. Let K be a field. A *linear recurrence sequence* (defined over K) is a sequence $f : \mathbb{N}_0 \to K$ for which there exists $m \in \mathbb{N}$ and $b_1, \dots, b_m \in K$ such that

$$(2.5.1.2) \qquad f(n) = \sum_{j=1}^{m} b_j f(n-j)$$

for all $n \geq m$.

LEMMA 2.5.1.3. *Let K be a field and let F be an extension of K. Then the collection of F-valued sequences that satisfy a linear recurrence over K forms an F-algebra.*

PROOF. We only prove that the product of two linear recurrence sequences is again a linear recurrence sequence; the proof that the sum of two such sequences is again linearly recurrent is similar. Let

$$S : F^{\mathbb{N}_0} \to F^{\mathbb{N}_0}$$

denote the shift map; i.e., given an F-valued sequence $f(n)$, we define

$$S \cdot f(n) = f(n+1).$$

Then $\{f(n)\}_{n\in\mathbb{N}_0}$ satisfies a linear recurrence over K if and only if the K-vector space V_f spanned by $\{S^i f\}_{i\in\mathbb{N}_0}$ is finite-dimensional.

Suppose now that $\{f(n)\}$ and $\{g(n)\}$ are two F-valued sequences that satisfy linear recurrences over K. Let W denote the K vector space spanned by all sequences that can be expressed as the product of an element of V_f and an element of V_g. Then we have a surjective map

$$V_f \otimes_K V_g \to W$$

and hence W is finite-dimensional. Furthermore

$$\{f(n)g(n)\}_{n\in\mathbb{N}_0} \in W$$

and W is evidently closed under the operator S. It follows that $\{f(n)g(n)\}_{n\in\mathbb{N}_0}$ is a linear recurrence sequence. $\qquad\square$

PROPOSITION 2.5.1.4. *Let K be a field and let $f : \mathbb{N}_0 \to K$. The following are equivalent:*

  (i) *$\{f(n)\}$ satisfies a linear recurrence;*
  (ii) *there exist a matrix $A \in M_m(K)$ (for some $m \in \mathbb{N}$) and vectors $v, w \in K^m$ such that*

$$f(n) = w^t A^n v \text{ for } n \geq 0;$$

(iii) *there exist polynomials $P(x), Q(x) \in K[x]$ with $Q(0) = 1$ such that*

$$\sum_{n \geq 0} f(n) x^n$$

is the power series expansion of $P(x)/Q(x)$ about $x = 0$;

(iv) *there exist $\ell \in \mathbb{N}$, $d \in \mathbb{N}_0$ and $c_{i,j}, \alpha_j \in \overline{K}$ for $0 \leq i \leq d$ and $1 \leq j \leq \ell$ such that (for n sufficiently large) we have*

$$(2.5.1.5) \qquad\qquad f(n) = \sum_{i=0}^{d} \sum_{j=1}^{\ell} c_{i,j} n^i \alpha_j^n.$$

PROOF. We first show that (i) implies (ii). For each $i = 1, \ldots, m$, we denote by e_i the i-th vector in the canonical basis for the vector space K^m; i.e., the entries in e_i are all equal to 0 except the i-th entry which equals 1. Suppose that $f(n)$ satisfies the linear recurrence

$$f(n) = \sum_{i=1}^{m} b_i f(n-i)$$

for some $b_1, \ldots, b_m \in K$. Let $v = [f(0), \ldots, f(m-1)]^t$ and let A be the $m \times m$ matrix whose i-th row is e_{i+1}^t for $i \in \{1, \ldots, m-1\}$ and whose m-th row is $[b_m, \ldots, b_1]$. Then the linear recurrence gives $A^n v = [f(n), \ldots, f(n+m-1)]$. Taking $w = e_1^t$ gives that $f(n) = w^t A^n v$.

We now show that (ii) implies (iii). Let $Q(x) = \det(I - Ax) \in K[x]$. Then

$$\sum_{n \geq 0} f(n) x^n \;=\; \sum_{n \geq 0} (w^t A^n v) x^n \;=\; w^t (I - Ax)^{-1} v.$$

The classical adjoint formula then shows that $(I - Ax)^{-1} = R(x)/Q(x)$, where $R(x) \in M_d(K)[x]$. Thus the generating sequence for $f(n)$ is given by the power series expansion of $P(x)/Q(x)$, where $P(x) = w^t R(x) v$.

Next suppose that (iii) holds. Then there exist polynomials $P(x), Q(x) \in K[x]$ with $Q(0) = 1$ such that the generating series for $f(n)$ is equal to $P(x)/Q(x)$. Using the division algorithm, we find polynomials $R, P_1 \in K[x]$ such that

$$P(x) = Q(x) \cdot R(x) + P_1(x)$$

and $\deg(P_1) < \deg(Q)$ (if Q is constant, then $P_1 = 0$). Then for all sufficiently large n, the coefficient of x^n in $P(x)/Q(x)$ is the same as the coefficient of x^n in $P_1(x)/Q(x)$. Therefore, from now on we assume $R = 0$ and $P = P_1$ has degree less than Q.

We write $Q(x) = (1 - \alpha_1 x) \cdots (1 - \alpha_\ell x)$ with $\alpha_1, \ldots, \alpha_\ell \in \overline{K}$, not necessarily distinct. Since the degree of $P(x)$ is strictly less than the degree of $Q(x)$, we can use partial fractions to express $P(x)/Q(x)$ as

$$\sum_{i=0}^{d} \sum_{j=1}^{\ell} \beta_{i,j} / (1 - \alpha_j x)^i$$

for some nonnegative integer d and $\beta_{i,j} \in \overline{K}$ for $i \in \{1, \ldots, d\}$ and $j \in \{1, \ldots, \ell\}$. This then gives

$$f(n) \;=\; \sum_{i=0}^{d} \sum_{j=1}^{\ell} \beta_{i,j} \binom{n+i}{i} \alpha_j^n,$$

which yields (iv).

Finally, we show that (iv) implies (i). Given $\alpha \in \overline{K}$, we let

$$c_0 + c_1 x + \cdots + x^m \in K[x]$$

denote its minimal polynomial. Let $g(n) = \alpha^n$. Then

$$g(n) = -\sum_{i-1}^{m} c_i g(n-i)$$

for $n \geq m$ and so we see that α^n satisfies a linear recurrence over K. Also, $h(n) = n$ satisfies the recurrence $h(n) - 2h(n-1) + h(n-2) = 0$. By Lemma 2.5.1.3, the collection of $\overline{K}$-valued sequences satisfying a linear recurrence over K forms a $\overline{K}$-algebra, and so we see that (iv) implies (i). □

REMARK 2.5.1.6. If in the definition of a linear recurrence sequence (Definition 2.5.1.2), we have $b_m \neq 0$ then in Proposition 2.5.1.4 we have that

$$\det(A) = (-1)^{m+1} b_m \neq 0$$

and so, A is invertible and therefore $\deg(P) < \deg(Q)$ in part (iii) of the conclusion of Proposition 2.5.1.4 which in turn yields in part (iv) of Proposition 2.5.1.4 that (2.5.1.5) holds for all $n \geq 0$ (not only for n sufficiently large).

Part (iv) of Proposition 2.5.1.4 gives rise to some natural classes of linear recurrences. We say that a linear recurrence sequence $f(n)$ is *non-degenerate* if the expression for $f(n)$ given in part (iv) of the proposition satisfies that α_j/α_k is not a root of unity when $\alpha_j \neq \alpha_k$ and α_k is nonzero. We say that $f(n)$ is *simple* if $c_{i,j} = 0$ whenever $i > 0$; that is,

$$f(n) = \sum_{j=1}^{d} c_j \alpha_j^n$$

for all n sufficiently large. In this case, we define the *length* of $f(n)$ to be the smallest natural number d for which there exists an expression for $f(n)$ of the form

$$\sum_{j=1}^{d} c_j \alpha_j^n$$

for all sufficiently large n.

With the notation as in part (iv) of the conclusion of Proposition 2.5.1.4, in Section 2.5 (see Proposition 2.5.3.1 and Theorem 2.5.4.1), we show that that there exists a prime number p, an embedding

$$\mathbb{Q}\left((c_{i,j})_{i,j}, (\alpha_j)_j\right) \hookrightarrow \mathbb{Q}_p,$$

and there exists $k \in \mathbb{N}$ such that for each $b \in \{0, 1, \ldots, k-1\}$, the map

$$n \mapsto f(nk + b) = \sum_{i=1}^{d} \sum_{j=1}^{\ell} c_{i,j}(kn+b)^i \alpha_j^b (\alpha_j^k)^n$$

can be interpolated by a p-adic analytic function. This allows one to conclude that the set of $n \in \mathbb{N}_0$ such that $f(n) = 0$ is a union of finitely many arithmetic progressions.

2.5.2. More on linear recurrence sequences. Let $\{a_n\}_{n\in\mathbb{N}_0} \subseteq \mathbb{C}$ be a linear recurrence sequence; more precisely, there exist complex numbers $c_0, \ldots, c_{k-1}$ such that for each $n \in \mathbb{N}$ we have

$$a_{n+k} = c_{k-1}a_{n+k-1} + \cdots + c_1 a_{n+1} + c_0 a_n.$$

As proven in Proposition 2.5.1.4 (iv), there exist polynomials

$$f_1, \ldots, f_m \in \mathbb{C}[z]$$

such that

- $\deg(f_i)$ is less than the order of multiplicity of the root r_i; and
- for each $n \in \mathbb{N}$ we have

(2.5.2.1) $$a_n = f_1(n)r_1^n + \cdots + f_m(n)r_m^n.$$

The numbers r_i are called the *characteristic roots* and they are associated to the *characteristic equation* for this sequence:

$$x^k - c_{k-1}x^{k-1} - \cdots - c_1 x - c_0 = 0.$$

EXAMPLE 2.5.2.2. Let $\{a_n\}_{n\geq 1}$ be the sequence defined by

$$a_1 = -3;\ a_2 = 0;\ a_3 = -15 \text{ and}$$

$$a_{n+3} = 3a_{n+1} - 2a_n.$$

The characteristic equation is

$$x^3 - 3x + 2 = 0$$

whose roots are $r_1 = 1$ (twice) and $r_2 = -2$. Thus we search for a formula

$$a_n = (An + B)r_1^n + Cr_2^n,$$

with $A, B, C \in \mathbb{C}$. We find that $A = -3$, $B = 2$ and $C = 1$; so

$$a_n = -3n + 2 + (-2)^n.$$

It is natural to ask what is the set of $n \in \mathbb{N}$ such that $a_n = 0$. In Example 2.5.2.2, this leads to solving the equation

$$-3n + 2 + (-2)^n = 0$$

and it is easy to check that the only solution is $n = 2$. However, in general the above question might be more challenging if there are more roots r_i of the characteristic equation which have the same largest absolute value. Furthermore, if the linear recurrence sequence is degenerate, then there could be even infinitely many $n \in \mathbb{N}$ such that $a_n = 0$. So, in order to solve this question for any linear recurrence sequence we will turn this question into an analysis problem.

2.5.3. An embedding lemma. With the notation for a linear recurrence sequence $\{a_n\}$ as in (2.5.2.1), assume that each r_i is a positive real number; then

$$F(x) := \sum_{i=1}^{m} f_i(x)r_i^x$$

is a real analytic function. So, the question is when $F(x) = 0$ (especially, for which values x which are positive integers). Still this does not solve the problem, but it provides the motivation for the Skolem's method which will solve the problem. Rather than work over $\mathbb{R}$, we will instead work over $\mathbb{Q}_p$; as we shall see, this has the advantage that the integers embed inside $\mathbb{Z}_p$, which is a compact set and so analytic functions on $\mathbb{Z}_p$ that vanish at infinitely many integers will necessarily vanish at all integers. Our main goal is to prove that the general term of a linear recurrence sequence can be parametrized by finitely many p-adic analytic functions. First we state an embedding lemma due to Lech [**Lec53**].

PROPOSITION 2.5.3.1. *Let K be a finitely generated extension of $\mathbb{Q}$, and let $S \subseteq K$ be a finite set. Then there exist infinitely many prime numbers p and for each such prime number p there exists an embedding*

$$\sigma : K \longrightarrow \mathbb{Q}_p$$

such that $\sigma(S) \subseteq \mathbb{Z}_p$.

Before proving this result, we need a basic result which gives that a non-constant integer polynomial has roots modulo p for infinitely many prime numbers p.

LEMMA 2.5.3.2. *Let $f(x) \in \mathbb{Q}[x]$ be a non-constant polynomial. Then there are infinitely many primes p for which $f(x)$ has a root modulo p.*

PROOF. By clearing denominators, we see that without loss of generality we may assume that $f(x) \in \mathbb{Z}[x]$. Suppose that the set of primes p for which $f(x)$ has a root modulo p is finite and let $S = \{p_1, \ldots, p_k\}$ denote the set of such primes. Then for every integer n, we have that all prime factors of $f(n)$ are in S. Let $N = p_1 \cdots p_k$. Pick a natural number a such that $f(a)$ is nonzero and let j be a natural number with the property that

$$|f(a)|_{p_i} > p_i^{-j} \text{ for } i = 1, \ldots, k.$$

Then for each integer n we have

$$f(a + N^j n) \equiv f(a) \pmod{N^j}.$$

In particular, we have

$$|f(a + N^j n)|_{p_i} = |f(a)|_{p_i} \text{ for } i = 1, \ldots, k.$$

Since the only prime factors of $f(a + N^j n)$ are in S, we see that

$$f(a + N^j n) = \pm f(a)$$

for every integer n. In particular, one of $f(x) + f(a)$ and $f(x) - f(a)$ has infinitely many roots, which gives that $f(x)$ is constant, a contradiction. $\square$

PROOF OF PROPOSITION 2.5.3.1. Let $t_1, \ldots, t_d$ be a transcendence basis for K as an extension of $\mathbb{Q}$ and let $F = \mathbb{Q}(t_1, \ldots, t_d)$. By the theorem of the primitive element, there is an element $\theta \in K$ such that $K = F[\theta]$ with θ algebraic over F. Let $P(t_1, \ldots, t_d) \in \mathbb{Z}[t_1, \ldots, t_d]$ be a nonzero polynomial such that

(2.5.3.3) $P(t_1, \ldots, t_d)s \in \mathbb{Z}[t_1, \ldots, t_d, \theta]$ for every $s \in S$.

Let

$$f(x) = x^d + c_{d-1}x^{d-1} + \cdots + c_0 \in F[x]$$

denote the minimal polynomial of θ over F and let

$$\Delta = \Delta(t_1, \ldots, t_d) \in F$$

denote the discriminant of $f(x)$. Since $f(x)$ is irreducible, we see that it has distinct roots and hence Δ is nonzero. We also let

$$I := \{0 \leq i \leq d - 1 \colon c_i \neq 0\}.$$

Then

$$G(t_1, \ldots, t_d) := P \cdot \Delta \cdot \prod_{i \in I} c_i$$

is a nonzero rational function. Since G is nonzero, there exist integers $a_1, \ldots, a_d$ such that $G(a_1, \ldots, a_d)$ is a nonzero rational number. Furthermore, using Hilbert's Irreducibility Theorem (see [**Lan83**, Chapter 9]) we may ensure that

$$\bar{f}(x) := \sum_{i=0}^{d} c_i(a_1, \ldots, a_d)x^i \in \mathbb{Q}[x].$$

(where we take $c_d = 1$) is irreducible. Then by Lemma 2.5.3.2 we can choose a prime number p satisfying the following conditions:

   (i) $\bar{f}(x)$ has a root modulo p;
   (ii) p does not divide $P(a_1, \ldots, a_d)$;
   (iii) $|c_i(a_1, \ldots, a_d)|_p = 1$ for $i \in I$;
   (iv) $|\Delta(a_1, \ldots, a_d)|_p = 1$

Since $\mathbb{Z}_p$ is uncountable, we can find algebraically independent elements $\varepsilon_1, \ldots, \varepsilon_d \in \mathbb{Z}_p$. We note that the map σ which sends t_i to $a_i + p\varepsilon_i$ for $i = 1, \ldots, d$ extends to an embedding of F into $\mathbb{Z}_p$. Now, let

$$f_0(x) = \sum_{i=0}^{d} c_i(a_1 + p\varepsilon_1, \ldots, a_d + p\varepsilon_d)x^i \in \mathbb{Z}_p[x].$$

Then for $b \in \mathbb{Z}$, we have $|f_0(b) - \bar{f}(b)|_p < 1$ and thus by (i), we have that there is some $a \in \mathbb{Z}$ such that $|f_0(a)|_p < 1$. Condition (iv) gives that $f_0(x)$ has no repeated roots mod p and so $|f_0'(a)| = 1$. Hence by Hensel's lemma (see Lemma 2.3.2.1), there is $\mu \in \mathbb{Z}_p$ such that $f_0(\mu) = 0$. Moreover f_0 is irreducible over $\sigma(F)$ and hence we can extend σ to K to give an embedding of K into $\mathbb{Q}_p$ by sending θ to μ. By condition (iii),

$$|\sigma(P(t_1, \ldots, t_d))|_p = 1,$$

and so by Equation (2.5.3.3), we have $\sigma(S) \subseteq \mathbb{Z}_p$.   □

One can also prove Proposition 2.5.3.1 using the Chebotarev Density Theorem (see Theorem 2.4.0.1); for more details, see, for example, [**Bel06**].

2.5.4. Zeros in linear recurrence sequences. Using Proposition 2.5.3.1 we can prove the following result which is due in this form to Lech [**Lec53**] (see also [**Sko34**, **Mah35**] for earlier versions of the same result under stronger hypothesis).

THEOREM 2.5.4.1 (Skolem-Mahler-Lech). *Let $\{a_n\}_{n \in \mathbb{N}_0} \subseteq \mathbb{C}$ be a linear recurrence sequence. Then there exist*

$$r, m_1, \ldots, m_r, k_1, \ldots, k_r \in \mathbb{N}_0$$

such that

$$\{n \in \mathbb{N}_0 : a_n = 0\} = \bigcup_{i=1}^{r} \{m_i + \ell k_i : \ell \in \mathbb{N}_0\}.$$

REMARKS 2.5.4.2.
(i) In the above result, if $r = 0$, then that means there exist no integers n such that $a_n = 0$.

(ii) Note that we allow $k_i = 0$ in which case the above corresponding set consists of a single element; if $k_i > 0$, then the set is an infinite arithmetic progression.

(iii) Given a decomposition of the set of n for which $a_n = 0$ as above with r minimal, the elements of the form m_i with $k_i = 0$ are called the *exceptional zeros* of the linear recurrence.

PROOF OF THEOREM 2.5.4.1. Since $\{a_n\}$ is a linear recurrence sequence, then there exist polynomials $f_i \in \mathbb{C}[z]$ and complex numbers r_i (see Proposition 2.5.1.4) such that

$$a_n = \sum_{i=1}^{\ell} f_i(n) r_i^n.$$

Without loss of generality we may assume each $r_i \neq 0$. Choose a prime number p and an embedding of all coefficients of each f_i, and also of all r_i in $\mathbb{Z}_p$ (according to Proposition 2.5.3.1). Furthermore, we may assume each r_i is mapped to a p-adic unit (by asking that also each inverse $1/r_i$ is mapped into $\mathbb{Z}_p$). Looking at the natural surjection

$$\mathbb{Z}_p \to \mathbb{Z}/p\mathbb{Z},$$

we see that

$$|r_i^{p-1} - 1|_p \leq p^{-1} \text{ for } i = 1, \dots, \ell.$$

In particular, if we let $N = 2$ if $p = 2$, and $N = p - 1$ if p is odd, then for each $i = 1, \dots, \ell$ we have

(2.5.4.3) $$\mid r_i^N - 1 \mid_p < p^{-1/(p-1)}.$$

Therefore, the function

$$z \mapsto \left(r_i^N\right)^z$$

is a p-adic analytic function on $\mathbb{Z}_p$ (see Lemma 2.3.4.2). Then for each $k = 1, \dots, N$, the function

$$g_k(z) := \sum_{i=1}^{\ell} f_i(Nz + k) r_i^k \cdot \left(r_i^N\right)^z$$

is a p-adic analytic function on $\mathbb{Z}_p$. Furthermore, for each nonnegative integer m, we have $a_{Nm+k} = g_k(m)$. On the other hand, for each $k = 1, \dots, N$, we know by Lemma 2.3.6.1 that either g_k is identically equal to 0 in which case

$$a_{Nm+k} = 0 \text{ for all } m \geq 0,$$

or g_k is not identically equal to 0 in which case there exist at most finitely many $z \in \mathbb{Z}_p$ such that $g_k(z) = 0$ (and thus there exist at most finitely many integers $m \geq 0$ such that $a_{Nm+k} = 0$). □

REMARKS 2.5.4.4. We make two important observations.

(1) The above proof also yields an upper bound for the common difference of the (infinite) arithmetic progressions from the conclusion of Theorem 2.5.4.1. Indeed, given a prime number p for which the conclusion of Proposition 2.5.3.1 applies so we can embed each r_i and each coefficient of f_i into $\mathbb{Z}_p$, then the common difference N of the infinite arithmetic progressions such that $a_{\ell+Nk} = 0$ for all $k \in \mathbb{N}_0$ is bounded above by 2 if $p = 2$, and by $p - 1$ if p is odd. More precisely, N is a divisor of $p - 1$ if p is odd (see (2.5.4.3)).

(2) Furthermore there exist effective bounds (see [**Sch99, Sch00**]) for the number of exceptional solutions $n \in \mathbb{N}_0$ such that $a_n = 0$ (for a linear reccurence sequence $\{a_n\}$). Also, Schmidt [**Sch99, Sch00**] found effective bounds for the number of infinite arithmetic progressions appearing in the conclusion of Theorem 2.5.4.1.

2.6. Heights

In this section we recall some standard terminology about heights; for more details, we refer the reader to [**BG06, HS00, Lan83, Sil07**].

2.6.1. Product formula fields. A *product formula field* K comes equipped with a standard set M_K of absolute values $|\cdot|_v$ which satisfy a *product formula*

$$(2.6.1.1) \qquad \prod_{v \in M_K} |x|_v^{N_v} = 1 \quad \text{for every } x \in K^*,$$

where $N \colon M_K \to \mathbb{N}$ and $N_v := N(v)$ (see [**Lan83**] for more details).

The typical examples of product formula fields appearing in the book are

(1) *global fields* K which are either number fields or function fields of transcendence degree 1 over a finite field.
(2) function fields K of transcendence degree 1 over any field F (not necessarily a finite field F).

Actually, any finitely generated field K of characteristic 0 can be made into a product formula field; if it has positive transcendence degree over $\mathbb{Q}$, then it can be viewed as a function field of transcendence degree 1 over another field. Also, in certain applications, it is convenient to view a finitely generated field K of higher (finite) transcendence degree over $\mathbb{Q}$ as a function field of an irreducible, projective variety V defined over $\mathbb{Q}$ (at the expense of replacing K by a fintie extension we may even assume that V is smooth and geometrically irreducible). Then one can endow K with a set of places corresponding to irreducible divisors of V, which makes K a product formula field; for more details, see [**BG06**].

We also note that once K is either a number field or a function field of transcendence degree 1 over another field F, then any finite extension L of K is also a product formula field (see [**Lan83**]). Indeed, each absolute value $|\cdot|_v$ from M_K admits between 1 and $[L : K]$ extensions to absolute values on L; we let M_L be the set of all extensions on L of the absolute values from M_K. For each $v \in M_K$ and for each $w \in M_L$ lying above v, we let

$$N_w := N_v \cdot [L_w : K_v],$$

where K_v and L_w are the corresponding completions of K and L with respect to v and w. Then it is easy to check that L equipped with the absolute values $|\cdot|_w$ from M_L and the corresponding positive integers N_w is a product formula field:

$$(2.6.1.2) \qquad \prod_{w \in M_L} |x|_w^{N_w} = 1 \text{ if } x \in L^*.$$

Furthermore, for each $v \in M_K$ we have that

$$(2.6.1.3) \qquad \sum_{\substack{w \in M_L \\ w|v}} N_w = [L : K] \cdot N_v.$$

2.6.2. Absolute values for function fields. In the case K is the rational function field $F(t)$ for some field F, the places v in M_K are of two types:

(a) either $v := v_f$ corresponds to a irreducible monic polynomial $f \in F[t]$ in which case for any nonzero polynomials $P, Q \in F[t]$ we have

$$\left| \frac{P(t)}{Q(t)} \right|_v := e^{\mathrm{ord}_f(Q) - \mathrm{ord}_f(P)},$$

where $\mathrm{ord}_f(g)$ for a polynomial $g \in F[t]$ is the nonnegative exponent corresponding to the irreducible polynomial $f(t)$ in the factorization of $g(t)$ as a product of powers of distinct irreducible polynomials. In this case, we also let

$$N_v := \deg(f).$$

(b) or $v := v_\infty$ in which case for any nonzero polynomials $P, Q \in F[t]$ we have

$$\left| \frac{P(t)}{Q(t)} \right|_v = e^{\deg(f) - \deg(g)}.$$

In this case we let $N_v := 1$.

It is immediate to see that from our definition, we have that the product formula (2.6.1.1) holds. Also, we have that $\log |x|_v \in \mathbb{Q}$ for each nonzero $x \in F(t)$ and for each absolute value $|\cdot|_v$ in $M_{F(t)}$. It is immediate to see that for any finite extension K of $F(t)$ the property that

$$\log |x|_v \in \mathbb{Q}$$

remains valid for each nonzero $x \in K$ and for each absolute value $|\cdot|_v$ extending an absolute value of $F(t)$.

2.6.3. The Weil height. If K is a product formula field, the logarithmic Weil height of $x \in \overline{K}$ is defined as (see [**Lan83**, p. 52])

$$(2.6.3.1) \qquad h(x) := \frac{1}{[K(x) : K]} \cdot \sum_{v \in M_K} \sum_{\substack{w | v \\ w \in M_{K(x)}}} \log \max\{|x|_w^{N_w}, 1\}.$$

It is convenient to extend the above notion of height to the entire projective line over $\overline{K}$ by letting the height of the point at infinity be $h(\infty) = 0$.

The following result is the classical Northcott Theorem (see [**HS00**, Theorem B.2.3]).

THEOREM 2.6.3.2. *Let $d \in \mathbb{N}$ and let $B \in \mathbb{R}$. There exist at most finitely many algebraic numbers x such that $h(x) \leq B$ and $[\mathbb{Q}(x) : \mathbb{Q}] \leq d$.*

Our next result is a special case of [**Lan83**, Prop. 1.8, p. 81].

PROPOSITION 2.6.3.3. *Let K be a product formula field, and let $f \in K[z]$ be a polynomial. Then there exists $c_f > 0$ such that*

$$|h(f(x)) - \deg(f) \cdot h(x)| \leq c_f$$

for all $x \in \overline{K}$.

2.6.4. Canonical heights. We recall that the degree of a rational function $\varphi \in K(z)$ is

$$\deg(\varphi) := \max\{\deg(f), \deg(g)\},$$

where $\varphi(z) = f(z)/g(z)$ for some polynomials $f, g \in K[z]$ which are coprime.

DEFINITION 2.6.4.1. Let K be a product formula field, let $\phi \in K(z)$ with $\deg(\phi) > 1$, and let $z \in \overline{K}$. The *canonical height* $\widehat{\mathrm{h}}_\phi(z) := \widehat{\mathrm{h}}_\phi(z)$ of z with respect to ϕ is

$$\widehat{\mathrm{h}}_\phi(x) := \lim_{n \to \infty} \frac{h(\phi^n(x))}{\deg(\phi)^n}.$$

It is immediate to see that always the canonical height is nonnegative. Call and Silverman [**CS93**, Thm. 1.1] proved the existence of the above limit, using boundedness of

$$|h(\phi(x)) - (\deg\phi)h(x)|$$

and a telescoping sum argument due to Tate. We will use the following properties of the canonical height.

PROPOSITION 2.6.4.2. *Let K be a number field, let $\varphi \in K(z)$ be a rational function of degree greater than 1, and let $x \in \overline{K}$. Then*

(a) *for each $n \in \mathbb{N}$, we have $\widehat{\mathrm{h}}_\varphi(\varphi^n(x)) = \deg(\varphi)^n \cdot \widehat{\mathrm{h}}_\varphi(x)$;*
(b) *$|h(x) - \widehat{\mathrm{h}}_\varphi(x)|$ is bounded by a function which does not depend on x;*
(c) *if K is a number field then x is preperiodic if and only if $\widehat{\mathrm{h}}_\varphi(x) = 0$.*

PROOF. Part (a) is clear; for (b) see [**CS93**, Thm. 1.1]; and for (c) see [**CS93**, Cor. 1.1.1]. □

Part (c) of Proposition 2.6.4.2 is not true if K is a function field with constant field F which is not contained in the algebraic closure of a finite field, since $\widehat{\mathrm{h}}_\phi(x) = 0$ whenever $x \in F$ and $\varphi \in F(z)$. But these are essentially the only counterexamples in the function field case; for more details, see [**Ben05**] for the case when φ is a polynomial and [**Bak09**] for the general case of rational maps (we also note that Chatzidakis and Hrushovski [**CH08a**, **CH08b**] proved a generalization of [**Ben05, Bak09**] for endomorphisms of arbitrary varieties defined over a function field).

2.6.5. Local canonical heights for polynomials. Let K be a product formula field endowed with a set M_K of absolute values which satisfy a product formula, let $\phi \in K[z]$ be a polynomial of degree greater than 1, and let $v \in M_K$. We let

$$(2.6.5.1) \qquad \widehat{\mathrm{h}}_{\phi,v}(x) := \lim_{n \to \infty} \frac{\log\max\{|\phi^n(x)|_v^{N_v}, 1\}}{\deg(\phi)^n}$$

be the canonical local height of $x \in K$ at v. Clearly, for all but finitely many $v \in M_K$, both x and all coefficients of ϕ are v-adic integers. Hence, for such $v \in M_K$, we have $\widehat{\mathrm{h}}_{\phi,v}(x) = 0$. It turns out that

$$(2.6.5.2) \qquad \widehat{\mathrm{h}}_\phi(x) = \sum_{v \in M_K} \widehat{\mathrm{h}}_{\phi,v}(x).$$

For a proof of the existence of the limit in (2.6.5.1), and of the equality in (2.6.5.2), see [**CS93**]. We note that Call and Silverman [**CS93**] defined local canonical heights also for rational maps of degree greater than 1; however, it is only in the case of polynomials that all local canonical heights are nonnegative (for more details, see [**Sil07**]).

2.6.6. Canonical heights in function fields. We assume now that K is a function field of transcendence degree 1 over a field F, i.e., K is a finite extension of $F(t)$ (where t is transcendental over F). We endow K with the set M_K of absolute values extending the absolute values of $F(t)$ (see Subsection 2.6.2). The following result is a nice observation which we will use later.

LEMMA 2.6.6.1. *For each $x \in K$, and for each $\phi \in K[z]$ with $\deg(\phi) > 1$, we have $\widehat{\mathrm{h}}_\phi(x) \in \mathbb{Q}$.*

PROOF. Since (2.6.5.2) is a finite sum, it suffices to prove that $\widehat{\mathrm{h}}_{\phi,v}(x) \in \mathbb{Q}$ for each $v \in M_K$. For some fixed $v \in M_K$, write

$$\phi(z) = \sum_{i=0}^{d} \delta_i z^i$$

with $\delta_i \in K$ and $\delta_d \neq 0$, and let

$$M_v := \max\left\{1, |\delta_d|_v^{-\frac{1}{d-1}}, \max_{0 \leq i < d} \left|\frac{\delta_i}{\delta_d}\right|_v^{\frac{1}{d-i}}\right\}.$$

The following claim is essentially [**GT08a**, Lemma 4.4].

CLAIM 2.6.6.2. *If $|x|_v > M_v$, then $\widehat{\mathrm{h}}_{\phi,v}(x) = \log|x|_v + \frac{\log|\delta_d|_v}{d-1}$.*

PROOF OF CLAIM 2.6.6.2. From the definition of M_v and the fact that $|x|_v > M_v$, we obtain that

$$|\delta_d x^d|_v > \max_{0 \leq i < d} |\delta_i x^i|_v.$$

Thus $|\phi(x)|_v = |\delta_d x^d|_v$; but since $|x|_v > M_v \geq |\delta_d|_v^{-1/(d-1)}$, it follows that $|\phi(x)|_v > |x|_v > 1$. An easy induction now shows that

$$|\phi^n(x)|_v = |\delta_d|_v^{\frac{d^n-1}{d-1}} \cdot |x|_v^{d^n} > 1 \text{ for all } n \in \mathbb{N}.$$

Hence

$$\widehat{\mathrm{h}}_{\phi,v}(x) = \lim_{n \to \infty} \frac{\log|\phi^n(x)|_v}{d^n} = \log|x|_v + \frac{\log|\delta_d|_v}{d-1},$$

as desired. □

Finally, if $\widehat{\mathrm{h}}_{\phi,v}(x) > 0$ then there exists $n \in \mathbb{N}$ such that $|\phi^n(x)|_v > M_v$. Thus,

$$(2.6.6.3) \qquad \widehat{\mathrm{h}}_{\phi,v}(x) = \frac{\widehat{\mathrm{h}}_{\phi,v}(\phi^n(x))}{d^n} = \frac{\log|\phi^n(x)|_v + \frac{\log|\delta_d|_v}{d-1}}{d^n}$$

is a rational number (see Subsection 2.6.2), which concludes the proof of Lemma 2.6.6.1. □

The following result about canonical heights of non-preperiodic points for non-*isotrivial* polynomials will also be used later.

DEFINITION 2.6.6.4. We say a polynomial $\phi \in K[z]$ is *isotrivial* over F if there exists a linear $\ell \in \overline{K}[z]$ such that $\ell \circ \phi \circ \ell^{-1} \in \overline{F}[z]$.

Benedetto proved that a non-isotrivial polynomial has nonzero canonical height at its non-preperiodic points [**Ben05**, Thm. B] (as mentioned before, Baker [**Bak09**] extended Benedetto's result to rational maps, and later Chatzidakis and Hrushovski [**CH08a**, **CH08b**] proved a far-reaching generalization to endomorphisms of higher dimensional varieties).

LEMMA 2.6.6.5. *Let $\phi \in K[z]$ with $\deg(\phi) \geq 2$, and let $x \in \overline{K}$. If ϕ is non-isotrivial over F, then $\widehat{\mathrm{h}}_\phi(x) = 0$ if and only if x is preperiodic for ϕ.*

We state one more preliminary result, which is proved in [**GTZ08**, Lemma 6.8].

LEMMA 2.6.6.6. *Let $\phi \in K[z]$ be isotrivial over F, and let ℓ be as in Definition 2.6.6.4. If $x \in \overline{K}$ satisfies $\widehat{\mathrm{h}}_\phi(x) = 0$, then $\ell(x) \in \overline{F}$.*

PROOF. Let $f := \ell^{-1} \circ \phi \circ \ell$; then
$$f^n(\ell^{-1}(x)) = \ell^{-1}(\phi^n(x)),$$
and so, by Proposition 2.6.3.3, $\widehat{\mathrm{h}}_\phi(x) = 0$ yields $\widehat{\mathrm{h}}_f(\ell^{-1}(x)) = 0$. Since $f \in \overline{F}[z]$, this means
$$|\ell^{-1}(x)|_v \leq 1 \text{ for every } v \in M_{K'(x)};$$
so $\ell^{-1}(x) \in \overline{F}$.   □

DEFINITION 2.6.6.7. With the notation as in Lemma 2.6.6.6, we call the pair (ϕ, x) isotrivial. Furthermore, if $E \subseteq K$ is any subfield, and there exists a linear polynomial $\ell \in \overline{K}[z]$ such that $\ell \circ \phi \circ \ell^{-1} \in \overline{E}[z]$ and $\ell(x) \in \overline{E}$, then we call the pair (ϕ, x) isotrivial over E.

2.6.7. Weil height on projective spaces. Let K be product formula field with the corresponding set M_K of absolute values, and let n be a positive integer. Then for each finite extension L of K, we let M_L be the corresponding set of absolute values on L extending the absolute values from M_K; also, we let N_w be the corresponding positive integers for the absolute values $|\cdot|_w$ for $w \in M_L$ as in the product formula (2.6.1.2). We define the Weil height of any point
$$[a_0 : x_1 : \cdots : a_n] \in \mathbb{P}^n(L), \text{ as follows}$$
(2.6.7.1)
$$h([a_0 : a_1 : \cdots : a_N]) = \frac{1}{[L:K]} \cdot \sum_{w \in M_L} N_w \log \max\{|a_0|_w, |a_1|_w, \ldots, |a_N|_w\}.$$

Using the product formula (2.6.1.2) we see that the definition of the Weil height is independent of the particular choice of coordinates for the representation of the point in $\mathbb{P}^n(L)$. Also, using (2.6.1.3) we get that the formula (2.6.7.1) is independent of the choice of the field L containing the coordinates a_i. Finally, it is immediate to see that if $n = 1$, the Weil height of $[1 : x]$ is the same as the Weil height of $x \in \overline{K}$ defined in (2.6.3.1).

2.6.8. Canonical heights associated to polarizable endomorphisms. Let K be a product formula field endowed with a set M_K of absolute values, let X be a projective variety defined over K, and let $f : X \longrightarrow X$ be a polarizable endomorphism with respect to the very ample line bundle $\mathcal{L}$ on X (see Subsection 2.2.4). In particular,
$$f^*(\mathcal{L}) = \mathcal{L}^{\otimes d} \text{ for some integer } d > 1.$$

As discussed in Subsection 2.1.14, there exists a natural embedding $\iota_{\mathcal{L}} : X \longrightarrow \mathbb{P}^N$ for some $N \in \mathbb{N}$. For each $x \in \overline{K}$, we denote by

$$h_{\mathcal{L}}(x) := h(\iota_{\mathcal{L}}(x)),$$

where $h(\cdot)$ is the usual Weil height on $\mathbb{P}^N(\overline{K})$ as defined in Subsection 2.6.7. Then, following Call and Silverman [**CS93**], we define the canonical height of $x \in X(\overline{K})$ as follows:

$$\widehat{h}_f(x) := \lim_{n \to \infty} \frac{h_{\mathcal{L}}(f^n(x))}{d^n}.$$

2.6.9. The case of abelian varieties. Let A be an abelian variety defined over a global field K. Let $\mathcal{L}$ be a very ample line bundle on A with respect to which the multiplication-by-n-map $[n]$ on A is polarized; more precisely,

$$[n]^*(\mathcal{L}) \xrightarrow{\sim} \mathcal{L}^{\otimes n^2}$$

for each nonzero integer n. Then we can define the Néron-Tate canonical height on A as in Subsection 2.6.8:

$$\widehat{h}(x) = \lim_{n \to \infty} \frac{h_{\mathcal{L}}([n]x)}{n^2}.$$

CHAPTER 3

The Dynamical Mordell-Lang problem

We start in this chapter by proving various reductions of the Dynamical Mordell-Lang Conjecture (Conjecture 1.5.0.1). We continue by presenting the connections of the Dynamical Mordell-Lang Conjecture (Conjecture 1.5.0.1) with the classical Mordell-Lang Conjecture, and also with the Denis-Mordell-Lang Conjecture. Essentially, the Dynamical Mordell-Lang Conjecture is the outcome of combining (at the level of general principles) these last two conjectures in a dynamical setting. We discuss two possible extensions of Conjecture 1.5.0.1 (see Questions 3.2.0.1 and 3.6.0.1), and also discuss various applications of the Dynamical Mordell-Lang Conjecture and of its extensions.

3.1. The Dynamical Mordell-Lang Conjecture

3.1.1. The statement revisited. We recall the Dynamical Mordell-Lang Conjecture (stated first in our book as Conjecture 1.5.0.1). We note that throughout this chapter, similar as in the rest of our book, unless otherwise stated, a subvariety is always closed.

CONJECTURE 3.1.1.1 (Dynamical Mordell-Lang Conjecture). *Let K be a field of characteristic 0, let X be a quasiprojective variety defined over $\mathbb{C}$, let Φ be an endomorphism of X, let $\alpha \in X(K)$, and let $V \subseteq X$ be a subvariety. Then the set of $n \in \mathbb{N}_0$ for which $\Phi^n(\alpha) \in V(K)$ is a union of finitely many arithmetic progressions.*

We observe that we state Conjecture 3.1.1.1 for an arbitrary field of characteristic 0, while Conjecture 1.5.0.1 was stated for varieties defined over the complex numbers. On the other hand, we see in Proposition 3.1.2.1 that it suffices to prove Conjecture 3.1.1.1 when $K = \mathbb{C}$.

We will state next various reductions and equivalent formulations of Conjecture 1.5.0.1.

3.1.2. Reductions. First we show that Conjecture 3.1.1.1 reduces to Conjecture 1.5.0.1.

PROPOSITION 3.1.2.1. *It suffices to prove Conjecture 3.1.1.1 when $K = \mathbb{C}$.*

PROOF. There exists a finitely generated subfield $K_0 \subseteq K$ such that
- X, V and Φ are all defined over K_0; and
- $\alpha \in X(K_0)$.

Since $\mathbb{C}$ is an algebraically closed field of infinite transcendence degree over $\mathbb{Q}$, then there exists an embedding
$$\iota : K_0 \longrightarrow \mathbb{C}.$$
Hence, Conjecture 3.1.1.1 reduces to proving Conjecture 1.5.0.1. $\qquad\square$

Actually, the proof of Proposition 3.1.2.1 yields the following useful statement.

PROPOSITION 3.1.2.2. *It suffices to prove Conjecture* 3.1.1.1 *when K is a finitely generated subfield of $\mathbb{C}$.*

So, from now on, in all of the following reductions we work under the hypothesis and with the notation from Conjecture 1.5.0.1. Before proceeding to our reductions we introduce a simple but useful notation for our later arguments.

DEFINITION 3.1.2.3. Let $S \subseteq \mathbb{C}$ and $c \in \mathbb{C}$. We let

$$S + c = c + S := \{c + x \colon x \in S\}$$

and

$$S \cdot c = c \cdot S := \{cx \colon x \in S\}.$$

PROPOSITION 3.1.2.4. *Let ℓ be a positive integer. If Conjecture* 1.5.0.1 *holds for X, V, Φ and $\Phi^\ell(\alpha)$, then it also holds for X, V, Φ and α.*

PROOF. Indeed, we let $S_\alpha := \{n \in \mathbb{N}_0 \colon \Phi^n(\alpha) \in V(\mathbb{C})\}$ and similarly, we let $S_{\Phi^\ell(\alpha)} := \{n \in \mathbb{N}_0 \colon \Phi^{n+\ell}(\alpha) \in V(\mathbb{C})\}$. Then apart from a finite set, we have

$$S_\alpha = S_{\Phi^\ell(\alpha)} + \ell,$$

as claimed. □

PROPOSITION 3.1.2.5. *Let k be a positive integer. Conjecture* 1.5.0.1 *holds for the endomorphism Φ^k of X if and only if it holds for the endomorphism Φ of X.*

PROOF. Assume first that Conjecture 1.5.0.1 holds for the endomorphism Φ of the variety X. Then for any subvariety $V \subseteq X$ and for any $\alpha \in X(\mathbb{C})$, the set

$$(3.1.2.6) \qquad S(V, \Phi, \alpha) := \{n \in \mathbb{N}_0 \colon \Phi^n(\alpha) \in V(\mathbb{C})\}$$

is a finite union of arithmetic progressions. Then the set

$$S_k := S(V, \Phi, \alpha) \cap \{nk \colon n \in \mathbb{N}_0\}$$

is also a finite union of arithmetic progressions since the intersection of two arithmetic progressions is another arithmetic progression. On the other hand, letting

$$(3.1.2.7) \qquad S(V, \Phi^k, \alpha) := \{n \in \mathbb{N}_0 \colon \Phi^{nk}(\alpha) \in V(\mathbb{C})\},$$

we have that

$$S(V, \Phi^k, \alpha) = \frac{1}{k} \cdot S_k,$$

and thus $S(V, \Phi^k, \alpha)$ is also a finite union of arithmetic progressions, as claimed.

Assume now that Conjecture 1.5.0.1 holds for Φ^k, i.e., for any starting point $\gamma \in X(\mathbb{C})$ and for any subvariety $V \subseteq X$, the set $S(V, \Phi^k, \gamma)$ defined as in (3.1.2.7) is a finite union of arithmetic progressions. Our goal is to prove that for each $\alpha \in X(\mathbb{C})$, the set $S(V, \Phi, \alpha)$ defined as in (3.1.2.6) is also a finite union of arithmetic progressions. The result follows since

$$S(V, \Phi, \alpha) = \bigcup_{\ell=0}^{k-1} \left(S\left(V, \Phi^k, \Phi^\ell(\alpha)\right) \cdot k + \ell \right),$$

as claimed. □

The following simple, but useful observation will be often used even without mentioning this proposition.

PROPOSITION 3.1.2.8. *It suffices to prove Conjecture* 1.5.0.1 *for irreducible subvarieties.*

PROOF. Since it suffices to consider the intersection of an orbit with each of the finitely many irreducible components of a subvariety, the conclusion follows. □

PROPOSITION 3.1.2.9. *Conjecture* 1.5.0.1 *holds if* α *is preperiodic.*

PROOF. Using Propositions 3.1.2.4 and 3.1.2.5, we may replace α by an iterate of it under Φ, and we may also replace Φ by an iterate. Therefore, we may assume that α is fixed by Φ. Then for any subvariety $V \subseteq X$, either $\alpha \in V(\mathbb{C})$ in which case

$$S := \{n \in \mathbb{N}_0 \colon \Phi^n(\alpha) \in V(\mathbb{C})\}$$

is the entire set $\mathbb{N}_0$, or $\alpha \notin V(\mathbb{C})$ in which case S is empty. □

In particular, Proposition 3.1.2.9 has the following consequences.

COROLLARY 3.1.2.10. *Assume there exists* $n \in \mathbb{N}$ *such that* Φ^n *is the identity map on* X. *Then Conjecture* 1.5.0.1 *holds.*

PROOF. In this case, each point is periodic, and thus, Proposition 3.1.2.9 finishes the proof. □

COROLLARY 3.1.2.11. *Conjecture* 1.5.0.1 *holds when* $\dim(V) = 0$.

PROOF. By Proposition 3.1.2.8, it suffices to prove the result when V is simply a point, say $V = \{\beta\}$. If there exist finitely many $n \in \mathbb{N}$ such that

$$\Phi^n(\alpha) = \beta,$$

then we are done. Otherwise, α (and actually, also β) are preperiodic points for Φ, in which case the conclusion follows from Proposition 3.1.2.9. □

The following proposition is useful sometimes in proving the Dynamical Mordell-Lang Conjecture for an endomorphism which is *conjugated* to the original endomorphism.

DEFINITION 3.1.2.12. Let Φ and Ψ be endomorphisms of the same variety X. We say that Φ and Ψ are *conjugated* if there exists an automorphism μ of X such that

$$\Psi = \mu^{-1} \circ \Phi \circ \mu.$$

PROPOSITION 3.1.2.13. *Let* X *be a quasiprojective variety defined over* $\mathbb{C}$. *Conjecture* 1.5.0.1 *holds for the endomorphism* Φ *of* X *if and only if it holds for an endomorphism* Ψ *conjugated to* Φ.

PROOF. Let μ be an automorphism of X such that

$$\Psi = \mu^{-1} \circ \Phi \circ \mu.$$

Then for each $\alpha \in X(\mathbb{C})$, and each subvariety $V \subset X$, we have that

$$\Phi^n(\alpha) \in V(\mathbb{C})$$

if and only if

$$\Psi^n\left(\mu^{-1}(\alpha)\right) \in W(\mathbb{C}),$$

where $W := \mu^{-1}(V)$. □

The following easy result will allow us to reformulate the Dynamical Mordell-Lang Conjecture. Before stating our reduction of Conjecture 1.5.0.1 we recall our definition for periodic closed subvarieties: V is periodic under Φ, if

$$\Phi^k(V) \subset V \Leftrightarrow V \subseteq \Phi^{-k}(V),$$

for some $k \in \mathbb{N}$; for more details, we refer the reader to Subsection 2.2.2.

PROPOSITION 3.1.2.14. *Let X be a quasiprojective variety, let Φ be an endomorphism of X, let $\alpha \in X$, let $k \in \mathbb{N}$ and $\ell \in \mathbb{N}_0$, and let V be the Zariski closure of the set $\mathcal{O}_{\Phi^k}(\Phi^\ell(\alpha))$. Then V is Φ-periodic.*

PROOF. We know that $\Phi^{-k}(V) \supseteq \mathcal{O}_{\Phi^k}(\Phi^\ell(\alpha))$ and since $\Phi^{-k}(V)$ is Zariski closed, we have $\Phi^{-k}(V) \supseteq V$ which yields that V is Φ-periodic.   □

3.1.3. Equivalent statements. We will show that the following two conjectures are equivalent to Conjecture 1.5.0.1.

CONJECTURE 3.1.3.1. *Let X be a quasiprojective variety defined over $\mathbb{C}$, let Φ be any endomorphism of X, let $\alpha \in X(\mathbb{C})$, and let $V \subseteq X$ be an irreducible subvariety of positive dimension. If the intersection $\mathcal{O}_\Phi(\alpha) \cap V(\mathbb{C})$ is Zariski dense in V, then V is Φ-periodic.*

CONJECTURE 3.1.3.2. *Let X be a quasiprojective variety defined over $\mathbb{C}$, let Φ be any endomorphism of X, let $\alpha \in X(\mathbb{C})$ be a non-preperiodic point, and let $V \subseteq X$ be a subvariety. If V contains no periodic positive dimensional subvariety intersecting $\mathcal{O}_\Phi(\alpha)$, then $V(\mathbb{C}) \cap \mathcal{O}_\Phi(\alpha)$ is finite.*

CONJECTURE 1.5.0.1 IMPLIES CONJECTURE 3.1.3.1. Suppose that V is irreducible, of positive dimension, and that

$$V(\mathbb{C}) \cap \mathcal{O}_\Phi(\alpha) \text{ is Zariski dense in } V.$$

By Conjecture 1.5.0.1, there exist finitely many pairs $(k_i, \ell_i) \in \mathbb{N}_0 \times \mathbb{N}_0$ (for $1 \leq i \leq r$) such that

$$V(\mathbb{C}) \cap \mathcal{O}_\Phi(\alpha) = \bigcup_{i=1}^{r} \mathcal{O}_{\Phi^{k_i}}(\Phi^{\ell_i}(\alpha)).$$

Since $V(\mathbb{C}) \cap \mathcal{O}_\Phi(\alpha)$ is Zariski dense in V and V is irreducible, there exists a pair $(k, \ell) := (k_i, \ell_i)$ such that V is the Zariski closure of $\mathcal{O}_{\Phi^k}(\Phi^\ell(\alpha))$. Since $\dim(V) > 0$, we have $k \geq 1$. By Proposition 3.1.2.14 we conclude that V is Φ-periodic, or, more precisely, that $\Phi^k(V) \subseteq V$.   □

CONJECTURE 3.1.3.1 IMPLIES CONJECTURE 3.1.3.2. Suppose that V contains no positive dimensional periodic subvariety intersecting $\mathcal{O}_\Phi(\alpha)$. We must show that $V(\mathbb{C}) \cap \mathcal{O}_\Phi(\alpha)$ is finite. Assume not. Then the Zariski closure V_0 of $V(\mathbb{C}) \cap \mathcal{O}_\Phi(\alpha)$ is positive dimensional. Furthermore, any positive dimensional irreducible component V_1 of V_0 has Zariski dense intersection with $\mathcal{O}_\Phi(\alpha)$. By Conjecture 3.1.3.1 we obtain that V_1 is periodic, which contradicts the fact that V contains no positive dimensional periodic subvariety intersecting $\mathcal{O}_\Phi(\alpha)$.   □

CONJECTURE 3.1.3.2 IMPLIES CONJECTURE 1.5.0.1. By Proposition 3.1.2.9, we may assume α is not preperiodic.

Let $\mathcal{V}$ be the set of all subvarieties V of X with the property that there exists an endomorphism Φ of X, and there exists a point $x \in X(\mathbb{C})$ such that the set

$$T := \{n \in \mathbb{N}_0 \colon \Phi^n(\alpha) \in V(\mathbb{C})\}$$

is not a finite union of arithmetic progressions. Assume that $\mathcal{V}$ is non-empty. Then we let $V \in \mathcal{V}$ be a subvariety of minimal dimension.

Thus there exists an endomorphism Φ of X and there exists $\alpha \in X(\mathbb{C})$ such that
$$T = \{n \in \mathbb{N}_0 \colon \Phi^n(\alpha) \in V(\mathbb{C})\}$$
is not a finite union of arithmetic progressions. In particular, we may assume V is irreducible; otherwise it would be a finite union of proper subvarieties and one of these subvarieties must be in $\mathcal{V}$ (since otherwise V is not in $\mathcal{V}$). Furthermore, T must be an infinite set since it is not a finite union of arithmetic progressions (of nonnegative ratio). Then assuming Conjecture 3.1.3.2 we conclude that V contains a positive dimensional subvariety V_0 satisfying:

- $V(\mathbb{C}) \cap \mathcal{O}_{\Phi}(\alpha) \neq \emptyset$; and
- $V_0 \subseteq \Phi^{-k}(V)$, for some $k \in \mathbb{N}$.

There are two cases:

Case 1. $V \subseteq \Phi^{-k}(V)$.

Then for each $i \in \mathbb{N}_0$, if $\Phi^i(\alpha) \in V(\mathbb{C})$, we also have that $\Phi^{i+k}(\alpha) \in V(\mathbb{C})$. In particular, for each $\ell \in \{0, \ldots, k-1\}$, the intersection of T with the (infinite) arithmetic progression $\{\ell + nk\}_{n \in \mathbb{N}_0}$ is either empty or it contains all sufficiently large integers from the arithmetic progression. So T is a finite union of arithmetic progressions.

Case 2. $W := V \cap \Phi^{-k}(V)$ is a proper subvariety of V.

Since V is irreducible, then $\dim(W) < \dim(V)$. For each $\ell \in \{0, \ldots, k-1\}$, we let W_ℓ be the Zariski closure of $\Phi^\ell(W)$. In particular, $W_0 = W$. Furthermore,
$$\dim(W_j) < \dim(V) \text{ for each } j.$$
By our assumption, $V_0 \subseteq V$ and $V_0 \subseteq \Phi^{-k}(V_0)$ and so, $V_0 \subseteq W_0$. Also V_0 meets $\mathcal{O}_{\Phi}(\alpha)$ and so there exists $a \in \mathbb{N}_0$ such that $\Phi^a(\alpha) \in V_0(\mathbb{C})$. Then
$$\Phi^{a+nk}(\alpha) \in V_0(\mathbb{C}) \subseteq W_0(\mathbb{C})$$
for all $n \in \mathbb{N}_0$. Furthermore, $\Phi^{a+j+nk}(\alpha) \in W_j(\mathbb{C})$ for each $j < k$. We let
$$T_j := \{n \in \mathbb{N}_0 \colon \Phi^{a+j+nk}(\alpha) \in (W_j \cap V)(\mathbb{C})\}$$
for each $j \in \{0, \ldots, k-1\}$. Since
$$\dim(W_j \cap V) \leq \dim(W_j) < \dim(V)$$
and since V has minimal dimension among the subvarieties contained in $\mathcal{V}$, we have $W_j \cap V \notin \mathcal{V}$ and so applying the conclusion of Conjecture 1.5.0.1 to each subvariety $W_j \cap V$ with respect to the action of Φ^k and the starting point $\Phi^{j+a}(\alpha)$, we conclude that T_j is a finite union of arithmetic progressions. Now, using that
$$\Phi^{a+j+nk}(\alpha) \in W_j(\mathbb{C})$$
for all $n \in \mathbb{N}_0$, we conclude that apart from a finite set, T equals
$$\bigcup_{j=0}^{k-1} (k \cdot T_j + a + j).$$

Since each T_j is a finite union of arithmetic progressions, we also have that T is a finite union of arithmetic progressions, contradicting the choice of V and thus proving Conjecture 1.5.0.1. $\qquad\square$

3.2. The case of rational self-maps

One can formulate the same question as in Conjecture 1.5.0.1 for any rational self-map.

QUESTION 3.2.0.1. *Let X be a quasiprojective variety defined over $\mathbb{C}$, let $V \subseteq X$ be a subvariety, let $\alpha \in X(\mathbb{C})$, and let Φ be a rational self-map on X. If for each $n \in \mathbb{N}_0$, the point $\Phi^n(\alpha)$ is contained in the domain of definition for Φ, is it true then that the set $S_V := \{n \in \mathbb{N}_0 \colon \Phi^n(\alpha) \in V(\mathbb{C})\}$ is a finite union of arithmetic progressions?*

So far there are very few partial results towards an answer to Question 3.2.0.1. We mention here Xie's result [**Xie14**, Theorem 1.1] for birational self-maps on surfaces (for more details, see Section 10.3), and also Theorem 11.1.0.7 which yields that if the set S_V has positive Banach density, then V contains a positive dimensional periodic subvariety (for more details, see Section 11.4 and [**BGT15b**]). In Subsection 3.2.1 we present a surprising application of a positive answer to Question 3.2.0.1 (see also [**BGT15b**]). This material is not referenced again, but it serves to illustrate the applicability of the types of problems that we study in this monograph.

3.2.1. Applications to ODE.
In [**Rub83**, Problem 16], Rubel posed the following question regarding the coefficients of a power series which is a solution to an ordinary differential equation. Note that in Question 3.2.1.1, the n-th derivative of a function $y(z)$ is denoted by $y^{(n)}(z)$.

QUESTION 3.2.1.1 (Rubel). *Let $r \in \mathbb{N}$ and let*

$$(3.2.1.2) \qquad\qquad y(z) = \sum_{n=0}^{\infty} a_n z^n$$

be a solution to a linear differential equation:

$$(3.2.1.3) \qquad C_r(z)y^{(r)}(z) + C_{r-1}(z)y^{(r-1)}(z) + \cdots + C_0(z)y(z) + D(z) = 0,$$

where the coefficients $C_i(z)$ are polynomials, and also $D(z)$ is a polynomial. Is it true that the set

$$S_{\mathbf{a}} := \{n \in \mathbb{N}_0 \colon a_n = 0\}$$

is a union of finitely many arithmetic progressions?

There was extensive work done on this problem and the best result known in this direction is that the set $S_{\mathbf{a}}$ of zeros for the coefficients of $f(z)$ is a union of at most finitely many arithmetic progressions along with a set of Banach density equal to 0 (see [**BGT15b**] and also Chapter 11). Previously, Bézivin [**Bez89**] showed that when 0 and ∞ are not irregular singular points for the linear ordinary differential equation (ODE) satisfied by $f(z)$, then the set $S_{\mathbf{a}}$ is a union of finitely many arithmetic progressions along with a set of natural density equal to 0. Later, Methfessel [**Met00**] was able to eliminate the technical condition of Bézivin regarding the corresponding ODE (and also proved an appropriate extension of this result to positive characteristic). On the other hand, note that a set of normal density equal to 0 is also of Banach density 0, but the converse is not true as shown by Example 11.1.0.2.

It is easy to see that if $y(z)$ is a power series as in (3.2.1.2) which satisfies a linear ODE as in (3.2.1.3), then the sequence $\{a_n\}_{n \in \mathbb{N}_0}$ satisfies a recurrence relation of the following form: there exist polynomials $P_0, \ldots, P_r$ (depending on the polynomials C_i), such that

$$(3.2.1.4) \qquad P_r(n)a_{n+r} + P_{r-1}(n)a_{n+r-1} + \cdots + P_0(n)a_n = 0,$$

for all $n > \deg(D)$. So, Question 3.2.1.1 asks whether the Skolem-Mahler-Lech principle (see Section 2.5) extends also to recurrence sequences satisfying (3.2.1.4) when the polynomials P_i are not constant. We describe in this Subsection 3.2.1 that a positive answer to Question 3.2.0.1 provides a positive answer to Question 3.2.1.1 as well.

Let K be a field of characteristic 0, let $r \geq 1$ be an integer, let

$$P_0(z), \ldots, P_r(z) \in K[z]$$

and let

$$\{a_n\}_{n \geq 0} \subset K$$

be a sequence satisfying the recurrence relation (3.2.1.4) for all $n \geq 0$. For each $i = 1, \ldots, r$ we let

$$Q_i(z) := \frac{P_i(z)}{P_r(z)}.$$

Choose a positive integer N such that no integer $n \geq N$ is a pole for any of the Q_i's, i.e., $P_r(n) \neq 0$ for all $n \geq N$. Next we define an r-by-r matrix $A(z)$ whose entries are rational functions in $K(z)$:

$$A(z) := \begin{pmatrix} 0 & 1 & 0 & 0 & \cdots & 0 \\ 0 & 0 & 1 & 0 & \cdots & 0 \\ \cdots & \cdots & \cdots & \cdots & \cdots & \cdots \\ 0 & 0 & \cdots & \cdots & 0 & 1 \\ -Q_0(z) & -Q_1(z) & \cdots & \cdots & \cdots & -Q_{r-1}(z) \end{pmatrix},$$

We define the following rational self-map on $\mathbb{A}^{r+1}$ defined over K:

$$(3.2.1.5) \qquad \Phi(z, v) := (z + 1, A(z) \cdot v),$$

for any $(z, v) \in \mathbb{A}^1 \times \mathbb{A}^r$.

Assume Question 3.2.0.1 has a positive answer for the rational map Φ. We let

$$v_0 := (a_N, a_{N+1}, \ldots, a_{N+r-1})$$

and we note that for each $n \geq 0$ we have that

$$\Phi^n(0, v_0) = (n, a_{N+n}, a_{N+n+1}, \ldots, a_{N+n+r-1}).$$

So applying the conclusion in Question 3.2.0.1 to $X := \mathbb{A}^{r+1}$, rational self-map Φ, subvariety $Y := \mathbb{A}^r \times \{0\} \subset X$, and starting point $\alpha := (0, v_0) \in X(K)$ we obtain that the set

$$(3.2.1.6) \qquad S_{\mathbf{a}} := \{n \geq 0 \colon a_n = 0\}$$

is a union of finitely many arithmetic progressions. This answers in the affirmative the Question 3.2.1.1 raised by Rubel [**Rub83**, Problem 16].

We also mention that when both P_0 and P_r in (3.2.1.4) are nonzero constants, then the rational self-map Φ from (3.2.1.5) is actually an automorphism of $\mathbb{A}^{r+1}$. In that case, Question 3.2.1.1 is known to have a positive answer (see [**BBY12**, Theorem 1.3] and more generally, see [**BGT10**] which solves the Dynamical Mordell-Lang Conjecture for all étale endomorphisms).

3.3. Known cases of the Dynamical Mordell-Lang Conjecture

In the present Section we discuss briefly the various cases when the Dynamical Mordell-Lang Conjecture is known to hold. We also note that, so far, there are no known counterexamples to Conjecture 1.5.0.1.

Each of the partial results towards Conjecture 1.5.0.1 imposes some restriction either on the ambient variety X, or on the endomorphism Φ of X, or on the subvariety V, or on the starting point α. Now, when X is fixed, and α is allowed to be an arbitrary point of X, then the strength of the restriction on Φ is usually inversely proportional to the strength of the restriction on V, as can be seen later in this section from the examples we discuss.

One could attempt to solve Conjecture 1.5.0.1, or at least to register the progress towards proving it by considering the classification of algebraic varieties and see for which varieties X the Dynamical Mordell-Lang Conjecture is known to hold. We wish to stress that we do not take this approach either here in this section, or later in the book. It is true that for many varieties X, there are no endomorphisms of infinite order (i.e., for each endomorphism Φ of X we have that $\Phi^n = \mathrm{id}_X$ for some $n \in \mathbb{N}$), and therefore, by Corollary 3.1.2.10, the Dynamical Mordell-Lang Conjecture holds in this case. On the other hand, if Φ has infinite order and also its ramification locus is nonempty (see Subsection 3.3.3 for the unramified case), Conjecture 1.5.0.1 is generally difficult, even when X has small dimension; for example, Conjecture 1.5.0.1 is still open for endomorphisms of $\mathbb{A}^3$. So, even though the geometry of X sometimes limits the complexity of the dynamics of its endomorphisms, we found that the endomorphism itself was more relevant in the Dynamical Mordell-Lang Conjecture than the ambient variety X. Also, due to the apparent simplicity of the question posed by the Dynamical Mordell-Lang Conjecture, many people searched for proving it in special cases when X is either $\mathbb{A}^N$, or $\mathbb{P}^N$, or $(\mathbb{P}^1)^N$, and Φ satisfies additional properties. We record in this section most of the known results on this problem.

3.3.1. Generic endomorphisms of projective spaces.
First we note that the Dynamical Mordell-Lang Conjecture was shown by Fakhruddin [**Fak14**] to hold for generic endomorphisms of $\mathbb{P}^N$. A *generic endomorphism* of degree d of $\mathbb{P}^n$ is a map

$$[X_0 : \cdots : X_n] \mapsto [F_0(X_0, \ldots, X_n) : \cdots : F_n(X_0, \ldots, X_n)],$$

where the homogeneous polynomials F_i have degree d and all their coefficients are algebraically independent. Also, Fakhruddin proved that for a generic endomorphism of $\mathbb{P}^N$ there exist no periodic proper subvarieties; so, generically one expects that the intersection from Conjecture 1.5.0.1 is finite.

3.3.2. Arbitrary endomorphisms of affine and projective spaces.
So, in light of the results presented in Subsection 3.3.1, it is natural that the next step in attacking the Dynamical Mordell-Lang Conjecture would be to relax the hypothesis in Fakhruddin's theorem [**Fak14**] and see if the conclusion can still be established. For example, one could consider first arbitrary endomorphisms of $\mathbb{P}^N$ for small N. This turns out to be a very difficult question already when $N = 2$. For example, for the case Φ is an endomorphism of $\mathbb{A}^2$, Xie [**Xiea, Xieb**] was able to prove the Dynamical Mordell-Lang Conjecture (see also Section 10.3) at the end of a proof over 100 pages long! Furthermore, it is unclear whether Xie's method can be extended beyond endomorphisms of $\mathbb{A}^2$. Also, it is worth

mentioning that Xie [**Xie14**] proved first the Dynamical Mordell-Lang Conjecture for all polynomial birational automorphisms of $\mathbb{A}^2$, and already that proof was itself a tour de force. So, we believe that proving the Dynamical Mordell-Lang Conjecture for all endomorphisms of $\mathbb{P}^N$ (for any positive integer N) would be a major result!

3.3.3. Étale endomorphisms.

Since we saw in Subsection 3.3.2 that it seems very hard to obtain a proof of the Dynamical Mordell-Lang Conjecture for *all* endomorphisms of $\mathbb{A}^N$ or $\mathbb{P}^N$ (even when N is small), the next thing one can consider is to restrict Φ to certain classes of endomorphisms, and the first thing one can ask is the case of automorphisms. The Dynamical Mordell-Lang Conjecture holds for automorphisms of $\mathbb{P}^N$ (as proven by Denis [**Den94**]); the proof essentially reduces to the classical Skolem-Mahler-Lech theorem (see Theorem 2.5.4.1) because one can easily find a p-adic analytic parametrization of the orbit in this case. Similarly, Cutkosy and Srinivas [**CS93**] gave proved the Dynamical Mordell-Lang conjecture for translation maps on algebraic groups using the classical Skolem-Mahler-Lech theorem (see also [**Zan09**, Chapter 4]). For automorphisms of $\mathbb{A}^N$, it is not at all clear that a p-adic analytic parametrization of the orbit can be found; however, the surprising result of Bell [**Bel06**] is that such a parametrization does exist! Actually, the p-adic methods (as introduced by Bell in [**Bel06**]) were key for most of the work that has been done towards proving the Dynamical Mordell-Lang Conjecture. The method of finding a p-adic analytic parametrization of the orbit of a point is called now the *p-adic arc lemma* and it will be the main topic of Chapter 4, but it will also be an important tool in most of the remaining chapters of our book. We call it the *p-adic arc lemma* since each time it can be used it generates a p-adic analytic submanifold of $X(\mathbb{C}_p)$ (we call it *arc*) containing the orbit of α under Φ.

Using a generalization of the original idea from [**Bel06**], the authors [**BGT10**] proved the Dynamical Mordell-Lang Conjecture for all étale endomorphisms of any variety; in particular, Conjecture 1.5.0.1 holds for any automorphism of any variety. At the moment when we write this book, the result of [**BGT10**] is the most general result towards the Dynamical Mordell-Lang Conjecture valid for all varieties X (and all subvarieties V). Chapter 4 is devoted to the result from [**BGT10**].

3.3.4. Endomorphisms of semiabelian varieties.

Since each endomorphism of a semiabelian variety is unramified (but not necessarily étale because it may not be flat), it is natural to ask whether the Dynamical Mordell-Lang conjecture holds when X is a semiabelian variety. Conjecture 1.5.0.1 does indeed hold for any endomorphism Φ of a semiabelian variety X, as proven in [**GT09**]. Once again, one constructs a p-adic analytic parametrization for the orbit of any point α under Φ; however, the method is much more direct than the method from [**BGT10**] due to the existence of the local analytic uniformization map for the semiabelian variety X. In Chapter 9 we discuss the proof from [**GT09**] in more depth.

Working from a slightly different angle, but still dealing with endomorphisms Φ of $\mathbb{P}^N$ which are induced by endomorphisms of $\mathbb{G}_m^N$, Silverman and Viray [**SV13**] considered the number of linear subvarieties $L \subseteq \mathbb{P}^N$ such that the intersection $\mathcal{O}_\Phi(\alpha) \cap L$ is "larger than expected" (in a manner made precise in [**SV13**]). When Φ is the d-th power map (for $d \geq 2$) and the coordinates of α (seen as a point in $\mathbb{A}^N$) are multiplicatively independent, Silverman and Viray prove that there exists a finite subset S, whose cardinality is bounded in terms of N, such that any $N + 1$ points in $\mathcal{O}_\Phi(\alpha) \setminus S$ are in linear general position in $\mathbb{P}^N$.

Working in a somewhat related direction to [**SV13**], Ostafe and Sha [**OS**] obtained quantitative bounds for the set

$$S(V, \Phi, \alpha) = \{n \in \mathbb{N}_0 \colon \Phi^n(\alpha) \in V(\mathbb{C})\},$$

when $\Phi \colon \mathbb{A}^N \longrightarrow \mathbb{A}^N$ restricts to certain endomorphisms of $\mathbb{G}_m^N$, V is a hypersurface in $\mathbb{A}^N$, and the coordinates of α are multiplicatively independent. As previously stated (see [**GT09**] and also Chapter 9), the Dynamical Mordell-Lang Conjecture holds for any endomorphism of a semiabelian varieties (in particular, of $\mathbb{G}_m^N$); hence $S(V, \Phi, \alpha)$ is known to be a finite union of arithmetic progressions without additional technical assumptions on Φ, V, and α. However, imposing extra hypotheses on the endomorphism Φ, on the subvariety V, and also on the starting point α allowed Ostafe and Sha [**OS**] to obtain uniform quantitative bounds for $\#S(V, \Phi, \alpha)$.

The case of endomorphisms of semiabelian varieties in the Dynamical Mordell-Lang Conjecture is connected with one of the deepest questions in Diophantine geometry: the classical Mordell-Lang conjecture (see Subsection 3.4.1). Actually, the connection between these two conjectures will be the topic of Section 3.4.

3.3.5. Split endomorphisms. Besides the cases discussed in Subsections 3.3.1 to 3.3.4, Conjecture 1.5.0.1 was studied in the following special case: $X = (\mathbb{P}^1)^N$ and

(3.3.5.1) $\Phi := (f_1, \ldots, f_N),$

where each $f_i \colon \mathbb{P}^1 \longrightarrow \mathbb{P}^1$ is a rational map. Sometimes, the maps f_i restrict to polynomial mappings on $\mathbb{A}^1$; in this case, we call them simply polynomials. Endomorphisms of the form (3.3.5.1) are called *split*.

The results for maps Φ as in (3.3.5.1) are presented in full detail in Chapters 5 and 7. Once again, the results show a balance between the hypotheses one needs to impose on the maps f_i, and the hypotheses imposed on the subvariety V. Just to give a taste of these results, we note three cases when Conjecture 1.5.0.1 was proven:

   (a) each f_i is a polynomial map over $\mathbb{C}$, and V is a line (see Theorem 5.1.0.2);
   (b) $f_1 = \cdots = \cdots f_N$ is a quadratic polynomial map over $\mathbb{C}$, and V is any subvariety (see Corollary 7.0.0.1);
   (c) $f_1 = \cdots = f_N$ is a rational map defined over $\overline{\mathbb{Q}}$ whose critical points are not periodic, and V is a curve defined over $\overline{\mathbb{Q}}$ (see Theorem 7.1.0.1).
   (d) each f_i is a polynomial map over $\overline{\mathbb{Q}}$, and V is a curve.

So, (a) is the most general result in terms of allowing different maps f_i and have no restriction on the f_i's (other than they are polynomial mappings, which is quite essential for the proof; see Chapter 5). On the other hand, the subvariety in case (a) is the most restrictive, since the results of Chapter 5 only hold for lines. In (b) we have a restrictive hypothesis on the maps f_i, but on the other hand the result holds for any subvariety V. At the expense of relaxing a bit the hypothesis on the maps f_i (but still assuming they are all equal, and furthermore defined over $\overline{\mathbb{Q}}$ instead of $\mathbb{C}$), Conjecture 1.5.0.1 is proven for all curves defined over $\overline{\mathbb{Q}}$. Finally, using his deep analysis of valuation rings of a polynomial ring in two variables over a field, Xie [**Xieb**, Theorem 0.3] established result (d) above (see also Section 10.3).

We believe that an extension of the arc lemma construction (see Subsection 3.3.3 and Chapter 4) *should* lead to a proof of the Dynamical Mordell-Lang Conjecture for all endomorphisms of $(\mathbb{P}^1)^N$; in other words, each time when we deal with *split* rational maps in Conjecture 1.5.0.1, the arc lemma should be the key to the proof. Chapter 7 details the various progress made in this direction. Also, we note that for the case of endomorphisms of $\mathbb{P}^N$ (so, no longer *split endomorphisms*), there is heuristic evidence, which we discuss in Chapter 8, suggesting that the arc lemma might not necessarily work for proving the Dynamical Mordell-Lang Conjecture, at least when $N > 5$.

3.4. The Mordell-Lang conjecture

In this section we discuss the classical Mordell-Lang conjecture, at times in parallel with the Dynamical Mordell-Lang Conjecture. First we start by presenting a special case of Conjecture 1.5.0.1 which leads us naturally towards the Mordell-Lang Conjecture.

3.4.1. A special case of the Dynamical Mordell-Lang Conjecture. We consider next a very special case of Conjecture 1.5.0.1 in which case we recover the cyclic case in Laurent's theorem [**Lau84**] (formerly known as the Mordell-Lang Conjecture for an algebraic torus, or the *full* Manin-Mumford Conjecture; for more details, see Subsection 3.4.2).

THEOREM 3.4.1.1 (Laurent [**Lau84**]). *Let N be a positive integer, and let Γ be a finitely generated subgroup of $\mathbb{G}_m^N(\mathbb{C})$. Then for each subvariety $V \subseteq \mathbb{G}_m^N$, the intersection*

$$V(\mathbb{C}) \cap \Gamma$$

is a finite union of cosets of subgroups of Γ.

In particular, Theorem 3.4.1.1 yields that if an irreducible subvariety $V \subseteq \mathbb{G}_m^N$ contains a Zariski dense set of points from a finitely generated subgroup of $\mathbb{G}_m^N(\mathbb{C})$, then V must be a translate of an algebraic subtorus of $\mathbb{G}_m^N$. The proof of this consequence is immediate since the Zariski closure of a coset of a subgroup of $\mathbb{G}_m^N$ is itself a coset of an algebraic subgroup of $\mathbb{G}_m^N$.

Essentialy, Theorem 3.4.1.1 shows that any possible algebraic relation simultaneously satisfied by an infinite subset of a finitely generated subgroup of an algebraic torus has to be of the form

$$x_1^{m_1} \cdots x_N^{m_N} = c,$$

for some integers $m_1, \ldots, m_N$ and a constant $c \in \mathbb{C}$. This type of rigidity for subvarieties of $\mathbb{G}_m^N$ which contain a Zariski dense set of points from a finitely generated subgroup of $\mathbb{G}_m^N(\mathbb{C})$ was part of the motivation for conjecturing the Dynamical Mordell-Lang Conjecture. Indeed, we have a similarly rigid description for the subvarieties V of a given variety X endowed with an endomorphism Φ such that V contains a Zariski dense set of points from a single orbit of a point of X (see Conjecture 3.1.3.1 and its equivalence to Conjecture 1.5.0.1 proven in Subsection 3.1.3).

In order to see better the connection between Theorem 3.4.1.1 and the Dynamical Mordell-Lang Conjecture, assume now that Γ is a cyclic subgroup of $\mathbb{G}_m^N$ spanned by the point

$$(a_1, \ldots, a_N) \in \mathbb{G}_m^N(\mathbb{C}).$$

Then the conclusion of Theorem 3.4.1.1 is that for any given subvariety $V \subseteq \mathbb{G}_m^N$, the set
$$S = \{n \in \mathbb{Z} \colon (a_1^n, \ldots, a_N^n) \in V(\mathbb{C})\}$$
is a finite union of double-sided arithmetic progressions. We see next that this conclusion would follow also from a positive answer to a special case of the Dynamical Mordell-Lang Conjecture.

So, with the above notation for N, and $a_1, \ldots, a_N$, let $X = \mathbb{A}^N$, let Φ be the endomorphism of $\mathbb{A}^N$ given by
$$\Phi(x_1, \ldots, x_N) := (a_1 x_1, \ldots, a_N x_N),$$
and let
$$\gamma := (1, \ldots, 1) \in \mathbb{A}^N(\mathbb{C}).$$
Then for any subvariety $V \subseteq \mathbb{A}^N$, Conjecture 1.5.0.1 yields that the set
$$S_1 := \{n \in \mathbb{N}_0 \colon \Phi^n(\gamma) \in V(\mathbb{C})\}$$
is a finite union of arithmetic progressions. Now, noting that Φ is an automorphism of $\mathbb{A}^N$, we also obtain (using Conjecture 1.5.0.1 for the endomorphism Φ^{-1} of $\mathbb{A}^N$) that
$$S_2 := \{n \in \mathbb{N}_0 \colon \Phi^{-n}(\gamma) \in V(\mathbb{C})\}$$
is a finite union of arithmetic progressions.

Let $\{a + rn\}_{n \in \mathbb{N}_0}$ be an infinite arithmetic progression contained either in S_1 or in S_2. We show next that
$$\{a + rn\}_{n < 0} \subset S_1 \cup S_2.$$
This suffices to show that $S_1 \cup S_2$ is a union of finitely many double-sided arithmetic progressions, thus proving Theorem 3.4.1.1. Without loss of generality, we assume $r > 0$ and therefore we have that
$$\{a + rn\}_{n \in \mathbb{N}_0} \subset S_1$$
and we prove that

(3.4.1.2) $$\{a + rn\}_{n < 0} \subset S_2.$$

The Zariski closure W of $\mathcal{O}_{\Phi^r}(\Phi^a(\gamma))$ is contained in V. Furthermore, according to Conjecture 3.1.3.1 (which is equivalent to Conjecture 1.5.0.1 as we showed in Subsection 3.1.3), we obtain that
$$\Phi^r(W) \subseteq W.$$
However, Φ^r is an automorphism, and so,

(3.4.1.3) $$\Phi^r(W) = W.$$

In particular (3.4.1.3) yields (3.4.1.2), as desired.

Hence, a positive answer to a special case of Conjecture 1.5.0.1 yields that for a cyclic subgroup $\Gamma \subset \mathbb{G}_m^N(\mathbb{C})$, and for any subvariety $V \subseteq \mathbb{G}_m^N$, the intersection
$$V(\mathbb{C}) \cap \Gamma$$
is a finite union of cosets of subgroups of Γ. This is a special case of the Mordell-Lang Conjecture which we discuss in more detail in Subsection 3.4.2.

3.4.2. Statement of the conjecture. Mordell's Conjecture (solved by Faltings [**Fal83**]) predicts that on any curve of genus greater than 1 defined over a number field K, there exist at most finitely many K-rational points. Lang extended Mordell's original conjecture to a question of describing the intersection between subvarieties of semiabelian varieties X and finitely generated subgroups Γ of $X(\mathbb{C})$. This generalization (stated below in Theorem 3.4.2.1) was proven by Vojta [**Voj96**] who extended Faltings' proof for the case of abelian varieties [**Fal91**].

THEOREM 3.4.2.1 (Vojta [**Voj96**]; originally Mordell-Lang Conjecture). *Let X be a semiabelian variety defined over $\mathbb{C}$, let $V \subseteq X$ be a subvariety, and let $\Gamma \subseteq X(\mathbb{C})$ be a finitely generated subgroup. Then*

$$\Gamma \cap V(\mathbb{C})$$

is a finite union of cosets of subgroups of Γ.

3.4.3. Interpretation of the conjecture. In particular, a reformulation of the Mordell-Lang Conjecture (see also the reformulation of the Dynamical Mordell-Lang Conjecture done in Conjecture 3.1.3.1) is that if V is irreducible and it contains a Zariski dense set of points in common with a finitely generated subgroup

$$\Gamma \subseteq X(\mathbb{C}),$$

then V is a translate of an algebraic subgroup, since V would have to contain a Zariski dense coset of a subgroup of Γ (see [**Hin88**, Lemme 10]). This interpretation of the Mordell-Lang Conjecture can be stated in terms of *special points* and *special varieties*, which is a central concept in arithmetic geometry (also present in the Manin-Mumford, Bogomolov, André-Oort, and Pink-Zilber conjectures). This principle predicts that the only irreducible subvarieties containing a Zariski dense set of special points, are special subvarieties. In the Mordell-Lang conjecture, the special points are the points on the finitely generated subgroup Γ and the special (irreducible) subvarieties are translates of algebraic subgroups of X by points in Γ. In the Dynamical Mordell-Lang conjecture, the special points are the points of the orbit $\mathcal{O}_\Phi(\alpha)$ while the special (irreducible) subvarieties are periodic subvarieties which intersect $\mathcal{O}_\Phi(\alpha)$.

3.4.4. The case of other algebraic groups. It is essential that one restricts the Mordell-Lang conjecture to semiabelian varieties since the same statement fails for a power of the additive group scheme. Indeed, the plane curve C given by the equation

$$y^2 - 2x^2 = 1$$

contains infinitely many points in common with the rank-2 subgroup $\mathbb{Z} \times \mathbb{Z}$ of $\mathbb{G}_a^2$, even though C is not a translate of an algebraic subgroup of $\mathbb{G}_a^2$. Also, the intersection of C with $\mathbb{Z} \times \mathbb{Z}$ is not a finite union of cosets of subgroups of $\mathbb{Z} \times \mathbb{Z}$ (there are infinitely many solutions to the above Pell's equation, but they do not have an additive group structure).

On the other hand, the Mordell-Lang principle holds for cyclic subgroups of arbitrary commutative algebraic groups G. If $\Gamma \subset G(\mathbb{C})$ is a cyclic subgroup generated by a point γ, then (using again Proposition 2.5.3.1) we can find a suitable embedding of Γ into $\mathbb{Z}_p$ and thus we have a p-adic parametrization of the group Γ (essentially, this is the Skolem-Chabauty method or the arc lemma as we call it, see [**Cha41**, **Sko34**] and also Chapter 4). More precisely, there exists $N \in \mathbb{N}$ and there exist p-adic analytic maps

$$f_\ell : \mathbb{Z}_p \longrightarrow G(\mathbb{Z}_p) \text{ for } \ell = 0, \ldots, N-1$$

such that for each $k \in \mathbb{Z}$ we have

$$f_\ell(k) = (kN + \ell) \cdot \gamma.$$

So, for any subvariety V of G, and for any polynomial H in the vanishing ideal of V, we have that for each $k \in \mathbb{N}_0$,

$$(Nk + \ell) \cdot \gamma \in V \text{ if and only if } (H \circ f_\ell)(k) = 0.$$

An application of Lemma 2.3.6.1 finishes the argument. We discuss this method in more detail in Chapter 4.

However, if G is not semiabelian, and $\Gamma \subset G(\mathbb{C})$ is a subgroup of rank larger than 1, then the Mordell-Lang principle fails. Even if $G = \mathbb{G}_a \times \mathbb{G}_m$, one may take Γ to be the subgroup generated by $(1,1)$ and $(0,2)$, and V to be the diagonal line given by the equation $x = y$ in G (seen as a subvariety of $\mathbb{A}^2$ with the usual affine coordinates x and y). Then

$$V(\mathbb{C}) \cap \Gamma = \{(2^k, 2^k) : k \in \mathbb{N}_0\},$$

which is not a union of cosets of subgroups of Γ.

3.4.5. The case of positive characteristic. If one attempts a naïve translation of the Mordell-Lang Conjecture in characteristic p there are obvious counterexamples arising from varieties defined over finite fields, as shown in the following example.

EXAMPLE 3.4.5.1. Let p be a prime number. Consider the line V given by the equation

$$x + y = 1$$

defined over the field $\mathbb{F}_p(t)$. Then V contains infinitely many points in common with the cyclic subgroup of $\mathbb{G}_m^2$ spanned by $(t, 1-t)$ even though the line itself is not *special*, i.e., it is not a coset of a one-dimensional subtorus of $\mathbb{G}_m^2$. In particular, this example works also as a counterexample to a verbatim translation of the Dynamical Mordell-Lang Conjecture in positive characteristic. If we let

$$\Phi : \mathbb{A}^2 \longrightarrow \mathbb{A}^2$$

be the endomorphism given by

$$\Phi(x, y) = (tx, (1-t)y),$$

then the intersection of the diagonal line $V \subset \mathbb{A}^2$ with the orbit $\mathcal{O}_\Phi((1,1))$ consists of all points of the form $\Phi^{p^n}((1,1))$.

The problem lies in the fact that the line V from Example 3.4.5.1 is defined over a finite field, and thus once V contains a point, then it contains the entire orbit of that point under the corresponding Frobenius map. Essentially the above example shows that besides algebraic subgroups, varieties defined over $\overline{\mathbb{F}}_p$ are also *special* for the Mordell-Lang problem in characteristic p. Hrushovski [**Hru96**] proved a characteristic p Mordell-Lang theorem which gives a full description of the special varieties in this case.

THEOREM 3.4.5.2 (Hrushovski [**Hru96**]). *Let X be a semiabelian variety defined over a field K of characteristic p, let $\Gamma \subseteq X(K)$ be a finitely generated subgroup, and let $V \subseteq X$ be a subvariety defined over K. If*

$$V(K) \cap \Gamma$$

is Zariski dense in V, then there exists a semiabelian variety X_0 defined over $\overline{\mathbb{F}}_p$, a subvariety $V_0 \subseteq X_0$ defined over $\overline{\mathbb{F}}_p$, and there exists an algebraic group homomorphism $\Psi : X \longrightarrow X_0$ defined over $\overline{K}$ such that

$$V = \gamma + \Psi^{-1}(V_0),$$

for some point $\gamma \in X(\overline{K})$.

The above positive characteristic counterexamples to a direct translation of the Mordell-Lang Conjecture also prevent an immediate translation of the Dynamical Mordell-Lang Conjecture in characteristic p. In Chapter 13 we will formulate a revised Dynamical Mordell-Lang Conjecture in positive characteristic, which addresses the issues discussed here.

In the next section we discuss another motivation for the Dynamical Mordell-Lang Conjecture which is a Mordell-Lang type conjecture for the additive group scheme in characteristic p. So, there exists a natural formulation of the Mordell-Lang problem which works both in characteristic p and also for the additive group scheme, as long as the statement is formulated in the context of Drinfeld modules (which are themselves families of polynomial dynamical systems acting on the affine line).

3.5. Denis-Mordell-Lang conjecture

The Dynamical Mordell-Lang Conjecture is also motivated by the Denis-Mordell-Lang Conjecture for Drinfeld modules (see [**Den92a**] and also our Chapter 12). A *Drinfeld module* Φ is a family of endomorphisms of $\mathbb{G}_a$ defined over a field K of characteristic p. More precisely, Φ is a ring homomorphism

$$\Phi : \mathbb{F}_p[t] \longrightarrow \operatorname{End}_K(\mathbb{G}_a)$$

such that $\Phi_a := \Phi(a)$ is separable for each $a \in \mathbb{F}_p[t]$ (and moreover, Φ_t is not linear). We can extend the action of Φ diagonally to $\mathbb{G}_a^g$ for any $g \in \mathbb{N}$. A subset $\Gamma \subseteq \mathbb{G}_a^g$ is called a $\Phi(\mathbb{F}_p[t])$-*submodule* if it is invariant under the action of Φ. Inspired by the classical Mordell-Lang conjecture (see Theorem 3.4.2.1), Denis [**Den92a**] conjectured the following statement for Drinfeld modules.

CONJECTURE 3.5.0.1. *Let g be a positive integer, let K be a field extension of $\mathbb{F}_p(t)$, let*

$$\Phi : \mathbb{F}_p[t] \longrightarrow \operatorname{End}_K(\mathbb{G}_a)$$

be a Drinfeld module, let $\Gamma \subset \mathbb{G}_a^g(K)$ be a finitely generated $\Phi(\mathbb{F}_p[t])$-submodule, and let $V \subseteq \mathbb{G}_a^g$ be any subvariety. Then the intersection

$$V(K) \cap \Gamma$$

is a finite union of cosets of $\Phi(\mathbb{F}_p[t])$-submodules of Γ.

Only few cases of Conjecture 3.5.0.1 are known (see [**Ghi05, Ghi10, GT08b**]). In [**GT08b**], Ghioca and Tucker proved the conjecture in the case when Γ is a cyclic submodule and K is a finite extension of $\mathbb{F}_p(t)$. That result was obtained by deriving a v-adic analytic parametrization of Γ for some suitably chosen place v of K (which does not lie over the place at infinity from $\mathbb{F}_p(t)$). The method is somewhat similar with the arc lemma described in the Chapter 4 for p-adically parametrizing orbits of points under an étale map (note that $\Phi'_a(z)$ is always a nonzero constant if $a \neq 0$, and thus Φ_a is also an étale map if $a \in \mathbb{F}_p[t] \setminus \{0\}$). However, the parametrization from [**Den92a**] is much easier to obtain since a Drinfeld module comes equipped with a global analytic uniformization map $\exp_\Phi$ corresponding to the place v_∞ at infinity from $\mathbb{F}_p(t)$ (very similar to the global analytic uniformization map for abelian varieties over $\mathbb{C}$). One can prove that for a place v other than v_∞, the map $\exp_\Phi$ is a local analytic isomorphism and that is sufficient to obtain the desired parametrization of the cyclic $\Phi(\mathbb{F}_p[t])$-module Γ. Then one concludes the argument using Theorem 2.3.6.1 by noting that a v-adic analytic map either has finitely many zeros in a compact set, or it is identically equal to 0 (for more details, see Chapter 12).

It is interesting to note that for Drinfeld modules Φ one does not need to allow for special subvarieties other than algebraic subgroups invariant under Φ. Essentially, a Drinfeld module plays in characteristic p the role of an abelian variety from characteristic 0; many of the classical conjectures in arithmetic geometry are formulated for both abelian varieties defined over fields in characteristic 0, and for Drinfeld modules. We will return to a positive characteristic version of the Dynamical Mordell-Lang conjecture in Chapter 13.

3.6. A more general Dynamical Mordell-Lang problem

It is natural to ask a more general question than the Dynamical Mordell-Lang Conjecture which would also generalize the classical Mordell-Lang Conjecture. In particular, one would like to understand the case when a quasiprojective variety X is endowed with the action of finitely many commuting endomorphisms.

QUESTION 3.6.0.1 (The Dynamical Mordell-Lang problem). *Let X be a quasiprojective variety defined over $\mathbb{C}$, let $\Phi_1, \ldots, \Phi_r$ be commuting endomorphisms of X, let $\alpha \in X(\mathbb{C})$, and let $V \subseteq X$ be a subvariety. Is it true that the set of tuples*

$$(n_1, \ldots, n_r) \in \mathbb{N}_0^r$$

for which $\Phi_1^{n_1} \cdots \Phi_r^{n_r}(\alpha) \in V$ is a union of at most finitely many sets of the form

$$\gamma + (H \cap \mathbb{N}_0^r),$$

where $\gamma \in \mathbb{N}_0^r$ and $H \subseteq \mathbb{Z}^r$ is a subgroup?

If X is a semiabelian variety, α is the identity of X, and each Φ_i is a translation map, then a positive answer to Question 3.6.0.1 is equivalent to the classical Mordell-Lang conjecture. However, Question 3.6.0.1 is much more general, and often it has a negative answer. For example, if $X = \mathbb{A}^2$, $r = 2$, $\alpha = (1, 2)$, and

$$\Phi_1(x, y) = (x + 1, y) \text{ while } \Phi_2(x, y) = (x, y^2),$$

then the set of $(n_1, n_2) \in \mathbb{N}_0^2$ such that $\Phi_1^{n_1} \Phi_2^{n_2}(\alpha) \in \Delta$, where Δ is the diagonal line $y = x$ in $\mathbb{A}^2$ is the set

$$\{(2^m, m) : m \in \mathbb{N}_0\}.$$

Clearly this set does not satisfy the description from Question 3.6.0.1. One might think the problem with the previous example lies in the fact that $\deg(\Phi_1) = 1$ and so the points in any orbit of Φ_1 are not sparse enough. However one can construct negative examples to the conclusion from Question 3.6.0.1 even for algebraic group endomorphisms of semiabelian varieties such that the endomorphisms in question do not restrict to automorphisms of any positive dimensional subvariety. The following example is from [**GTZ11b**].

EXAMPLE 3.6.0.2. Let Φ and Ψ be the endomorphisms of $\mathbb{G}_m^3$ given by

$$\Phi(x, y, z) = \left(x^2 y^{-1},\, y^2 z^{-2},\, z^2\right)$$

and

$$\Psi(x, y, z) = \left(x^2 y^2,\, y^2 z^4,\, z^2\right).$$

Then

$$\Phi\Psi(x, y, z) = \Psi\Phi(x, y, z) = \left(x^4 y^2 z^{-4},\, y^4 z^4,\, z^4\right).$$

Moreover, for any $m, n \in \mathbb{N}_0$ we have that $\Phi^m \Psi^n(x, y, z)$ equals

$$\left(x^{2^{m+n}}\, y^{n2^{m+n} - m2^{m+n-1}}\, z^{n(n-1)2^{m+n} + m(m-1)2^{m+n-2} - mn2^{m+n}},\right.$$
$$\left. y^{2^{m+n}}\, z^{n2^{m+n+1} - m2^{m+n}},\, z^{2^{m+n}}\right).$$

Let V be the subvariety of $\mathbb{G}_m^3$ defined by $x = 1$, and let $\alpha := (1, 1/3, 9)$. Then, for $m, n \in \mathbb{N}_0$, the point $\Phi^m \Psi^n(\alpha)$ lies in $V(\mathbb{C})$ if and only if

$$\left(\frac{1}{3}\right)^{n2^{m+n} - m2^{m+n-1}} 9^{n(n-1)2^{m+n} + m(m-1)2^{m+n-2} - mn2^{m+n}} = 1,$$

or in other words

$$2^{m+n-1}(-2n + m + 4n(n - 1) + m(m - 1) - 4mn) = 0.$$

This last equation is equivalent to

$$(2n - m)^2 = 6n,$$

whose solutions in nonnegative integers are

$$n = 6k^2 \text{ and } m = 12k^2 \pm 6k$$

with $k \in \mathbb{N}_0$. But the set of solutions

$$\{(12k^2 \pm 6k,\, 6k^2) : k \in \mathbb{N}_0\}$$

is not a union of cosets of semigroups of $\mathbb{N}_0^2$ (for instance, because there are arbitrarily large gaps between consecutive values of the second coordinate).

One can prove that Φ and Ψ do not induce a degree-one map on any subvariety of $\mathbb{G}_m^3$ (see [**GTZ11b**, Lemma 6.1]) thus showing that Question 3.6.0.1 may fail in a non-trivial way.

On the other hand, Question 3.6.0.1 is answered in the affirmative in [**GTZ11b**] for several interesting cases.

THEOREM 3.6.0.3. *Let G be a semiabelian variety defined over $\mathbb{C}$, let $\alpha \in G(\mathbb{C})$, let $V \subseteq G$ be a closed subvariety, and let $\Phi_1, \ldots, \Phi_r$ be commuting algebraic group endomorphisms of G. Assume that either*

  (a) *The Jacobian at the origin of each endomorphism Φ_i is diagonalizable; or*
  (b) *V is a connected algebraic subgroup of G of dimension one; or*
  (c) *$G = A^j$ for $0 \leq j \leq 2$, where A is a one-dimensional semiabelian variety.*

Then the set E of tuples

$$(n_1, \ldots, n_r) \in \mathbb{N}^r \ \text{such that} \ \Phi_1^{n_1} \ldots \Phi_r^{n_r}(\alpha) \in V(\mathbb{C})$$

is the union of finitely many sets of the form

$$u + (\mathbb{N}^r \cap H)$$

with $u \in \mathbb{N}^r$ and H a subgroup of $\mathbb{Z}^r$.

The steps in the proof of Theorem 3.6.0.3 are as follows (for more details, see [**GTZ11b**]).

(1) Let S be the commutative semigroup generated by all the Φ_i, and denote by $\mathcal{O}_S(\alpha)$ the orbit of α under S. Then one can reduce to the case

$$V(\mathbb{C}) \cap \mathcal{O}_S(\alpha) \ \text{is Zariski dense in} \ V.$$

The point here is that if W is the Zariski closure of $V(\mathbb{C}) \cap \mathcal{O}_S(\alpha)$, then

$$V(\mathbb{C}) \cap \mathcal{O}_S(\alpha) = W(\mathbb{C}) \cap \mathcal{O}_S(\alpha).$$

(2) Furthermore, replacing V with one of its irreducible components, we may also assume V is irreducible.

(3) Since each Φ_i is integral over $\mathbb{Z}$ (seen as a subring of the endomorphism ring of G), one observes that $\mathcal{O}_S(\alpha)$ is contained in a finitely generated subgroup Γ of $G(\mathbb{C})$. Indeed, if each Φ_i satisfies a monic equation over $\mathbb{Z}$ of degree at most g, then $\mathcal{O}_S(\alpha)$ is contained in the subgroup Γ which is spanned by all the points

$$\Phi_1^{m_1} \cdots \Phi_r^{m_r}(\alpha) \ \text{for all} \ 0 \le m_i \le g - 1.$$

(4) Using Theorem 3.4.2.1, one knows that if the irreducible subvariety V contains a Zariski dense set of points from Γ then V is a translate of an algebraic subgroup Y of G by a point in Γ.

(5) We focus now on part (a) of Theorem 3.6.0.3; parts (b) and (c) are done similarly. We apply the p-adic logarithmic map $\log_p$ associated to the semiabelian variety G (for a suitable prime p; for more details, see also Chapter 9) and thus the theorem is equivalent to showing that if $A_1, \ldots, A_r$ (where each A_i is the Jacobian of Φ_i at the identity of G) are commuting matrices in $M_{n,n}(\mathbb{C})$ (where $n = \dim(G)$), if $\beta \in \mathbb{C}^n$ (where $\beta = \log_p(\alpha)$) and if L is a translate of a linear subvariety of $\mathbb{C}^n$, then the set of

$$(n_1, \ldots, n_r) \in \mathbb{N}_0^r \ \text{such that} \ A_1^{n_1} \cdots A_r^{n_r}(\beta) \in L$$

is a finite union of sets of the form

$$\gamma + (H \cap \mathbb{N}_0^r),$$

where $\gamma \in \mathbb{N}_0^r$ and $H \subseteq \mathbb{Z}^r$ is a subgroup. The existence of a suitable prime p and of the method of taking the p-adic logarithm will be discussed further in Chapter 9. The difficulty lies in showing that one can reduce to the case where α is in the domain of definition for the p-adic logarithmic map—this step requires several clever combinatorial arguments (for details see [**GTZ11b**, Lemma 3.2]).

(6) Finally, using mainly linear algebra, the previous restatement of the problem is proven.

It is worth pointing out that there are counterexamples to any possible weakening of conditions (a)—(c) from Theorem 3.6.0.3. Also, Scanlon and Yasufuku [**SY14**] showed that essentially any set of tuples $(n_1, \ldots, n_r) \in \mathbb{N}_0^r$ which is the solution set of finitely many polynomial-exponential equations

$$P_1\left(n_1, \ldots, n_r, c_1^{n_1}, \ldots, c_r^{n_r}\right) = \cdots = P_s\left(n_1, \ldots, n_r, c_1^{n_1}, \ldots, c_r^{n_r}\right) = 0$$

(for any given $c_i \in \overline{\mathbb{Q}}$, and any polynomials P_i with algebraic coefficients) may be obtained in the above intersection once one does not assume that the endomorphisms Φ_i have diagonalizable Jacobian. However, it would be interesting to prove a similar criterion for non-commuting endomorphisms of semiabelian varieties.

Besides the classical Mordell-Lang conjecture (proven by Faltings and Vojta) and the above Theorem 3.6.0.3, the only other case when Question 3.6.0.1 has been proven in the affirmative for a non-cyclic semigroup of endomorphisms of a quasiprojective variety X is the case when $X = \mathbb{A}^r$, V is a line, and each Φ_i is given by the action on the i-th coordinate of a one-variable polynomial f_i of degree larger than 1. We will explain in detail this case in Chapter 5.

CHAPTER 4

A geometric Skolem-Mahler-Lech Theorem

In this chapter we give a geometric interpretation of Theorem 2.5.4.1, which leads naturally to the formulation of the Dynamical Mordell-Lang Conjecture. Then we prove Conjecture 1.5.0.1 for all étale endomorphisms. The material for this chapter overlaps with the papers [**Bel06**] and [**BGT10**].

4.1. Geometric reformulation

The Skolem-Mahler-Lech theorem yields a geometric statement with direct consequences to solving a special case of Conjecture 1.5.0.1. The point is that for each linear polynomial $f(z)$, we can find a parametrization of the n-th iterate of any $c \in \mathbb{C}$ under f. Indeed, we let $f(z) = az + b$, and so

(1) if $a \neq 1$, then

$$(4.1.0.1) \qquad f^n(c) = a^n c + b \cdot \frac{a^n - 1}{a - 1};$$

(2) if $a = 1$, then

$$(4.1.0.2) \qquad f^n(c) = c + nb.$$

Hence if $f_1, \ldots, f_\ell \in \mathbb{C}[z]$ are linear polynomials, if $c_1, \ldots, c_\ell \in \mathbb{C}$, and $F \in \mathbb{C}[z_1, \ldots, z_\ell]$, then there exist $r_1, \ldots, r_m \in \mathbb{C}^*$ and there exist $g_1, \ldots, g_m \in \mathbb{C}[z]$ such that for each $n \in \mathbb{N}$ we have

$$(4.1.0.3) \qquad F\left(f_1^n(c_1), \ldots, f_\ell^n(c_\ell)\right) = \sum_{i=1}^{m} g_i(n) r_i^n.$$

In other words, using the equivalences from Proposition 2.5.1.4, we see that

$$F\left(f_1^n(c_1), \ldots, f_\ell^n(c_\ell)\right)$$

is the n-th term in a linear recurrence sequence. Hence the Skolem-Mahler-Lech theorem (see Theorem 2.5.4.1) yields that if $\Phi : \mathbb{A}^\ell \longrightarrow \mathbb{A}^\ell$ is the automorphism given by

$$\Phi(z_1, \ldots, z_\ell) = (f_1(z_1), \ldots, f_\ell(z_\ell))$$

then the set of integers n such that $\Phi^n(c_1, \ldots, c_\ell)$ lies on the hypersurface

$$F(z_1, \ldots, z_\ell) = 0$$

is a finite union of arithmetic progressions (with the understanding, as always, that an arithmetic progression may consist of a single number). The following more general statement follows identically since the intersection of two arithmetic progressions is another arithmetic progression (or the empty set).

THEOREM 4.1.0.4. *Let $f_1, \ldots, f_\ell \in \mathbb{C}[z]$ be linear polynomials, let $\Phi : \mathbb{A}^\ell \longrightarrow \mathbb{A}^\ell$ be the automorphism given by*

$$\Phi(z_1, \ldots, z_\ell) = (f_1(z_1), \ldots, f_\ell(z_\ell)),$$

let $(c_1, \ldots, c_\ell) \in \mathbb{C}$, and let $V \subseteq \mathbb{A}^\ell$ be a subvariety. Then the set of integers n such that $\Phi^n(c_1, \ldots, c_\ell) \in V(\mathbb{C})$ is a finite union of two-sided arithmetic progressions.

As with Remark 2.5.4.4, we note that the common differences of the (infinite) arithmetic progressions from the conclusion of Theorem 4.1.0.4 divide $p - 1$, where p is an odd prime number for which we can find an embedding into $\mathbb{Z}_p$ of each r_i and of each coefficient of g_i from (4.1.0.3). So, Theorem 4.1.0.4 is another instance of the *arc lemma* which is the common thread in the present Chapter, and as said before, it is the most important idea behind the attempts made so far on the Dynamical Mordell-Lang Conjecture. Furthermore, there exist effective bounds (see [**Sch99, Sch00**]) for the number of iterates (in case there are at most finitely many such iterates) which land on a subvariety V (as in Theorem 4.1.0.4).

4.2. Automorphisms of affine varieties

Denis [**Den94**] extended Theorem 4.1.0.4 to any automorphism of $\mathbb{P}^\ell$. The key is once again that the automorphisms are linear and therefore one finds easily the general form of the n-th iterate of a point α under Φ. Then, as shown in Proposition 2.5.3.1, one finds a suitable prime p and p-adic analytic parametrizations of the orbit. Note that in each case one splits $\mathbb{N}_0$ into finitely many arithmetic progressions of some suitable common difference $N \in \mathbb{N}$ and for each $j \in \{0, \ldots, N-1\}$ one finds p-adic analytic functions $F_0, \ldots, F_\ell$ such that for all $m \in \mathbb{N}_0$ we have

$$\Phi^{Nm+j}(\alpha) = [F_0(m) : \cdots : F_\ell(m)].$$

Hence, for any polynomial $H \in \mathbb{C}[z_0, \ldots, z_\ell]$ vanishing on the variety V, and for any $m \in \mathbb{N}_0$, we have that

$$\Phi^{Nm+j}(\alpha) \in V \Leftrightarrow H(F_0(m), \ldots, F_\ell(m)) = 0.$$

Thus the integer m is a zero of the p-adic analytic function

$$z \mapsto H(F_0(z), \ldots, F_\ell(z)).$$

Then Lemma 2.3.6.1 finishes our proof. Denis [**Den94**] also proved that assuming the set of $n \in \mathbb{N}_0$ such that $\Phi^n(\alpha) \in V(\mathbb{C})$ is *very dense* (in a precise sense defined in [**Den94**]), V must contain a positive dimensional periodic subvariety, and thus there exists an infinite arithmetic progression of integers n such that $\Phi^n(\alpha) \in V(\mathbb{C})$.

Later, Bell [**Bel06**] showed that the analogous result holds when Φ is an automorphism of any affine variety.

THEOREM 4.2.0.1 (Bell [**Bel06**]). *Let X be an affine variety defined over $\mathbb{C}$, let Φ be an automorphism of X, let $\alpha \in X(\mathbb{C})$, and let $V \subseteq X$ be a subvariety. Then the set*

$$\{n \in \mathbb{Z} : \Phi^n(\alpha) \in V\}$$

is a finite union of two-sided arithmetic progressions.

The novelty of the result from [**Bel06**] lies in the fact that there exists *no* general form for the n-th iterate of a point α under an automorphism of $\mathbb{A}^N$. In fact, the set of all automorphisms of $\mathbb{A}^N$ is very large and one cannot employ the exact same strategy as above, i.e., find the general form of the n-th iterate and then split the entire orbit of a generic point into finitely many pieces and then find suitable p-adic parametrizations of each piece. However, it is quite surprising that (at least in spirit) the same general strategy can be implemented: one is still able to split the orbit of a *single point* into finitely many pieces (each corresponding to some arithmetic progression for the order n of the iterates), and then find p-adic analytic parametrizations of each piece. But all this is done without having a general form of the n-th iterate, i.e., changing the starting point of the orbit likely changes the splitting of the orbit of that point and also the parametrization one needs to employ for that point. Also, just as in the case of automorphisms of the projective space, one essentially obtains a p-adic analytic continuation to $\mathbb{Z}_p$ of the n-th iterate of α under Φ, i.e., locally (restricting to some suitable arithmetic progressions $\{Nm + j\}_{m \in \mathbb{N}_0}$) there exists a well-defined p-adic analytic map

$$z \mapsto \Phi^{Nz+j}(\alpha) \text{ for } z \in \mathbb{Z}_p.$$

So, the $\mathbb{Z}_p$-manifold consisting of all these points

$$\Phi^{Nz+j} \text{ for } z \in \mathbb{Z}_p$$

is a p-adic analytic *arc* which gives the name of *arc lemma* to this p-adic method for parametrizing the orbit of a point.

The proof of Theorem 4.2.0.1 is a consequence of the following steps:

(1) using a result of Srinivas [**Sri91**], reduce to the case $X = \mathbb{A}^N$ (by showing that any automorphism of an affine variety extends to an automorphism of an affine space of suitably large dimension);

(2) choose a suitable prime number p and embed *everything* into $\mathbb{Z}_p$, i.e., find a prime p such that Φ, V and α can be viewed as being defined over $\mathbb{Z}_p$, and moreover the determinant of the Jacobian of Φ is a p-adic unit;

(3) show that there exists a positive integer k, and there exist p-adic analytic functions $f_{i,j}$ convergent on $\mathbb{Z}_p$ (for each $i = 0, \ldots, k - 1$ and for each $j = 1, \ldots, N$) such that

$$\Phi^{km+i}(\alpha) = (f_{i,1}(m), \ldots, f_{i,N}(m))$$

for all $m \in \mathbb{N}$; and

(4) use a compactness argument for zeros of p-adic analytic series such as our Lemma 2.3.6.1.

Clearly, Step (3) is the hardest, as the other steps are almost identical with the work previously done for automorphisms of the projective space; for example, Step (2) is essentially a consequence of the result we have proved in Proposition 2.5.3.1. The common feature for all automorphisms of affine varieties that in exploited in [**Bel06**] is that their Jacobians are constant (this is used both in Step (2), but more importantly in Step (3)). The next lemma is proven in [**Bel06**] and it constitutes the first move for proving Step (3).

LEMMA 4.2.0.2. *Let* $\Phi := (F_1, \ldots, F_N) : \mathbb{Z}_p^N \longrightarrow \mathbb{Z}_p^N$ *be a surjective polynomial map whose Jacobian has constant determinant which is a p-adic unit. Then there exists a positive integer k such that*

$$\Phi^k := (H_1, \ldots, H_N)$$

has the following two properties:

(i) $H_i(z_1, \ldots, z_N) \equiv z_i \pmod{p}$ *for all* $z_1, \ldots, z_N \in \mathbb{Z}_p$; *and*

(ii) *for each $\overline{z} := (z_1, \ldots, z_N) \in \mathbb{Z}_p^N$, the Jacobian of Φ^k at $\overline{z}$ is of the form $I_N + p \cdot M_{\overline{z}}$, for some matrix $M_{\overline{z}}$ with entries in $\mathbb{Z}_p$.*

PROOF. The first part of Lemma 4.2.0.2 is easy since Φ induces a bijective action on $\mathbb{F}_p^N$, denoted $\overline{\Phi}$. We let j be the order of $\overline{\Phi}$ and thus, we have

$$\Phi^{jn}(z) \equiv z \pmod{p},$$

for each $z \in \mathbb{Z}_p^N$ and for each $n \in \mathbb{N}$. Now we let m denote the order of the finite group $\mathrm{GL}_N(\mathbb{F}_p)$. We have then

$$J(\Phi^{jm}, \overline{z}) = J(\Phi^j, \overline{z}) \cdot J(\Phi^j, \Phi^j(\overline{z})) \cdots J(\Phi^j, \Phi^{j(m-1)}(\overline{z})).$$

By our choice of j, we have that

$$J(\Phi^j, \Phi^{ji}(\overline{z})) \equiv J(\Phi^j, \overline{z}) \pmod{p}$$

for all i. So,

$$J(\Phi^{jm}, \overline{z}) \equiv J(\Phi^j, \overline{z})^m \equiv I_n \pmod{p},$$

which yields that $k = j \cdot m$ works for Lemma 4.2.0.2.   □

Lemma 4.2.0.2 allows one to obtain the following result.

LEMMA 4.2.0.3. *Let $\Psi := (H_1, \ldots, H_N) : \mathbb{Z}_p^N \longrightarrow \mathbb{Z}_p^N$ be a polynomial map satisfying:*

(i) *$H_i(z_1, \ldots, z_N) \equiv z_i \pmod{p}$ for all $z_1, \ldots, z_N \in \mathbb{Z}_p$; and*
(ii) *for each $\overline{z} := (z_1, \ldots, z_N) \in \mathbb{Z}_p^N$, the Jacobian of Ψ at $\overline{z}$ is of the form $I_N + p \cdot M_{\overline{z}}$, for some matrix $M_{\overline{z}}$ with entries in $\mathbb{Z}_p$.*

Then for each given point $\overline{z_0} \in \mathbb{Z}_p^N$, there exist $g_1, \ldots, g_N \in \mathbb{Q}_p[[x]]$ which are analytic on $\mathbb{Z}_p$ such that

(1) *$(g_1(0), \ldots, g_N(0)) = \overline{z_0}$; and*
(2) *$g_i(z+1) = H_i(g_1(z), \ldots, g_N(z))$ for each $i = 1, \ldots, N$ and for each $z \in \mathbb{Z}_p$.*

We defer the proof of Lemma 4.2.0.3 to the next Section where we will prove a more general result valid for all unramified maps (not necessarily automorphisms). The above lemma shows that

$$\Psi^n(\overline{z_0}) = (g_1(n), \ldots, g_N(n)),$$

which is the desired *p*-adic analytic parametrization for our orbit.

4.3. Étale maps

The *p*-adic parametrization method from Section 4.2 was extended in [**BGT10**] to étale endomorphisms of any quasiprojective variety; for more details on étale endomorphisms, see Subsection 2.1.11 and also [**Har77**, Chapter III, Section 10].

THEOREM 4.3.0.1 ([**BGT10**]). *Let X be a quasiprojective variety defined over $\mathbb{C}$, let Φ be an étale endomorphism of X, let $\alpha \in X(\mathbb{C})$, and let $V \subseteq X$ be any subvariety. Then the set of $n \in \mathbb{N}$ such that $\Phi^n(\alpha) \in V(\mathbb{C})$ is a union of finitely many arithmetic progressions.*

The key to proving Theorem 4.3.0.1 is to find (again) a *p*-adic analytic parametrization of the orbit $\mathcal{O}_\Phi(\alpha)$. We present the proof of Theorem 4.3.0.1 in the Section 4.4. We note that Theorem 4.3.0.1 (though following along the same general principles outlined above) is stronger than Theorem 4.2.0.1. So, the progression in the above results, which are all treated by the same classical Skolem's method, is as follows:

    (a) in the Skolem-Mahler-Lech theorem regarding the distribution of zeros in a linear recurrence sequence, and in the case of automorphisms of the projective space, one has an explicit formula for the n-th element in the sequence (or the n-th iterate in the case of automorphisms of $\mathbb{P}^N$);

    (b) in the case of automorphisms of $\mathbb{A}^N$, one does not have an explicit formula for the n-th iterate of an automorphism of $\mathbb{A}^N$ under an automorphism Φ, but one knows that the Jacobian of Φ has constant (nonzero) determinant;

    (c) in Theorem 4.3.0.1, the Jacobian of the étale map Φ does not have constant determinant, but it has no ramification.

Hypothesis (c) is the weakest one can allow in order to still guarantee that Skolem's method can be applied. In Chapter 8 we discuss heuristic evidence that supports a random probabilistic model, which predicts that for endomorphisms Φ of $\mathbb{P}^N$ for $N > 5$, Skolem's method can not be used to find p-adic analytic parametrizations for the orbit of a point under Φ.

There are interesting consequences of Theorem 4.3.0.1 in the case when Φ is an automorphism. In [**KRS05**], the following question was raised in the context of understanding when so-called naïve blow-up algebras are noetherian.

QUESTION 4.3.0.2. *Let Φ be an automorphism of an irreducible (quasi)projective variety X defined over $\mathbb{C}$, and let $\alpha \in X(\mathbb{C})$ such that $\overline{\mathcal{O}}_\Phi(\alpha)$ is Zariski dense in X (where $\overline{\mathcal{O}}_\Phi(\alpha)$ is the set of all $\Phi^n(\alpha)$ where $n \in \mathbb{Z}$). Is it true that every infinite subset of $\mathcal{O}_\Phi(\alpha)$ is also Zariski dense in X?*

We note that Question 4.3.0.2 is actually *equivalent* to the Dynamical Mordell-Lang Conjecture for automorphisms of quasiprojective varieties; therefore, Theorem 4.3.0.1 yields a positive answer to Question 4.3.0.2.

PROPOSITION 4.3.0.3. *Question 4.3.0.2 is equivalent with Conjecture 1.5.0.1 when Φ is an automorphism.*

PROOF. **Assume first that Conjecture 1.5.0.1 holds when Φ is an automorphism, and we prove Question 4.3.0.2.**

So, under the hypothesis that $\overline{\mathcal{O}}_\Phi(\alpha)$ is Zariski dense in X, then we have to prove that any infinite subset of $\overline{\mathcal{O}}_\Phi(\alpha)$ is Zariski dense in X.

Assume there exists a proper, closed subvariety V of X containing an infinite subset S of $\overline{\mathcal{O}}_\Phi(\alpha)$; without loss of generality we may assume that V contains infinitely many points from the orbit $\mathcal{O}_\Phi(\alpha)$ (otherwise, we can replace Φ by Φ^{-1}). We know by the Dynamical Mordell-Lang Conjecture applied to the automorphism Φ that there exists an infinite arithmetic progression $\{\ell + kn\}_{n \in \mathbb{N}_0}$ (for some given $k, \ell \in \mathbb{N}$) such that $\Phi^{\ell+kn}(\alpha) \in V(\mathbb{C})$. In particular, V is fixed under the action of Φ^k (see Proposition 3.1.2.14) and thus

$$\Phi^{\ell+kn}(\alpha) \in V(\mathbb{C}) \text{ for all } n \in \mathbb{Z}.$$

In particular, we obtain that

$$\overline{\mathcal{O}}_\Phi(\alpha) \subseteq \bigcup_{i=0}^{k-1} \Phi^i(V).$$

This contradicts the fact that $\mathcal{O}_\Phi(\alpha)$ is Zariski dense in X, and that

$$\dim(V) < \dim(X),$$

since X is irreducible, and V is a proper subvariety.

Assume now that Question 4.3.0.2 holds, and we prove that Conjecture 1.5.0.1 holds for automorphisms.

So, $\Phi : X \longrightarrow X$ is an automorphism, and $V \subseteq X$ is a subvariety; clearly, we may assume that V is a proper subvariety.

If $\mathcal{O}_\Phi(\alpha)$ is Zariski dense in X, then we know from Question 4.3.0.2 that V contains only finitely many points from $\mathcal{O}_\Phi(\alpha)$ and thus, in this case, Conjecture 1.5.0.1 holds.

If $\mathcal{O}_\Phi(\alpha)$ is not Zariski dense, then we let $Y \subset X$ be the Zariski closure of $\mathcal{O}_\Phi(\alpha)$. The variety Y may be reducible, so we let Y_i for $i = 1, \dots, \ell$ be all the irreducible components of Y. Clearly, each Y_i is periodic under Φ, i.e., there exists $N \in \mathbb{N}$ such that for each $i = 1, \dots, \ell$ we have

$$\Phi^N(Y_i) = Y_i.$$

In particular, this means that for each $i = 1, \dots, \ell$, the set

$$S(Y_i, \Phi, \alpha) := \{m \in \mathbb{N}_0 : \Phi^m(\alpha) \in Y_i\}$$

is a union of finitely many arithmetic progressions of common difference N. In particular, this yields that the Zariski closure of $\mathcal{O}_{\Phi^N}(\alpha)$ is an irreducible subvariety Z of same dimension as each Y_i. Hence each $S(Y_i, \Phi, \alpha)$ is an arithmetic progression of common difference a divisor of N (possibly smaller than N). Indeed, for each

$$m, n \in S(Y_i, \Phi, \alpha) \text{ and } n > m,$$

we have that

$$\Phi^{n-m}(Y_i) = Y_j$$

for some $j \in \{1, \dots, \ell\}$ and moreover,

$$\dim(\Phi^{n-m}(Y_i) \cap Y_i) = \dim(Y_i),$$

because $\Phi^n(Z) \subseteq \Phi^{n-m}(Y_i) \cap Y_i$. In particular, we also get that

(4.3.0.4) $$S(Y_i, \Phi, \alpha) \cap S(Y_j, \Phi, \alpha) = \emptyset, \text{ if } i \neq j,$$

since otherwise we would get that $S(Y_i, \Phi, \alpha) \cap S(Y_j, \Phi, \alpha)$ contains an arithmetic progression of common difference N and thus

$$\dim(Y_i \cap Y_j) = \dim(Y_i) = \dim(Y_j),$$

contradiction. Now, we claim that

$$S(V, \Phi, \alpha) := \{n \in \mathbb{N}_0 : \Phi^n(\alpha) \in V\}$$

is a finite union of arithmetic progressions of common difference either 0 or N. Indeed, if $Y_i \subseteq V$, then

$$S(Y_i, \Phi, \alpha) \subseteq S(V, \Phi, \alpha),$$

while if Y_i is not contained in V, then we claim that

$$S(V, \Phi, \alpha) \cap S(Y_i, \Phi, \alpha) \text{ is finite.}$$

This follows immediately from the fact that the sets $S(Y_i, \Phi, \alpha)$ are disjoint (see (4.3.0.4)) and from Question 4.3.0.2 applied to the endomorphism Φ^N of Y_i and the proper subvariety

$$V \cap Y_i \subsetneq Y_i,$$

which contains only finitely many points in common with $\mathcal{O}_{\Phi^N}(\Phi^{m_i}(\alpha))$, where $m_i \in \mathbb{N}_0$ such that $\Phi^{m_i}(\alpha) \in Y_i$. $\square$

4.4. Proof of the Dynamical Mordell-Lang Conjecture for étale maps

Theorem 4.3.0.1 is a generalization of Theorem 4.2.0.1 both in its conclusion and also in the method for its proof. In order to prove Theorem 4.3.0.1, one uses a p-adic parametrization of the given orbit of Φ, and then the conclusion is obtained by inferring the discreteness of the set of all zeros of a non-trivial p-adic analytic function. The construction of the p-adic parametrization of the orbit is done using the geometric information about the map (essentially, that it has no ramification) and also using a local p-adic argument. The rest of this section follows closely the arguments from [**BGT10**], with the exception of Subsection 4.4.4 which presents a generalization of Theorem 4.3.0.1.

4.4.1. The geometric argument. First we note that it suffices to prove Theorem 4.3.0.1 when Φ is an unramified endomorphism of a smooth, irreducible quasiprojective variety X. Indeed, assuming the conclusion holds under the latter hypothesis, one can descend to this case for any étale endomorphism of any quasiprojective variety X. Since an étale map permutes the irreducible components of X, at the expense of replacing Φ by an iterate of it, we may assume Φ fixes each irreducible component, and thus we may assume that X is irreducible. Secondly, if α is in the smooth locus of X, then so is the entire orbit of α since Φ is étale and thus it induces an isomorphism between the tangent spaces of α and of each $\Phi^n(\alpha)$ (for $n \in \mathbb{N}$). Now, if α is not a smooth point for X, then $\mathcal{O}_\Phi(\alpha)$ is contained in the complement of the smooth locus of X, which is a lower dimensional subvariety Y. Hence a simple inductive argument on the dimension of the ambient space finishes the proof. So, it suffices to prove the following result.

THEOREM 4.4.1.1. *Let* $\Phi : X \longrightarrow X$ *be an unramified endomorphism of an irreducible smooth quasiprojective variety defined over* $\mathbb{C}$. *Then for any subvariety* V *of* X, *and for any point* $\alpha \in X(\mathbb{C})$ *the intersection* $V(\mathbb{C}) \cap \mathcal{O}_\Phi(\alpha)$ *is a union of at most finitely many orbits of the form* $\mathcal{O}_{\Phi^N}(\Phi^\ell(\alpha))$ *for some* $N, \ell \in \mathbb{N}_0$.

We note that since X is smooth, Φ is étale since the induced morphism on the tangent space at each space is an isomorphism as shown in [**Sha74**, Theorem 5, page 145]. Also, it is worth pointing out that all we use in our proof is the fact that Φ is unramified at the points of the orbit $\mathcal{O}_\Phi(\alpha)$ (see Theorem 4.4.4.1).

Before proceeding to the proof of Theorem 4.4.1.1, we note the following important corollary of it.

COROLLARY 4.4.1.2. *Conjecture 1.5.0.1 holds for any endomorphism of a semi-abelian variety* X *defined over* $\mathbb{C}$, *i.e., given a semiabelian variety* X *defined over* $\mathbb{C}$, *an endomorphism* Φ *of* X, *a subvariety* $V \subseteq X$, *and a point* $\alpha \in X(\mathbb{C})$, *the set of all* $n \in \mathbb{N}_0$ *such that*

$$\Phi^n(\alpha) \in V(\mathbb{C})$$

is a union of finitely many arithmetic progressions.

PROOF. The result follows from Theorem 4.4.1.1 since any semiabelian variety X is smooth, while any endomorphism Φ of X is unramified (see [**Iit76**, Theorem 2]). $\square$

We proceed to proving Theorem 4.4.1.1. First we make several geometric reductions, then we construct the p-adic analytic parametrization of the orbit and finally we finish our proof in Section 4.4.3.

Our first observation (which is the exact same observation used in proving the equivalence of Conjectures 1.5.0.1 and 3.1.1.1) is that there exists a finitely generated subfield $K \subset \mathbb{C}$ such that X, Φ and V are defined over K, and also $\alpha \in X(K)$. Furthermore, at the expense of replacing K by a finitely generated extension, we may assume there exists an embedding of X into $\mathbb{P}^M$ as an open subset of a projective subvariety.

The following result can be viewed as a geometric generalization of Skolem's embedding lemma (see Proposition 2.5.3.1) and it will be used several times throughout our book.

PROPOSITION 4.4.1.3. *Let M be a positive integer, let K be a finitely generated subfield of $\mathbb{C}$, let $X \subseteq \mathbb{P}^M$ be an open subset of a projective subvariety defined over K, let $V \subseteq X$ be a subvariety defined over K, let $\alpha \in X(K)$, and let*

$$\Phi : X \longrightarrow X$$

be an endomorphism defined over K. Then there exists a finitely generated $\mathbb{Z}$-algebra $R \subset K$ whose fraction field is K, there exists a $\mathrm{Spec}(R)$-scheme

$$\pi : \mathcal{X} \longrightarrow \mathrm{Spec}(R)$$

whose generic fiber is isomorphic to X, and there exists a dense open subset U of $\mathrm{Spec}(R)$ such that the following properties hold:

(1) *the scheme*

$$\pi|_U : \mathcal{X}_U \longrightarrow U$$

   is quasiprojective, where $\pi|_U$ and $\mathcal{X}_U$ are the corresponding restrictions of π and of $\mathcal{X}$ above the subset U of $\mathrm{Spec}(R)$.

(2) *if X is smooth and geometrically irreducible, then $\mathcal{X}|_U$ is smooth and geometrically irreducible.*

(3) *Φ extends to an endomorphism Φ_U of $\mathcal{X}_U$. Furthermore, if Φ is unramified, then Φ_U is unramified.*

(4) *α extends to a section $U \longrightarrow \mathcal{X}_U$.*

(5) *there exists a $\mathrm{Spec}(R)$-subscheme $\mathcal{V} \subseteq \mathcal{X}$ whose generic fiber is isomorphic to V.*

PROOF OF PROPOSITION 4.4.1.3. For any homogeneous ideal $\mathfrak{c} \in K[z_0, \ldots, z_M]$, we denote by $Z(\mathfrak{c})$ the Zariski closed subset of $\mathbb{P}^M$ on which the ideal $\mathfrak{c}$ vanishes. Then there exist homogeneous ideals $\mathfrak{a}, \mathfrak{b} \in K[z_0, \ldots, z_M]$ such that

$$X = Z(\mathfrak{a}) \setminus Z(\mathfrak{b}).$$

We choose generators $F_1, \ldots, F_m$ and $G_1, \ldots, G_n$ for $\mathfrak{a}$ and $\mathfrak{b}$, respectively. Let R be a finitely generated $\mathbb{Z}$-algebra containing the coefficients of the F_i, G_i, of the polynomials defining the variety V, and of the polynomials defining the morphism Φ and such that $\alpha \in \mathbb{P}^M(R)$. Let

$$\mathcal{X} \subseteq \mathbb{P}^M_{\mathrm{Spec}(R)}$$

be the model for X over $\mathrm{Spec}\, R$ defined by $Z(\mathfrak{a}') \setminus Z(\mathfrak{b}')$ where $\mathfrak{a}'$ and $\mathfrak{b}'$ are the homogeneous ideals in $R[z_0, \ldots, z_M]$ defined by $F_1, \ldots, F_m$ and $G_1, \ldots, G_n$, respectively. Similarly, let $\mathcal{V}$ be the model of V over $\mathrm{Spec}(R)$.

We cover X by a finite set $(Y_i)_{1 \leq i \leq \ell}$ of open subsets such that Φ restricted to each Y_i is represented by polynomials $P_{i,j}$ for $j \in \{0, \ldots, M\}$. Let $\mathcal{B}$ be the closed subset of $\mathcal{X}$ which is the zero set of the polynomials

$$P_{i,j} \text{ for } i \in \{1, \ldots, \ell\} \text{ and } j \in \{0, \ldots, M\}.$$

Since Φ is a well-defined morphism on the generic fiber X, we conclude that $\mathcal{B}$ does not intersect the generic fiber of

$$\mathcal{X} \longrightarrow \mathrm{Spec}(R).$$

Therefore $\mathcal{B}$ is contained in the pullback under

$$\mathbb{P}^M_{\mathrm{Spec}(R)} \longrightarrow \mathrm{Spec}(R)$$

of a proper closed subset E_1 of $\mathrm{Spec}(R)$. Similarly, let $\mathcal{C}$ be the closed subset defined by the intersection of $Z(\mathfrak{b}')$ with the Zariski closure of α in $\mathbb{P}^M_{\mathrm{Spec}(R)}$. Since $\alpha \in X$, we have that $\mathcal{C}$ is contained in the pullback under

$$\mathbb{P}^M_{\mathrm{Spec}(R)} \longrightarrow \mathrm{Spec}(R)$$

of a proper closed subset E_2 of $\mathrm{Spec}(R)$. Let

$$U' = \mathrm{Spec}\, R \setminus (E_1 \cup E_2),$$

let $\mathcal{X}'$ be the restriction of $\mathcal{X}$ above U', and let $\Phi_{U'}$ be the base extension of Φ to an endomorphism of $\mathcal{X}'$.

Assuming X is smooth, there is an open subset of $\mathcal{X}'$ on which the restriction of the projection map to

$$\mathcal{X}' \longrightarrow U'$$

is smooth, by [**AK70**, Remark VII.1.2, page 128]. Similarly, assuming Φ is unramified, there exists an open subset of $\mathcal{X}'$ on which $\Phi_{U'}$ is unramified by [**AK70**, Proposition VI.4.6, page 116] or [**GW10**, Appendix E]. Also, [**vdDS84**, Theorem (2.10)] shows that the condition of being geometrically irreducible is a first order property which is thus inherited by fibers above a dense open subset of $\mathrm{Spec}(R)$. Since each of these open sets contains the generic fiber, the complement of their intersection must be contained in the pullback under

$$\mathcal{X}' \longrightarrow U'$$

of a proper closed subset E_3 of $\mathrm{Spec}(R)$. Let

$$U := U' \setminus E_3;$$

then letting $\mathcal{X}_U$ be the restriction of $\mathcal{X}'$ above U yields the model and the endomorphism Φ_U (which is the restriction of $\Phi_{U'}$ above U) with the desired properties. $\square$

The following result is an easy consequence of Proposition 2.5.3.1 (see also [**Bel06**, Lemma 3.1] and [**Lec53**]).

PROPOSITION 4.4.1.4. *There exists a prime $p \geq 3$, an embedding of R into $\mathbb{Z}_p$, and a $\mathbb{Z}_p$-scheme $\mathcal{X}_{\mathbb{Z}_p}$ such that*

(1) *$\mathcal{X}_{\mathbb{Z}_p}$ is smooth and quasiprojective over $\mathbb{Z}_p$, and its generic fiber equals X;*
(2) *both the generic and the special fiber of $\mathcal{X}_{\mathbb{Z}_p}$ are geometrically irreducible;*
(3) *Φ extends to an unramified endomorphism $\Phi_{\mathbb{Z}_p}$ of $\mathcal{X}_{\mathbb{Z}_p}$; and*
(4) *α extends to a section $\mathrm{Spec}\,\mathbb{Z}_p \longrightarrow \mathcal{X}_{\mathbb{Z}_p}$.*

PROOF. Let U be the open subset of $\operatorname{Spec}(R)$ defined as in the conclusion of Proposition 4.4.1.3. So, let $x \in R$ be a nonzero element such that

$$\operatorname{Spec}\left(R\left[\frac{1}{x}\right]\right) \subseteq U.$$

Proposition 2.5.3.1 yields the existence of a prime number p (actually infinitely many such primes) and there exists an embedding

$$\iota : R \longrightarrow \mathbb{Z}_p$$

such that $\iota(x) \in \mathbb{Z}_p^*$. Then

$$\mathcal{X}_{\mathbb{Z}_p} := \mathcal{X}_U \times_U \operatorname{Spec}(\mathbb{Z}_p)$$

has the desired property. Also, Φ extends to an endomorphism $\Phi_{\mathbb{Z}_p}$ of $\mathcal{X}_{\mathbb{Z}_p}$ and α extends to a section $\operatorname{Spec}(\mathbb{Z}_p) \longrightarrow \mathcal{X}(\mathbb{Z}_p)$. □

Note that the hypothesis $p \geq 3$ will be used in Subsection 4.4.2 when we prove the construction of the p-adic analytic parametrization of the orbit. On the other hand, Proposition 2.5.3.1 yields that there are infinitely many primes p which satisfy the conditions given in Proposition 4.4.1.4.

For the sake of simplifying the notation, we let

$$\mathcal{X} := \mathcal{X}_{\mathbb{Z}_p}$$

and Φ denote the $\mathbb{Z}_p$-endomorphism $\Phi_{\mathbb{Z}_p}$ of $\mathcal{X}_{\mathbb{Z}_p}$ constructed in Proposition 4.4.1.4. Also, we use α to denote the section

$$\operatorname{Spec}(\mathbb{Z}_p) \longrightarrow \mathcal{X}(\mathbb{Z}_p)$$

induced in Proposition 4.4.1.4; i.e. $\alpha \in \mathcal{X}(\mathbb{Z}_p)$. Finally, we use $\mathcal{V} = \mathcal{V}_{\mathbb{Z}_p}$ to denote the $\mathbb{Z}_p$-scheme which is the Zariski closure in $\mathcal{X}$ of the subvariety V of X.

Since the special fiber $\overline{\mathcal{X}}$ of $\mathcal{X}$ has finitely many $\mathbb{F}_p$-points, some iterate of α under Φ is in a periodic residue class modulo p. At the expense of replacing α by a suitable iterate under Φ, we may assume that the residue class of α is Φ-periodic, say of period N (note that replacing α by one of its iterates under Φ will not change the conclusion of Theorem 4.4.1.1, as proven in Proposition 3.1.2.4). Also, at the expense of replacing Φ by Φ^N (which also does not change the conclusion of Theorem 4.4.1.1, as proven in Proposition 3.1.2.5) we may also assume that the residue class of α is fixed by Φ.

Let x be the reduction of α modulo p, i.e., the intersection of the corresponding section

$$\operatorname{Spec}(\mathbb{Z}_p) \longrightarrow \mathcal{X}$$

with the special fiber of $\mathcal{X}$. The next result gives the local ring structure for $\mathcal{X}$ at x.

PROPOSITION 4.4.1.5. *Let $\mathcal{O}_{\mathcal{X},x}$ be the local ring of x as a point on $\mathcal{X}$, let $\widehat{\mathcal{O}}_{\mathcal{X},x}$ be the completion of $\mathcal{O}_{\mathcal{X},x}$ at its maximal ideal $\mathfrak{m}$, and let $\widehat{\mathfrak{m}}$ be the maximal ideal in $\widehat{\mathcal{O}}_{\mathcal{X},x}$. Then there are elements $T_1, \dots, T_g$ of $\widehat{\mathcal{O}}_{\mathcal{X},x}$ such that*

$$\widehat{\mathcal{O}}_{\mathcal{X},x} = \mathbb{Z}_p[[T_1, \dots, T_g]].$$

PROOF. This result is proven in [**BGT10**, Proposition 2.1]. It is an application of the Cohen structure theorem (see [**Mat86**, Section 29] or [**Bou06**, Chapter IX]). □

There is a one-to-one correspondence between the points in $\mathcal{X}(\mathbb{Z}_p)$ that reduce to x and the primes $\mathfrak{p}$ in $\mathcal{O}_{\mathcal{X},x}$ such that

$$\mathcal{O}_{\mathcal{X},x}/\mathfrak{p} \cong \mathbb{Z}_p.$$

For each such prime $\mathfrak{p}$, its completion $\hat{\mathfrak{p}}$ in $\widehat{\mathcal{O}}_{\mathcal{X},x}$ has the property that

$$\mathfrak{p}\widehat{\mathcal{O}}_{\mathcal{X},x} = \hat{\mathfrak{p}};$$

for more details, see [**Mat86**, Theorem 8.7]. Furthermore, $\hat{\mathfrak{p}}$ is a prime ideal in $\widehat{\mathcal{O}}_{\mathcal{X},x}$ with residue domain $\mathbb{Z}_p$ since the sequence

$$0 \longrightarrow \hat{\mathfrak{p}} \longrightarrow \widehat{\mathcal{O}}_{\mathcal{X},x} \longrightarrow \mathbb{Z}_p \longrightarrow 0$$

is exact; this follows from the fact that $\widehat{\mathcal{O}}_{\mathcal{X},x}$ is flat over $\mathcal{O}_{\mathcal{X},x}$ ([**Mat86**, Theorem 8.8]) along with the fact that the quotient

$$\mathcal{O}_{\mathcal{X},x}/\mathfrak{p} \cong \mathbb{Z}_p$$

is complete with respect to the $\mathfrak{m}$-adic topology. Thus, if $\mathfrak{q}$ is any prime in $\widehat{\mathcal{O}}_{\mathcal{X},x}$ with residue domain $\mathbb{Z}_p$ then $\mathfrak{q}$ must be the completion of $\mathfrak{q} \cap \mathcal{O}_{\mathcal{X},x}$, because

$$\dim \mathcal{O}_{\mathcal{X},x} = \dim \widehat{\mathcal{O}}_{\mathcal{X},x},$$

by [**AM69**, Corollary 11.19]. Hence, we have a one-to-one correspondence between the points in $\mathcal{X}(\mathbb{Z}_p)$ that reduce to x and the primes $\mathfrak{q}$ in $\widehat{\mathcal{O}}_{\mathcal{X},x}$ such that

$$\widehat{\mathcal{O}}_{\mathcal{X},x}/\mathfrak{q} \cong \mathbb{Z}_p.$$

Note that the primes $\mathfrak{q}$ in $\widehat{\mathcal{O}}_{\mathcal{X},x}$ for which $\widehat{\mathcal{O}}_{\mathcal{X},x}/\mathfrak{q} \cong \mathbb{Z}_p$ are simply the ideals of the form

$$(T_1 - pz_1, \ldots, T_g - pz_g)$$

where the z_i are in $\mathbb{Z}_p$. For each $\mathbb{Z}_p$-point β in $\mathcal{X}$ such that $r(\beta) = x$, we write

$$\iota(\beta) = (\beta_1, \ldots, \beta_g)$$

where β corresponds to the prime ideal

$$(T_1 - p\beta_1, \ldots, T_g - p\beta_g)$$

in $\widehat{\mathcal{O}}_{\mathcal{X},x}$. Note that

$$\iota^{-1} : \mathbb{Z}_p^g \longrightarrow \mathcal{X}(\mathbb{Z}_p)$$

induces an analytic bijection between $\mathbb{Z}_p^g$ and the analytic neighborhood of $\mathcal{X}(\mathbb{Z}_p)$ consisting of points β such that $r(\beta) = x$. The next result is [**BGT10**, Proposition 2.2].

PROPOSITION 4.4.1.6. *There are power series*

$$F_1, \ldots, F_g \in \mathbb{Z}_p[[z_1, \ldots, z_g]]$$

such that:

    (1) *each F_i converges on $\mathbb{Z}_p^g$;*
    (2) *for each $\beta \in \mathcal{X}(\mathbb{Z}_p)$ such that $r(\beta) = x$, we have*

(4.4.1.7) $$\iota(\Phi(\beta)) = (F_1(\beta_1, \ldots, \beta_g), \ldots, F_g(\beta_1, \ldots, \beta_g)); \text{ and}$$

    (3) *each F_i is congruent to a linear polynomial mod p (in other words, all the coefficients of terms of degree greater than one are divisible by p).*

PROOF. The map Φ induces a ring homomorphism

$$\Phi^* : \widehat{\mathcal{O}}_{\mathcal{X},x} \longrightarrow \widehat{\mathcal{O}}_{\mathcal{X},x}$$

that sends the maximal ideal $\widehat{\mathfrak{m}}$ in $\widehat{\mathcal{O}}_{\mathcal{X},x}$ to itself. For each i, there is a power series

$$H_i \in \mathbb{Z}_p[[T_1,\ldots,T_g]]$$

such that $\Phi^* T_i = H_i$. Furthermore, since $\Phi^* T_i$ must be in the maximal ideal of $\widehat{\mathcal{O}}_{\mathcal{X},x}$, the constant term in H_i must be in $p\mathbb{Z}_p$. Then, for any

$$(\alpha_1,\ldots,\alpha_g) \in p\mathbb{Z}_p,$$

we have

$$(\Phi^*)^{-1}(T_1 - \alpha_1,\ldots,T_g - \alpha_g) = (T_1 - H_1(\alpha_1,\ldots,\alpha_g),\ldots,T_g - H_g(\alpha_1,\ldots,\alpha_g))$$

since

$$(T_1 - H_1(\alpha_1,\ldots,\alpha_g),\ldots,T_g - H_g(\alpha_1,\ldots,\alpha_g))$$

is a prime ideal of coheight equal to one, and

$$H_i(T_1,\ldots,T_g) - H_i(\alpha_1,\ldots,\alpha_g)$$

is in the ideal

$$(T_1 - \alpha_1,\ldots,T_g - \alpha_g)$$

for each i. Thus, if β corresponds to the prime ideal

$$(T_1 - p\beta_1,\ldots,T_g - p\beta_g)$$

then $\Phi(\beta)$ corresponds to the prime ideal

$$(T_1 - H_1(p\beta_1,\ldots,p\beta_g),\ldots,T_g - H_g(p\beta_1,\ldots,p\beta_g)).$$

Hence, letting

$$F_i(T_1,\ldots,T_g) := \frac{1}{p} H_i(pT_1,\ldots,pT_g)$$

gives the desired map. Since

$$H_i \in \mathbb{Z}_p[[T_1,\ldots,T_g]],$$

it follows that F_i must converge on $\mathbb{Z}_p$ and that all the coefficients of terms of degree greater than one of F_i are divisible by p. Since the constant term in H_i is divisible by p, we conclude that

$$F_1,\ldots,F_g \in \mathbb{Z}_p[[T_1,\ldots,T_g]],$$

as desired. $\qquad\square$

Switching to vector notation, we write

$$\vec{\beta} := (\beta_1,\ldots,\beta_g) \in \mathbb{Z}_p^g,$$

and we let

$$\mathcal{F}(\vec{\beta}) := (F_1(\beta_1,\ldots,\beta_g),\ldots,F_g(\beta_1,\ldots,\beta_g)).$$

From Proposition 4.4.1.6, we see that there is a $g \times g$ matrix L with coefficients in $\mathbb{Z}_p$ and a constant $\vec{C} \in \mathbb{Z}_p^g$ such that

(4.4.1.8) $$\mathcal{F}(\vec{\beta}) = \vec{C} + L(\vec{\beta}) + \text{ higher order terms}$$

Note that since all of the higher order terms are divisible by p, we also have

(4.4.1.9) $$\mathcal{F}(\vec{\beta}) \equiv \vec{C} + L(\vec{\beta}) \pmod{p}.$$

This will allow us in Subsection 4.4.2 to construct the p-adic analytic parametrization of the orbit.

PROPOSITION 4.4.1.10. *Let L be as in (4.4.1.8). Then L is invertible modulo p.*

PROOF. Let $\mathcal{O}_{\overline{\mathcal{X}},x}$ denote the local ring of x on $\overline{\mathcal{X}}$ and let $\overline{\mathfrak{m}}$ denote its maximal ideal. Since Φ is unramified, the map

$$\Phi^* : \mathcal{O}_{\overline{\mathcal{X}},x} \longrightarrow \mathcal{O}_{\overline{\mathcal{X}},x}$$

sends $\overline{\mathfrak{m}}$ surjectively onto itself (see [**BG06**, Appendix B.2]). Thus in particular it induces an isomorphism on the $\mathbb{F}_p$-vector space $\overline{\mathfrak{m}}/\overline{\mathfrak{m}}^2$. Completing $\mathcal{O}_{\overline{\mathcal{X}},x}$ at $\overline{\mathfrak{m}}$, we then get an induced isomorphism

$$\sigma : \widehat{\overline{\mathfrak{m}}}/\widehat{\overline{\mathfrak{m}}}^2 \longrightarrow \widehat{\overline{\mathfrak{m}}}/\widehat{\overline{\mathfrak{m}}}^2,$$

where $\widehat{\overline{\mathfrak{m}}}$ is the maximal ideal in the completion of $\mathcal{O}_{\overline{\mathcal{X}},x}$ at $\overline{\mathfrak{m}}$. This isomorphism is obtained by taking the map

$$\Phi^* : \widehat{\mathfrak{m}}/\widehat{\mathfrak{m}}^2 \longrightarrow \widehat{\mathfrak{m}}/\widehat{\mathfrak{m}}^2$$

and reducing mod p, where $\widehat{\mathfrak{m}}$ is the maximal ideal of $\widehat{\mathcal{O}}_{\mathcal{X},x}$. Writing σ as a linear transformation with respect to the basis $\{T_1, \ldots, T_g\}$ for $\widehat{\overline{\mathfrak{m}}}/\widehat{\overline{\mathfrak{m}}}^2$, we obtain the dual of the reduction of L mod p. Thus, if Φ^* induces an isomorphism on $\overline{\mathfrak{m}}/\overline{\mathfrak{m}}^2$, then the reduction mod p of L itself must be invertible.   □

PROPOSITION 4.4.1.11. *There exists a positive integer M such that*

$$\mathcal{F}^M(\vec{\beta}) \equiv \vec{\beta} \pmod{p}$$

for each $\vec{\beta} \in \mathbb{Z}_p^g$.

PROOF. Since L is invertible modulo p, it follows that the reduction modulo p of the affine map

$$\vec{\beta} \mapsto \vec{C} + L(\vec{\beta})$$

induces an automorphism of $\mathbb{F}_p^g$. Therefore, there exists a positive integer M such that

(4.4.1.12) $$\mathcal{F}^M(\vec{\beta}) \equiv \vec{\beta} \pmod{p},$$

for all $\vec{\beta} \in \mathbb{Z}_p$.   □

Now we are ready to prove the p-adic analytic parametrization of the orbit.

4.4.2. The p-adic argument. The results of this subsection generalize some of the work of Rivera-Letelier [**RL03**] for parametrizing p-adic orbits of rational maps (for more information see Chapter 6).

We construct a p-adic analytic function

$$U : \mathbb{Z}_p \longrightarrow \mathbb{Z}_p^n \text{ such that } U(z+1) = \mathcal{F}(U(z)),$$

where $\mathcal{F}$ is constructed as in Subsection 4.4.1 for a closed point $x \in \mathcal{X}$ and an unramified endomorphism Φ of the n-dimensional smooth $\mathbb{Z}_p$-scheme $\mathcal{X}$. For this, we generalize the construction from [**Bel06**], and thus provide the key analytical result (see our Theorem 4.4.2.1) which will be used in the proof of Theorem 4.4.1.1. We use an argument of Poonen [**Poo14**] that simplified and extended the proofs from [**BGT10**, Section 3].

THEOREM 4.4.2.1. *Let n be a positive integer, let $p > 2$ be a prime number and let*

$$\varphi_1, \ldots, \varphi_n \in \mathbb{Z}_p[[x_1, \ldots, x_n]]$$

be convergent power series on $\mathbb{Z}_p^n$ such that for each $i = 1, \ldots, n$ we have

$$\varphi_i(x_1, \ldots, x_n) \equiv x_i \pmod{p}.$$

Let

$$(\omega_1, \ldots, \omega_n) \in \mathbb{Z}_p^n$$

be an arbitrary point. Then there exist p-adic analytic functions

$$f_1, \ldots, f_n \in \mathbb{Q}_p[[z]]$$

such that for each $i = 1, \ldots, n$ we have

    (1) *f_i is convergent for $|z|_p \leq 1$;*
    (2) *$f_i(0) = \omega_i$;*
    (3) *$|f_i(z)|_p \leq 1$ for $|z|_p \leq 1$; and*
    (4) *$f_i(z+1) = \varphi_i(f_1(z), \ldots, f_n(z))$.*

A particular case of our result, when $n = 1$ and φ_1 is a rational p-adic function, is proven in [**RL03**] (for more details, see our Chapter 6).

PROOF OF THEOREM 4.4.2.1. Let

$$F : \mathbb{Z}_p[[x_1, \ldots, x_n]]^n \to \mathbb{Z}_p[[x_1, \ldots, x_n]]^n$$

be given by

$$F(h_1, \ldots, h_n) = (\varphi_1(h_1, \ldots, h_n), \ldots, \varphi_n(h_1, \ldots, h_n)).$$

Since

$$\phi_i \equiv x_i \pmod{p}$$

for all i, we see that F is indeed well-defined (i.e., its image lies inside $\mathbb{Z}_p[[x_1, \ldots, x_n]]^n$). We define the operator

$$\Delta : \mathbb{Z}_p[[x_1, \ldots, x_n]]^n \to \mathbb{Z}_p[[x_1, \ldots, x_n]]^n$$

given by

$$\Delta(h_1, \ldots, h_n) = F(h_1, \ldots, h_n) - (h_1, \ldots, h_n).$$

Since $\varphi_i(x_1, \ldots, x_n) \equiv x_i \pmod{p}$ we see that in fact

$$\Delta : \mathbb{Z}_p[[x_1, \ldots, x_n]]^n \longrightarrow p\mathbb{Z}_p[[x_1, \ldots, x_n]]^n$$

and hence Δ^j maps $\mathbb{Z}_p[[x_1, \ldots, x_n]]^n$ into $p^j \mathbb{Z}_p[[x_1, \ldots, x_n]]^n$ for each $j \geq 1$. Moreover, since

$$\varphi_1, \ldots, \varphi_n \in \mathbb{Z}_p[[x_1, \ldots, x_n]]$$

are convergent power series on $\mathbb{Z}_p^n$, we see that for every $j \geq 0$ we have $\Delta^j(x_1, \ldots, x_n)$ converges at the point $(\omega_1, \ldots \omega_n)$ and is in $p^j \mathbb{Z}_p^n$. We now define

$$(f_1(z), \ldots, f_n(z)) = \sum_{j=0}^{\infty} \binom{z}{j} \cdot \left(\Delta^j(x_1, \ldots, x_n) \right)(\omega_1, \ldots, \omega_n),$$

where $\left(\Delta^j(x_1, \ldots, x_n) \right)(\omega_1, \ldots, \omega_n)$ simply denotes the n-tuple of p-adic integers obtained by evaluating the n-tuple of convergent power series $\left(\Delta^j(x_1, \ldots, x_n) \right)$ all at the point $(\omega_1, \ldots, \omega_n)$. Thus for $i \in \{1, \ldots, n\}$ there exist constants $c_{j,i} \in p^n \mathbb{Z}_p$ such that

$$f_i(z) = \sum_{j=0}^{\infty} c_{j,i} \binom{z}{j}$$

and so by Equation (2.3.5.4) we have that

$$|c_{j,i}/j!|_p < p^{-j+j/(p-1)} \to 0 \text{ as } j \to \infty,$$

since $p > 2$. Thus we have that each $f_i(z)$ is a p-adic analytic theorem by Theorem 2.3.5.3. By construction,

$$(f_1(0), \ldots, f_n(0)) = (\omega_1, \ldots, \omega_n)$$

and $|f_i(z)|_p \le 1$ for $|z|_p \le 1$. It only remains to show that

$$f_i(z+1) = \varphi_i(f_1(z), \ldots, f_n(z)) \text{ for } i \in \{1, \ldots, n\}.$$

To see this, we note that

$$(f_1(z+1), \ldots, f_n(z+1)) = \sum_{j=0}^{\infty} \binom{z+1}{j} \cdot \left(\Delta^j(x_1, \ldots, x_n) \right) (\omega_1, \ldots, \omega_n).$$

Since $\binom{z+1}{j} = \binom{z}{j} + \binom{z}{j-1}$ and the above sum is absolutely convergent we see that

$$(f_1(z+1), \ldots, f_n(z+1))$$

is equal to

$$\sum_{j=0}^{\infty} \binom{z}{j} \cdot \left(\Delta^j(x_1, \ldots, x_n) \right) (\omega_1, \ldots, \omega_n) + \sum_{j=1}^{\infty} \binom{z}{j-1} \cdot \left(\Delta^j(x_1, \ldots, x_n) \right) (\omega_1, \ldots, \omega_n).$$

Absolute convergence allows us to rearrange once more, and we see that this expression is equal to

$$\sum_{j=0}^{\infty} \binom{z}{j} \cdot \left(\left(\Delta^j(x_1, \ldots, x_n) \right) (\omega_1, \ldots, \omega_n) + \left(\Delta^{j+1}(x_1, \ldots, x_n) \right) \right) (\omega_1, \ldots, \omega_n).$$

Finally, we note that $\Delta^j + \Delta^{j+1} = F \circ \Delta^j$ and so we see that

$$F(f_1(z), \ldots, f_n(z)) = (f_1(z+1), \ldots, f_n(z+1)),$$

thus proving the last remaining claim. The result now follows. $\qquad\square$

We note that the argument used in the proof of Theorem 4.4.2.1 fails if $p = 2$, because $|2^k/k!|_2$ does not tend to zero. In fact, one can construct explicit examples which show that the conclusion to the statement of Theorem 4.4.2.1 does not hold if one eliminates the hypothesis that p be at least 3. For example, if we take $n = 1$ and let $\phi(z) = -z$ then

$$\phi(z) \equiv z \pmod{2}$$

but $\phi^n(1) = (-1)^n$ and so there cannot exist a 2-adic analytic function $f(z)$ such that

$$f(n) = \phi^n(1)$$

since we would then have $f(n) = 1$ for all even natural numbers n. Lemma 2.3.6.1 would then say that $f(z)$ must be identically 1, since the zeros of $f(z) - 1$ are dense in $2\mathbb{Z}_2$. But this is a contradiction, since $f(1)$ must be -1. Finally, we note that one can generalize Theorem 4.4.2.1 to the case when K is a finite extension of $\mathbb{Q}_p$, also by replacing the congruences modulo p by congruences modulo an uniformizer

π of K; the convergence of the corresponding power series is guaranteed once $p-1$ is larger than the ramification index of $K/\mathbb{Q}_p$. More generally the following result holds.

THEOREM 4.4.2.2. *Let n be a positive integer, let $(K_v, |\cdot|_p)$ be a finite extension of $(\mathbb{Q}_p, |\cdot|_p)$ (where p is a prime number and the p-adic absolute value is normalized so that $|p|_p = 1/p$), let π be a uniformizer of the ring $\mathcal{O}_v$ of integers in K_v such that $|\pi|_p^e = 1/p$ for some positive integer e, and let*

$$\varphi_1, \ldots, \varphi_n \in \mathcal{O}_v[[x_1, \ldots, x_n]]$$

be convergent power series on $\mathcal{O}_v^n$ such that for each $i = 1, \ldots, n$ we have

$$\varphi_i(x_1, \ldots, x_n) \equiv x_i \pmod{\pi^{\left[\frac{e}{p-1}\right]+1}}.$$

Let

$$(\omega_1, \ldots, \omega_n) \in \mathcal{O}_v^n$$

be an arbitrary point. Then there exist p-adic analytic functions

$$f_1, \ldots, f_n \in K_v[[z]]$$

such that for each $i = 1, \ldots, n$ we have

    (1) *f_i is convergent for $|z|_p \leq 1$;*
    (2) *$f_i(0) = \omega_i$;*
    (3) *$|f_i(z)|_p \leq 1$ for $|z|_p \leq 1$; and*
    (4) *$f_i(z+1) = \varphi_i(f_1(z), \ldots, f_n(z))$.*

PROOF. The proof is identical with the proof of Theorem 4.4.2.1. The exponent $g := \left[\frac{e}{p-1}\right] + 1$ yields that each coefficient $c_{j,i}$ corresponding to the power series

$$f_i(z) = \sum_{j=0}^{\infty} c_{j,i} \binom{z}{j}, \text{ for each } i = 1, \ldots, n,$$

satisfies the inequality

$$|c_{j,i}|_p \leq p^{-jg/e+j/(p-1)} \to 0, \text{ as } j \to \infty,$$

which yields the convergence of the power series $f_i(z)$. $\square$

4.4.3. Finishing the proof for étale maps. Now we can finish the proof of Theorem 4.3.0.1. We continue with the geometric reductions from Subsection 4.4.1. So, let

$$\alpha \in \mathcal{X}(\mathbb{Z}_p)$$

be a point such that its reduction x modulo p is fixed by Φ (which extends as an endomorphism of the $\mathbb{Z}_p$-scheme $\mathcal{X}$). Furthermore, we have a local p-adic analytic isomorphism

$$\iota : \mathcal{U}_x \longrightarrow \mathbb{Z}_p^g,$$

where $\mathcal{U}_x$ is the set of all $\beta \in \mathcal{X}(\mathbb{Z}_p)$ whose reduction modulo p equals x. We let

$$\iota(\beta) := (\beta_1, \ldots, \beta_g)$$

for any $\beta \in \mathcal{U}_x$. Then there exists (as proven in Subsection 4.4.1) a p-adic analytic map

$$\mathcal{F} : \mathbb{Z}_p^g \longrightarrow \mathbb{Z}_p^g$$

such that

$$\iota(\Phi(\beta)) = \mathcal{F}(\beta_1, \ldots, \beta_g).$$

Furthermore, as proven in Proposition 4.4.1.11, there exists a positive integer M such that
$$\mathcal{F}^M(z_1, \ldots, z_g) \equiv (z_1, \ldots, z_g) \pmod{p}.$$
Finally, we let $\alpha_{i,j} \in \mathbb{Z}_p$ such that for all $j = 0, \ldots, M-1$ we have
$$\iota(\Phi^j(\alpha)) = (\alpha_{1,j}, \ldots, \alpha_{g,j}).$$
Then using Theorem 4.4.2.1 we conclude that there exist p-adic analytic functions
$$f_{i,j} : \mathbb{Z}_p \longrightarrow \mathbb{Z}_p$$
such that

(1) $f_{i,j}(0) = \alpha_{i,j}$ for each $i = 1, \ldots, g$ and for each $j = 0, \ldots, M-1$; and
(2) $(f_{1,j}(z+1), \ldots, f_{g,j}(z+1)) = \mathcal{F}^M(f_{1,j}(z), \ldots, f_{g,j}(z))$ for each $z \in \mathbb{Z}_p$ and for each $j = 0, \ldots, M-1$.

Hence for each $j = 0, \ldots, M-1$ we have that
$$\iota\left(\Phi^{nM+j}(\alpha)\right) = (f_{1,j}(n), \ldots, f_{g,j}(n)).$$
So, for any function H in the vanishing ideal of $\mathcal{V}$, we obtain that
$$\Phi^{Mn+j}(\alpha) \in \mathcal{V}(\mathbb{Z}_p)$$
if and only if n is a zero of the p-adic analytic function
$$z \mapsto \left(H \circ \iota^{-1}\right)(f_{1,j}(z), \ldots, f_{g,j}(z)).$$
Another application of Lemma 2.3.6.1 concludes our proof of Theorem 4.4.1.1.

4.4.4. The Dynamical Mordell-Lang Conjecture holds if the orbit avoids the ramification locus. The exact same strategy as in the proof of Theorem 4.4.1.1 yields the following result.

THEOREM 4.4.4.1. *Let p be a prime number, let $(K_v, |\cdot|_p)$ be a finite extension of $(\mathbb{Q}_p, |\cdot|_p)$, let $\mathcal{O}_v$ be the ring of integers of K_v, let $\mathcal{X}$ be an $\mathcal{O}_v$-scheme whose generic fiber is a variety X defined over K_v. Let Φ be an endomorphism of $\mathcal{X}$ defined over $\mathcal{O}_v$, and we let φ be the induced endomorphism of X. Let α be a section of $\mathcal{X} \longrightarrow \operatorname{Spec}(\mathcal{O}_v)$, and we let x the intersection of α with the generic fiber of $\mathcal{X}$.*

Assume that there is an N such that for all $n \geq N$ we have the following:

(i) *$\Phi^n(\alpha)$ lies in the smooth locus of $\mathcal{X}$; and*
(ii) *the ramification locus of Φ does not intersect $\Phi^n(\alpha)$.*

Then for each subvariety V of X defined over K_v, the set of all $n \in \mathbb{N}_0$ such that $\varphi^n(x) \in V$ is a finite union of arithmetic progressions.

PROOF. Since the residue field of K_v is finite, at the expense of replacing Φ by a suitable Φ^m (for some positive integer m), and also replacing α by $\Phi^N(\alpha)$, we may assume the residue class $\overline{x}$ of α (i.e., the intersection of α with the special fiber of $\mathcal{X} \longrightarrow \operatorname{Spec}(\mathcal{O}_v)$) is fixed by Φ. Then conditions (i)–(ii) allow us to argue identically as in Subsection 4.4.1 to prove the Proposition 4.4.4.2. In order to state the next result, we let $g := \dim(X)$ and we also let π be a uniformizer of $\mathcal{O}_v$.

PROPOSITION 4.4.4.2. *There are power series*
$$F_1, \ldots, F_g \in \mathcal{O}_v[[z_1, \ldots, z_g]]$$
such that:

(a) *each F_i converges on $\mathcal{O}_v^g$;*

(b) *there exists a p-adic analytic morphism ι mapping all points $\beta \in \mathcal{X}(\mathcal{O}_v)$ to points $\iota(\beta) = (\beta_1, \ldots, \beta_g) \in \mathcal{O}_v^g$ such that*

(4.4.4.3) $$\iota(\Phi(\alpha)) = (F_1(\alpha_1, \ldots, \alpha_g), \ldots, F_g(\alpha_1, \ldots, \alpha_g)),$$

where $\iota(\alpha) = (\alpha_1, \ldots, \alpha_g)$; and

(c) *each F_i is congruent to a linear polynomial mod π (in other words, all the coefficients of terms of degree greater than one are divisible by π), and moreover*

$$(F_1(x_1, \ldots, x_g), \ldots, F_g(x_1, \ldots, x_g)) \equiv C + L(x_1, \ldots, x_g) \pmod{\pi},$$

where $C \in \mathcal{O}_v^g$ and L is a linear, invertible map $\mathcal{O}_v^g \longrightarrow \mathcal{O}_v^g$.

Then applying [**BGT15a**, Proposition 2.1], at the expense of replacing again Φ by an iterate (note that this does not affect the conclusion of the Dynamical Mordell-Lang Conjecture, as shown by Proposition 3.1.2.5), we can improve condition (c) of Proposition 4.4.4.2 to the following congruence relation:

$$F_i(x_1, \ldots, x_g) \equiv x_i \pmod{\pi^{\left\lceil \frac{e}{p-1} \right\rceil + 1}},$$

for each $i = 1, \ldots, g$, where e is the ramification index of $K_v/\mathbb{Q}_p$. Then Theorem 4.4.2.2 yields the technical ingredient to finish the proof of Theorem 4.4.4.1 arguing identically as in Subsection 4.4.3. $\square$

CHAPTER 5

Linear relations between points in polynomial orbits

In this chapter we prove Conjecture 1.5.0.1 for all affine lines $L \subset \mathbb{A}^N$ under the action of endomorphisms Φ of $\mathbb{A}^N$ of the form

$$\Phi(x_1, \ldots, x_N) - (f_1(x_1), \ldots, f_N(x_N)),$$

where the f_i's are one-variable polynomials of degree larger than 1 with complex coefficients. In particular, the results of this chapter refer to the *split case* of the Dynamical Mordell-Lang Conjecture. The results presented in this chapter are taken from papers [**GTZ08**] and [**GTZ12**]. For some results we include complete proofs, and in other cases we only sketch the proofs appearing in [**GTZ08**] and [**GTZ12**].

5.1. The main results

The first result of this chapter is the case of Conjecture 1.5.0.1 for all affine lines under the coordinatewise action of one-variable polynomials; this result is an immediate consequence of the main theorem of [**GTZ12**].

THEOREM 5.1.0.1. *Let $N \in \mathbb{N}$, let $f_1, \ldots, f_N \in \mathbb{C}[z]$ be polynomials of degree larger than 1, let $\alpha \in \mathbb{A}^N(\mathbb{C})$ and let $L \subset \mathbb{A}^N$ be a line defined over $\mathbb{C}$. We let the endomorphism*

$$\Phi : \mathbb{A}^N \longrightarrow \mathbb{A}^N \text{ be defined by}$$

$$\Phi(x_1, \ldots, x_N) = (f_1(x_1), \ldots, f_N(x_N)).$$

Then the set

$$S(L, \Phi, \alpha) := \{n \in \mathbb{N}_0 \colon \Phi^n(\alpha) \in L(\mathbb{C})\}$$

is a union of finitely many arithmetic progressions.

We prove Theorem 5.1.0.1 as a consequence of Theorem 5.1.0.2 (which was proven in [**GTZ12**]). Before stating Theorem 5.1.0.2, we recall the notation

$$\gamma + U := \{\gamma + x \colon x \in U\}$$

for any typle $\gamma \in \mathbb{N}_0^N$ and any set $U \subset \mathbb{N}_0^N$, where the addition of any two tuples in $\mathbb{N}_0^N$ is done coordinatewise.

THEOREM 5.1.0.2 ([**GTZ12**]). *Let $N \in \mathbb{N}$, let $\alpha \in \mathbb{A}^N(\mathbb{C})$, let $f_1, \ldots, f_N \in \mathbb{C}[z]$ satisfy $\deg(f_i) > 1$ for $i = 1, \ldots, N$, and let L be a complex line in $\mathbb{A}^N$. Let S be the semigroup generated by the maps*

$$\Phi_i \colon \mathbb{A}^N \to \mathbb{A}^N \text{ with } 1 \leq i \leq N,$$

where

$$\Phi_i(x_1, \ldots, x_N) = (x_1, \ldots, x_{i-1}, f_i(x_i), x_{i+1}, \ldots, x_N).$$

Then the intersection of $\mathcal{O}_S(\alpha)$ with L is $\mathcal{O}_T(\alpha)$, where $\mathcal{T}$ is the union of finitely many cosets of cyclic subsemigroups of S. More precisely, there exist $\ell \in \mathbb{N}_0$, tuples $\gamma_1, \ldots, \gamma_\ell \in \mathbb{N}_0^N$ and cyclic subgroups $H_i \subseteq \mathbb{Z}^N$ (for $i = 1, \ldots, \ell$) such that for each tuple $(n_1, \ldots, n_N) \in \mathbb{N}_0^N$ we have that

$$\Phi_1^{n_1} \cdots \Phi_N^{n_N}(\alpha) \in L(\mathbb{C})$$

if and only if

$$(n_1, \ldots, n_N) \in \bigcup_{i=1}^{\ell} \gamma_i + \left(H_i \cap \mathbb{N}_0^N\right).$$

In particular, if the intersection $\mathcal{O}_S(\alpha) \cap L(\mathbb{C})$ is infinite, then there are $m_1, \ldots, m_N \in \mathbb{N}_0$ not all equal to 0 such that

$$(f_1^{m_1}, \ldots, f_N^{m_N})(L) = L.$$

5.1.1. Proof of Theorem 5.1.0.1 and of its extensions. We show how to deduce Theorem 5.1.0.1 as a consequence of Theorem 5.1.0.2.

PROOF OF THEOREM 5.1.0.1. With the notation as in Theorem 5.1.0.2, we have that

$$(5.1.1.1) \quad \left\{(n_1, \ldots, n_N) \in \mathbb{N}_0^N : \Phi_1^{n_1} \cdots \Phi_N^{n_N}(\alpha) \in L(\mathbb{C})\right\} = \bigcup_{i=1}^{\ell} \gamma_i + \left(H_i \cap \mathbb{N}_0^N\right).$$

Intersecting the right hand-side of (5.1.1.1) with the diagonal subset Δ_N of $\mathbb{N}_0^N$, i.e. with the set

$$\Delta_N := \{(n, \ldots, n) : n \in \mathbb{N}_0\},$$

we obtain the conclusion of Theorem 5.1.0.1. To see this, note that for each

$$i = 1, \ldots, \ell,$$

we have that $H_i \cap \Delta_N$ is a set of the form

$$\{(a_i n, \ldots, a_i n) : n \in \mathbb{N}_0\},$$

for some $a_i \in \mathbb{N}_0$ (because H_i is a cyclic subgroup of $\mathbb{Z}^N$), and therefore the projection of the set

$$(\gamma_i + H_i) \cap \Delta_N$$

on any of the coordinates is an arithmetic progression, as desired.   □

As proven by Xie [**Xieb**], the Dynamical Mordell-Lang Conjecture holds for any curve defined over $\overline{\mathbb{Q}}$ under the coordinatewise action of one-variable polynomials defined over $\overline{\mathbb{Q}}$ (see also Section 5.10 and Section 10.3). It is expected that a specialization argument (similar to the one employed in [**BGKT12**]) would work to extend Xie's result (see Theorem 5.10.0.6) to all affine curves under the coordinatewise action of one-variable polynomials defined over $\mathbb{C}$. In particular, in Theorem 5.1.0.1 it is not essential that the polynomials f_i have degrees larger than 1. We prove the extension of Theorem 5.1.0.1 to the case of all polynomials defined over $\overline{\mathbb{Q}}$.

THEOREM 5.1.1.2. *Let $N \in \mathbb{N}$, let $\alpha \in \mathbb{A}^N(\overline{\mathbb{Q}})$, let $L \subset \mathbb{A}^N$ be a line defined over $\overline{\mathbb{Q}}$, let $f_1, \ldots, f_N \in \overline{\mathbb{Q}}[z]$ be arbitrary polynomials, and let $\Phi : \mathbb{A}^N \longrightarrow \mathbb{A}^N$ be defined as follows*

$$\Phi(x_1, \ldots, x_N) := (f_1(x_1), \ldots, f_N(x_N)).$$

Then the set

$$S(L, \Phi, \alpha) := \{n \in \mathbb{N}_0 \colon \Phi^n(\alpha) \in L(\overline{\mathbb{Q}})\}$$

is a finite union of arithmetic progressions.

As previously stated, Theorem 5.1.1.2 is a special case of Xie's result for all curves; see Theorem 5.10.0.6. We decided to include Theorem 5.1.1.2 since the ideas appearing in its proof also appear later in a couple of places in the more technical proof of Theorem 5.1.0.2.

PROOF OF THEOREM 5.1.1.2. We argue by induction on N; the case $N = 1$ is obvious.

Now, if L does not project dominantly on one of the axes of $\mathbb{A}^N$, then we take the projection of L on the remaining $(N-1)$ coordinates of $\mathbb{A}^N$ and apply the inductive hypothesis. So, from now on, we assume that L projects dominantly onto each coordinate axis of $\mathbb{A}^N$.

Now, letting

$$\alpha := (\alpha_1, \ldots, \alpha_N),$$

if one of the α_i is preperiodic under the action of f_i, then the intersection

$$L(\overline{\mathbb{Q}}) \cap \mathcal{O}_\Phi(\alpha)$$

consists of only finitely many points, and then Theorem 5.1.1.2 holds with an argument similar to the proof of Proposition 3.1.2.9. So, in particular, we may assume that each f_i is non-constant.

Now, if each $\deg(f_i) > 1$ then we are done by Theorem 5.1.0.1, while if each $\deg(f_i) = 1$ then we are done by Theorem 4.1.0.4. So, we may assume from now on that there exists at least one polynomial, say f_1, which is linear, and at least one polynomial, say f_2, which has degree larger than 1. We prove that in this case $S(L, \Phi, \alpha)$ is finite.

Clearly, if suffices to assume $N = 2$ (after taking the projection of L on the first two coordinate axes) and prove $S(L, \Phi, \alpha)$ is finite. So, L is a line of the form

$$x_2 = ax_1 + b,$$

for some $a \in \overline{\mathbb{Q}}^*$ and $b \in \overline{\mathbb{Q}}$. Hence there exists a positive constant c_1 such that for each point $(x_1, x_2) \in L(\overline{\mathbb{Q}})$, we have the following inequality between the Weil heights of the coordinates of the point on L (see Proposition 2.6.3.3)

$$h(x_2) \le h(x_1) + c_1.$$

So, for each $n \in \mathbb{N}_0$ such that $\Phi^N(\alpha) \in L(\overline{\mathbb{Q}})$, letting $\alpha := (\alpha_1, \alpha_2)$ we have

$$(5.1.1.3) \qquad h\left(f_2^n(\alpha_2)\right) \le h\left(f_1^n(\alpha_1)\right) + c_1.$$

Using the form of the n-th iterate of a linear polynomial (see (4.1.0.1) and (4.1.0.2)), we conclude that there exist positive constants c_2 and c_3 such that

$$(5.1.1.4) \qquad h\left(f_1^n(\alpha_1)\right) \le nc_2 + c_3,$$

for each $n \in \mathbb{N}_0$. On the other hand, using Proposition 2.6.4.2 (a)—(b), we obtain that there exists a positive constant c_4 such that

$$(5.1.1.5) \qquad h\left(f_2^n(\alpha_2)\right) \ge \deg(f_2)^n \cdot \widehat{\mathrm{h}}_{f_2}(\alpha_2) - c_4,$$

for each $n \in \mathbb{N}_0$, where $\widehat{\mathrm{h}}_{f_2}(\alpha_2)$ is the canonical height of α_2 with respect to the polynomial $f_2 \in \overline{\mathbb{Q}}[z]$ (for more details, see Subsection 2.6.4). According to Proposition 2.6.4.2 (c), we have that $\widehat{\mathrm{h}}_{f_2}(\alpha_2) > 0$ since we assumed that α_2 is not preperiodic under the action of f_2. Then, combining (5.1.1.3), (5.1.1.4) and (5.1.1.5), we obtain that

$$(5.1.1.6) \qquad \deg(f_2)^n \cdot \widehat{\mathrm{h}}_{f_2}(\alpha_2) \le nc_2 + c_1 + c_3 + c_4.$$

Using in (5.1.1.6) the fact that $\deg(f_2) \ge 2$ and that $\widehat{\mathrm{h}}_{f_2}(\alpha_2) > 0$, we conclude that n must be bounded, i.e., $S(L, \Phi, \alpha)$ is finite, which concludes the proof of Theorem 5.1.1.2. $\square$

Employing a specialization argument similar to the one we will present in Section 5.6, one can extend the result of Theorem 5.1.1.2 for all lines and polynomials defined over $\mathbb{C}$.

5.1.2. Question 3.6.0.1 for polynomial orbits. With the notation as in Theorem 5.1.0.2, letting

$$\alpha := (\alpha_1, \dots, \alpha_N), \text{ we have}$$

$$\mathcal{O}_S(\alpha_1, \dots, \alpha_N) = \{(f_1^{m_1}(\alpha_1), \dots, f_N^{m_N}(\alpha_N)) : m_1, \dots, m_N \in \mathbb{N}_0\}.$$

We state next a special case of Question 3.6.0.1 that is relevant for Theorem 5.1.0.2.

QUESTION 5.1.2.1. *Let* $N \in \mathbb{N}$, *let* $\alpha_1, \dots, \alpha_N \in \mathbb{C}$, *let* $V \subset \mathbb{A}^N$ *be a subvariety, and let* $f_1, \dots, f_N \in \mathbb{C}[z]$ *be polynomials of degree greater than 1. Is it true that the set*

$$S(V, f_1, \dots, f_N, \alpha) := \{(n_1, \dots, n_N) \in \mathbb{N}_0^N : (f_1^{n_1}(\alpha_1), \dots, f_N^{n_N}(\alpha_N)) \in V(\mathbb{C})\}$$

is a union of finitely many sets of the form

$$\gamma + \left(H \cap \mathbb{N}_0^N \right),$$

for some tuples $\gamma \in \mathbb{N}_0^N$ *and subgroups* $H \subseteq \mathbb{Z}^N$?

Theorem 5.1.0.2 solves a special case of Question 5.1.2.1 when V is a line. Theorem 5.1.0.2 can be interpreted as showing that the only linear relations between points in the orbits of the polynomials f_i come from a linear relation between the polynomials themselves. More generally, Question 5.1.2.1 asks for all *algebraic* relations between points in the orbits of the polynomials f_i. We expect that Question 5.1.2.1 is true.

It is crucial that the polynomials f_i be non-linear in Theorem 5.1.0.2 (and similarly in Question 5.1.2.1). Indeed, with the notation as in Theorem 5.1.0.2, if

$$N = 2, \ f_1(z) = z + 1, \ f_2(z) = z^2, \ \alpha_1 = 0 \text{ and } \alpha_2 = 2,$$

then the orbit $\mathcal{O}_S(\alpha_1, \alpha_2)$ intersects the diagonal line of the affine plane in the points

$$\left(2^{2^n}, 2^{2^n} \right) = \Phi_1^{2^{2^n}} \Phi_2^n(\alpha_1, \alpha_2).$$

Many similar counterexamples can be constructed when at least one of the polynomials f_i is linear. So, in other words, Question 5.1.2.1 (and in particular, an extension of Theorem 5.1.0.2) does not necessarily have a positive answer when an endomorphism

$$\Phi_i : \mathbb{A}^N \longrightarrow \mathbb{A}^N$$

defined as in Theorem 5.1.0.2 is actually an automorphism. Informally, the problem with the case when one of the f_i is linear is that the height of the iterates of α_i under the action of the linear f_i grows much slower than the height of the iterates of an α_j under the action of a non-linear polynomial f_j (see Section 2.6 for more details on heights). However, there are examples when Question 3.6.0.1 fails even when all polynomials f_i are linear. For example, if

$$N = 2, \ f_1(z) = z + 1, \ f_2(z) = 2z, \ \alpha_1 = 0 \ \text{and} \ \alpha_2 = 1,$$

then the orbit $\mathcal{O}_S(\alpha_1, \alpha_2)$ intersects the diagonal line of the affine plane in the points

$$(2^n, 2^n) = \Phi_1^{2^n} \Phi_2^n(\alpha_1, \alpha_2).$$

On the other hand, the original Dynamical Mordell-Lang Conjecture (see Conjecture 1.5.0.1) is already known to hold when the dynamical system is given by an automorphism (see Theorem 4.3.0.1).

5.2. Intersections of polynomial orbits

Theorem 5.1.0.2 is the main result of [**GTZ12**]; in the special case

$$\deg(f_1) = \cdots = \deg(f_N),$$

the result was proven in [**GTZ08**], and actually, the general result from [**GTZ12**] was proved by first reducing it to the special case when all polynomials have the same degree. In this section we prove Theorem 5.1.0.2 as a consequence of a result (see Theorem 5.2.0.1) regarding intersections of orbits under polynomial maps.

THEOREM 5.2.0.1. *Let $f, g \in \mathbb{C}[x]$ be polynomials of degrees larger than 1, and let $\alpha, \beta \in \mathbb{C}$. If $\mathcal{O}_f(\alpha) \cap \mathcal{O}_g(\beta)$ is infinite, then there exist positive integers m and n such that $f^m = g^n$.*

Theorem 5.2.0.1 was proven in [**GTZ12**] by reducing it to the case when

$$\deg(f) = \deg(g)$$

which was proved in [**GTZ08**]. It is essential that both f and g have degrees larger than 1, since otherwise, if $f(z) = z + 1$ and if g is a polynomial with all its coefficients positive integers, then $\mathcal{O}_f(0) \cap \mathcal{O}_g(0)$ is always infinite even though no iterate of f equals an interate of g. Again, the problem lies in the fact that the points in the orbit of f are not sparse enough, or, equivalently, that the height of the iterates under f does not grow exponentially.

Next we show how to deduce Theorem 5.1.0.2 from Theorem 5.2.0.1.

5.2.1. Proof of Theorem 5.1.0.2. Assume first that the intersection is infinite; otherwise the statement follows similarly to the proof of Proposition 3.1.2.9.

We proceed by induction on N. If $N = 1$, then the statement is trivial. Assume now that the statement holds for all $N < M$, and we prove it for $N = M$.

If all points of L take the same value, say a_i, on the i-th coordinate (for some $i = 1, \ldots, M$), then we take $\tilde{L}_i$ to be the projection of L on the remaining $M - 1$ coordinates and then reduce the problem to the intersection of $\tilde{L}_i \subseteq \mathbb{A}^{M-1}$ with the orbit of a point under the action of a semigroup generated by $M - 1$ one-variable polynomials, each acting on a different coordinate of $\mathbb{A}^{M-1}$. Thus the result follows by the inductive hypothesis.

Assume from now on that L projects dominantly onto each coordinate of $\mathbb{A}^M$. Let $\alpha = (\alpha_1, \ldots, \alpha_M) \in \mathbb{C}^M$; so we know that

$$L(\mathbb{C}) \cap (\mathcal{O}_{f_1}(\alpha_1) \times \cdots \times \mathcal{O}_{f_M}(\alpha_M)) \text{ is infinite.}$$

For each $i = 2, \ldots, M$, we let π_i denote the projection of $\mathbb{A}^M$ onto the first and i-th coordinates and we let $L_i := \pi_i(L)$. Since L_i projects dominantly onto each of the two coordinates, there exists a linear polynomial σ_i such that L_i is given by the equation $X_i = \sigma_i(X_1)$. Now, for any $k, \ell \in \mathbb{N}$ such that $\left(f_1^k(\alpha_1), f_i^\ell(\alpha_i)\right) \in L_i$ we have

$$\left(\sigma_i \circ f_1 \circ \sigma_i^{-1}\right)^k (\sigma_i(\alpha_1)) = f_i^\ell(\alpha_i).$$

Thus by Theorem 5.2.0.1 there exist $m_i, n_i \in \mathbb{N}$ such that $\left(\sigma_i \circ f_1 \circ \sigma_i^{-1}\right)^{m_i} = f_i^{n_i}$. Without loss of generality, we may assume m_i (and thus also n_i) is minimal with this property. Let D_1 be the least common multiple of all the m_i's, and let

$$D_i = \frac{n_i D_1}{m_i}$$

for each $i > 1$. Then

$$\left(\sigma_i \circ f_1 \circ \sigma_i^{-1}\right)^{D_1} = f_i^{D_i}, \text{ i.e. } f_i^{D_i} \circ \sigma_i = \sigma_i \circ f_1^{D_1}.$$

Since L is defined by the $M-1$ equations $X_i = \sigma_i(X_1)$, it follows that L is invariant under $(f_1^{D_1}, \ldots, f_N^{D_N})$.

Hence, for each $(k_1, \ldots, k_M) \in \mathbb{N}_0^N$ such that $(f_1^{k_1}(\alpha_1), \ldots, f_M^{k_M}(\alpha_M)) \in L(\mathbb{C})$, we also have that for all $m \in \mathbb{N}$,

$$(f_1^{k_1 + m D_1}(\alpha_1), \ldots, f_M^{k_M + m D_M}(\alpha_M)) \in L(\mathbb{C}).$$

For each $\ell \in \{1, \ldots, D_1\}$, let U_ℓ be the set of tuples $(k_1, \ldots, k_M) \in \mathbb{N}_0^N$ such that

$$k_1 \equiv \ell \pmod{M_1} \text{ and } (f_1^{k_1}(\alpha_1), \ldots, f_M^{k_M}(\alpha_M)) \in L(\mathbb{C}).$$

If U_ℓ is non-empty, pick $(s_1, \ldots, s_M) \in U_\ell$ for which s_1 is minimal. Then U_ℓ contains

$$V_\ell := \{(s_1 + j D_1, \ldots, s_M + j D_M) : j \in \mathbb{N}_0\}.$$

On the other hand, since L projects injectively onto each coordinate of $\mathbb{A}^M$, we conclude that $U_\ell = V_\ell$. Therefore, the set Z_ℓ of all

$$(f_1^{k_1}(\alpha_1), \ldots, f_M^{k_M}(\alpha_M)) \in L(\mathbb{C}) \text{ and } (k_1, \ldots, k_M) \in U_\ell$$

is the orbit of α under $\langle \rho_1^{D_1} \cdots \rho_M^{D_M} \rangle \rho_1^{k_1} \cdots \rho_M^{k_M}$.

5.2.2. Ritt's classification of polynomials with a common iterate. In Subsection 5.2.1 we proved that Theorem 5.2.0.1 yields Theorem 5.1.0.2, while in Subsection 5.1.1 we proved that Theorem 5.1.0.2 yields Theorem 5.1.0.1. So, most of the remaining part of this chapter is devoted to proving Theorem 5.2.0.1. But first, in this subsection we describe which are the polynomials satisfying the conclusion of Theorem 5.2.0.1, i.e. polynomials $f_1, f_2 \in \mathbb{C}[z]$ such that

$$f_1^{m_1} = f_2^{m_2} \text{ for some } m_1, m_2 \in \mathbb{N}.$$

Ritt [**Rit20**] proved the following result.

THEOREM 5.2.2.1 (Ritt [**Rit20**]). *Let* $f_1, f_2 \in \mathbb{C}[z]$ *with*

$$d_i := \deg(f_i) > 1 \ for \ i = 1, 2.$$

Then there exist $m_1, m_2 \in \mathbb{N}$ *such that*

$$f_1^{m_1} = f_2^{m_2}$$

if and only if there exists a polynomial $g \in z^r \mathbb{C}[z^s]$ *for some* $r \in \mathbb{N}_0$ *and* $s \in \mathbb{N}$, *there exist positive integers* n_1 *and* n_2 *such that*

$$n_1 m_1 = n_2 m_2,$$

and there exist $\epsilon_1, \epsilon_2, \beta \in \mathbb{C}$ *satisfying*

$$\epsilon_i^s = 1 \ and \ \epsilon_i^{\frac{d_i^{m_i}-1}{d_i-1}} = 1$$

for each $i = 1, 2$, *such that*

$$f_1(z) = -\beta + \epsilon_1 g^{n_1}(z + \beta)$$

and

$$f_2(z) = -\beta + \epsilon_2 g^{n_2}(z + \beta).$$

Using Theorem 5.2.2.1 we obtain the following consequence of Theorem 5.2.0.1.

COROLLARY 5.2.2.2. *Let* $\alpha_1, \alpha_2 \in \mathbb{C}$, *and let* $f_1, f_2 \in \mathbb{C}[z]$ *be non-linear polynomials. If*

$$\mathcal{O}_{f_1}(\alpha) \cap \mathcal{O}_{f_2}(\alpha_2) \ is \ infinite,$$

then there exists $h \in \mathbb{C}[z]$, $n_1, n_2 \in \mathbb{N}$, *and* $\ell_1, \ell_2 \in \mathbb{C}[z]$ *of degree 1 such that*

$$f_1 = \ell_1 \circ h^{n_1} \ and \ f_2 = \ell_2 \circ h^{n_2}.$$

PROOF. The result follows applying the conclusion of Theorem 5.2.2.1 to the polynomials f_1 and f_2 which have a common iterate according to Theorem 5.2.0.1. The conclusion of Corollary 5.2.2.2 holds with

$$h(z) := -\beta + g(z + \beta)$$

and

$$\ell_i(z) = (-\beta + \epsilon_i \beta) + \epsilon_i z$$

for $i = 1, 2$, where g, β, ϵ_i are as in the conclusion of Theorem 5.2.2.1. □

5.3. A special case

We first prove Theorem 5.2.0.1 in the special case when $\deg(f) = \deg(g)$, and in addition, $f, g \in \overline{\mathbb{Q}}[z]$ and $\alpha, \beta \in \overline{\mathbb{Q}}$.

THEOREM 5.3.0.1. *Let* K *be a number field, let* $\alpha, \beta \in K$, *and let* $f, g \in K[z]$ *satisfy* $\deg(f) = \deg(g) > 1$. *If* $\mathcal{O}_f(\alpha) \cap \mathcal{O}_g(\beta)$ *is infinite, then* $f^k = g^k$ *for some* $k \in \mathbb{N}$.

The hypothesis that $\deg(f) = \deg(g) = d > 1$ in Theorem 5.3.0.1 allows us to reduce Theorem 5.3.0.1 to the following special case of the Dynamical Mordell-Lang Conjecture.

THEOREM 5.3.0.2. *Let K be a number field, let $\alpha, \beta \in K$, and let $f, g \in K[z]$ satisfy $\deg(f) = \deg(g) > 1$. Let $\Delta \subset \mathbb{A}^2$ be the diagonal line, and let $\Phi : \mathbb{A}^2 \longrightarrow \mathbb{A}^2$ be defined as*

$$(5.3.0.3) \qquad \qquad \Phi(x, y) = (f(x), g(y)).$$

If the intersection

$$\mathcal{O}_\Phi(\alpha, \beta) \cap \Delta \text{ is infinite,}$$

then there exists $k \in \mathbb{N}$ such that

$$\Phi^k(\Delta) = \Delta, \text{ i.e. } f^k = g^k.$$

Theorem 5.3.0.2 is the case of Conjecture 1.5.0.1 for the diagonal line in the affine plane under the action of an endomorphism Φ as in (5.3.0.3) (see the equivalence between the Dynamical Mordell-Lang Conjecture and Conjecture 3.1.3.1 proven in Subsection 3.1.3).

5.3.1. Theorem 5.3.0.2 yields Theorem 5.3.0.1.

PROOF OF THEOREM 5.3.0.1. So, we assume Theorem 5.3.0.2 holds and we prove Theorem 5.3.0.1.

Let $d := \deg(f) = \deg(g)$. Since both $\mathcal{O}_f(\alpha)$ and $\mathcal{O}_g(\beta)$ are infinite, we conclude that α and β are not respectively preperiodic under the actions of f and g. We claim that $|m - n|$ is uniformly bounded for all pairs

$$(m, n) \text{ such that } f^m(\alpha) = g^n(\beta).$$

Indeed, using Proposition 2.6.4.2 (a)—(b), there exists a positive constant c such that for all $z \in \overline{\mathbb{Q}}$ we have

$$(5.3.1.1) \qquad \qquad |\widehat{\mathrm{h}}_f(z) - \widehat{\mathrm{h}}_g(z)| < c.$$

Using inequality (5.3.1.1) for $z = f^m(\alpha) = g^n(\beta)$, and also using that

$$\widehat{\mathrm{h}}_f(f^m(\alpha)) = d^m \cdot \widehat{\mathrm{h}}_f(\alpha) \text{ and } \widehat{\mathrm{h}}_g(g^n(\beta)) = d^n \cdot \widehat{\mathrm{h}}_g(\beta),$$

we conclude that

$$|d^m \widehat{\mathrm{h}}_f(\alpha) - d^n \widehat{\mathrm{h}}_g(\beta)| < c.$$

Furthermore, both $\widehat{\mathrm{h}}_f(\alpha)$ and $\widehat{\mathrm{h}}_g(\beta)$ are nonzero since neither α nor β is preperiodic under the action of f, respectively of g (see Proposition 2.6.4.2 (c)). Hence $|m - n|$ is uniformly bounded as claimed. Therefore, without loss of generality, we may assume that there exists $\ell \in \mathbb{N}_0$ such that

$$f^m(\alpha) = g^{m+\ell}(\beta)$$

for infinitely many $m \in \mathbb{N}$. Thus, the diagonal line Δ in the affine plane contains infinitely many points in common with the orbit $\mathcal{O}_{(f,g)}(\alpha, g^\ell(\beta))$ and hence by Theorem 5.3.0.2, there exists $k \in \mathbb{N}$ such that $f^k = g^k$. $\qquad \square$

5.4. Proof of Theorem 5.3.0.2

In Subsection 5.3.1 we proved that Theorem 5.3.0.2 yields Theorem 5.3.0.1. In this section we prove Theorem 5.3.0.2, using the arguments from [**GTZ08**].

We continue with the notation from Theorem 5.3.0.2; in particular we assume there exist infinitely many $k \in \mathbb{N}$ such that

$$f^k(\alpha) = g^k(\beta).$$

Also, we work under the assumption that

$$\deg(f) = \deg(g) = d \geq 2,$$

and that there exists a number field K containing α, β and all coefficients of f and g.

5.4.1. A useful theorem of Bilu and Tichy. Now, we let S be the finite set of places of K containing all archimedean places, and all non-archimedean places v such that at least one of α, β, or one of the coefficients of f and g is not integral at v. Then it is immediate to see that for all $k \in \mathbb{N}_0$, we have that

$$f^k(\alpha) \text{ and } g^k(\beta) \text{ are } S\text{-integers of } K.$$

Knowing that there exist infinitely many $k \in \mathbb{N}$ such that $f^k(\alpha) = g^k(\beta)$, we conclude that for *each* $k \in \mathbb{N}$, the plane curve C_k given by the equation

$$f^k(X) - g^k(Y) = 0$$

contains infinitely many S-integral points in K. Theorem 2.3.7.1 yields then the existence of an irreducible component $C_{k,0}$ of C_k which is a *Siegel curve*, i.e. $C_{k,0}$ has genus 0 and at most two points at infinity. Moreover, Bilu and Tichy [**BT00**] proved a strengthening (see Theorem 5.4.1.2) of Siegel's Theorem which gives a complete list of the polynomials f and g for which the curve $C := C_1$ given by the equation

$$f(X) - g(Y) = 0$$

contains a Siegel curve. We note from the beginning that we state Theorem 5.4.1.2 with the extra hypothesis that the polynomials f and g have the same degrees which simplifies the classification obtained by Bilu and Tichy [**BT00**]. In order to state this classification result we introduce the notation for a special class of polynomials of two variables denoted $D_n(X, Y)$ which may be viewed as a homogenized version of the Chebyshev polynomials (see also Definition 5.7.0.1).

DEFINITION 5.4.1.1. For each positive integer n we define the polynomial $D_n(X, Y)$ be the unique polynomial in $\mathbb{Z}[X, Y]$ such that

$$D_n(U + V, UV) = U^n + V^n.$$

It is immediate to see that $D_n(X, 0) = X^n$ (for all $n \in \mathbb{N}$) and, for $\alpha \in \mathbb{C}$ we have

$$\alpha^n D_n(X, 1) = D_n(\alpha X, \alpha^2).$$

THEOREM 5.4.1.2 (Bilu-Tichy [**BT00**]). *Let K be a number field, and let S be a finite set of places of K containing all the archimedean places of K. Let $F, G \in K[z]$ be polynomials of same degree and assume*

$$F(x) = G(y)$$

has infinitely many solutions in the ring of S-integers of K. Then

$$F = E \circ F_1 \circ a \text{ and } G = E \circ G_1 \circ b,$$

where $E, a, b \in K[z]$ with $\deg(a) = \deg(b) = 1$, and (F_1, G_1) or (G_1, F_1) are one of the following pairs:

1. (z, z);
2. $(z^2, c \circ z^2)$ *with $c \in K[z]$ linear;*
3. $(D_2(z, \alpha)/\alpha, D_2(z, \beta)/\beta)$ *with $\alpha, \beta \in K^*$;*
4. $(D_n(z, \alpha), -D_n(z \cos(\pi/n), \alpha))$ *with $\alpha \in K$, where $n \in \mathbb{N}$ is such that $\cos(2\pi/n)$ is in K.*

In the proof from [**GTZ08**] we only need the coarser information that F_1 and G_1 assume only finitely many shapes modulo composition with linear polynomials. We note that for $n \geq 3$ we have $[\mathbb{Q}(\cos(2\pi/n)) : \mathbb{Q}] = \phi(n)/2$ and hence degree considerations make it straightforward to find restrictions on the set of n for which $\cos(2\pi/n) \in K$, where K is a number field. In particular, there are only finitely many such n.

COROLLARY 5.4.1.3. *Let K, S, F, G satisfy the hypotheses of Theorem 5.4.1.2. Then*

$$F = E \circ H \circ a \text{ and } G = E \circ c \circ H \circ b$$

for some $E \in K[z]$, some linear $a, b, c \in K[z]$, and some $H = D_n(X, \alpha)$ with $\alpha \in \{0, 1\}$ and $n \in \mathbb{N}$ satisfying $\cos(2\pi/n) \in K$. In particular, for fixed K, there are only finitely many possibilities for H (even if we vary S, F, G).

Note very importantly that we can apply Theorem 5.4.1.2 and Corollary 5.4.1.3 not only to our polynomials f and g, but also to any pair of polynomials (f^k, g^k) with $k \in \mathbb{N}$ since each curve C_k contains infinitely many S-integral points of K.

5.4.2. Ritt's theorem regarding the decomposition of polynomials. The proof of Theorem 5.3.0.2 uses the following fundamental result of Ritt [**Rit20**] (for more recent results on polynomial decomposition, see [**MS14, MZ**]).

THEOREM 5.4.2.1 (Ritt [**Rit20**]). *Let K be a field of characteristic zero. If $A, B, C, D \in K[z] \setminus K$ satisfy $A \circ B = C \circ D$ and $\deg(B) = \deg(D)$, then there is a linear $\ell \in K[z]$ such that $A = C \circ \ell$ and $B = \ell \circ D$.*

The following technical result follows as a consequence of Theorem 5.4.2.1.

LEMMA 5.4.2.2. *Let K be a field of characteristic zero. Suppose that*

$$F, H, E, \tilde{E} \in K[z] \setminus K \text{ and linear } a, b, c, \tilde{a}, \tilde{b}, \tilde{c} \in K[z]$$

satisfy

$$F = E \circ H \circ a$$
$$G = E \circ c \circ H \circ b$$
$$F^k = \tilde{E} \circ H \circ \tilde{a}$$
$$G^k = \tilde{E} \circ \tilde{c} \circ H \circ \tilde{b}$$

for some integer $k > 1$. Then there is a linear $e \in K[z]$ such that $F^{k-1} = G^{k-1} \circ e$.

PROOF. See [**GTZ08**, Lemma 3.1]. □

Lemma 5.4.2.2 yields the first relation between the iterates of f and of g.

LEMMA 5.4.2.3. *Let K be a number field, let S be a finite set of places of K containing all archimedean places of K, and let $f, g \in K[z]$ with*

$$\deg(f) = \deg(g) > 1.$$

Suppose that for every $k \in \mathbb{N}$ the equation

$$f^k(X) = g^k(Y)$$

has infinitely many solutions in the ring of S-integers of K. Then there exists $r \in \mathbb{N}$ such that, for both $n = 1$ and infinitely many other values $n \in \mathbb{N}$, there is a linear $\ell_n \in K[z]$ such that

$$f^{rn} = g^{rn} \circ \ell_n.$$

PROOF. First we show that there exists $r \in \mathbb{N}$ such that

$$f^r = g^r \circ \ell \text{ for some linear } \ell \in \overline{K}[X].$$

By Corollary 5.4.1.3, for each m we have

$$f^{2^m} = E_m \circ H_m \circ a_m \text{ and } g^{2^m} = E_m \circ c_m \circ H_m \circ b_m$$

with $E_m \in \overline{K}[z]$, linear $a_m, b_m, c_m \in \overline{K}[z]$, and some $H_m \in \overline{K}[z]$ which comes from a finite set of polynomials. Thus

$$H_m = H_s \text{ for some } m \text{ and } s \text{ with } m < s.$$

Applying Lemma 5.4.2.2 with $F = f^{2^m}$ and $G = g^{2^m}$ and $k = 2^{s-m}$, it follows that there is a linear $\ell \in \overline{K}[z]$ such that

$$F^{k-1} = G^{k-1} \circ \ell, \text{ and thus}$$

$$f^r = g^r \circ \ell \text{ for } r = 2^s - 2^m.$$

Suppose there are only finitely many $n \in \mathbb{N}$ for which there is a linear $\ell_n \in \overline{K}[z]$ with $f^{rn} = g^{rn} \circ \ell_n$. Let N be an integer exceeding each of these finitely many integers n. We get a contradiction by applying the argument from the previous paragraph with (f^{rN}, g^{rN}) in place of (f, g). $\square$

5.4.3. The key result in the proof of Theorem 5.3.0.2. The following result (see [**GTZ08**, Proposition 3.3]) is key to the proof of Theorem 5.3.0.2.

PROPOSITION 5.4.3.1. *Let K be a field of characteristic zero, and let $F, \ell \in K[z]$ satisfy $\deg(F) = d > 1 = \deg(\ell)$. Suppose that, for infinitely many $n \in \mathbb{N}$, there is a linear $\ell_n \in K[z]$ such that $F^n = (F \circ \ell)^n \circ \ell_n$. Then either*

(1) *$F^k = (F \circ \ell)^k$ for some $k \in \mathbb{N}$; or*
(2) *$F(z) = v^{-1} \circ \epsilon z^d \circ v$ and $\ell = v^{-1} \circ \delta z \circ v$ for some linear $v \in K[z]$ and some $\epsilon, \delta \in K^*$.*

Before proving Proposition 5.4.3.1, we state two technical lemmas which are used in its proof.

LEMMA 5.4.3.2. *If H is a polynomial defined over a field K of characteristic 0, such that $u \circ H \circ v$ has at least two nonzero monomial terms for all linear polynomials u and v, then there exist at most finitely many linear polynomials a and b such that $H \circ b = a \circ H$. In fact, the number of solutions is at most the size of the largest group of roots of unity in K of order less than $\deg(H)$.*

PROOF. This is proven in [**GTZ08**, Lemma 4.1] as a consequence of an old theorem of Hällström [**Hal57**]. $\square$

LEMMA 5.4.3.3. *For any linear polynomial ℓ, and for any polynomial H defined over a field K of characteristic 0, if there exist linear polynomials $u, v \in K[z]$ such that $H(z) = u \circ z^d \circ v$ (for some $d > 1$), then the following statements are equivalent:*

(1) *the equation*

$$(5.4.3.4) \qquad\qquad H \circ \ell \circ H \circ b = a \circ H \circ H$$

has infinitely many solutions in linears $a, b \in K[z]$;

(2) $H(z) = v^{-1} \circ \epsilon z^d \circ v$ *and* $\ell(z) = v^{-1} \circ \delta z \circ v$ *for some* $\epsilon, \delta \in K^*$.

PROOF. This is proven in [**GTZ08**, Lemma 4.3] as an application of Theorem 5.4.2.1. $\square$

PROOF OF PROPOSITION 5.4.3.1. We have

$$(5.4.3.5) \qquad\qquad (F \circ \ell)^n \circ \ell_n = F^n$$

for every n in some infinite subset $\mathcal{M}$ of $\mathbb{N}$. For $n \in \mathcal{M}$, we apply Theorem 5.4.2.1 to the identity (5.4.3.5) with

$$B = F \circ \ell \circ \ell_n \text{ and } D = F,$$

and conclude that there is a linear $u_n \in K[z]$ such that

$$F \circ \ell \circ \ell_n = u_n \circ F.$$

By Lemma 5.4.3.2, if F is not obtained from a monomial by composing with linears on both sides then $\{\ell_n : n \in \mathcal{M}\}$ is finite.

Next, for $n \in \mathcal{M}$ with $n > 1$, apply Lemma 5.4.2.1 to (5.4.3.5) with

$$B = (F \circ \ell)^2 \circ \ell_n \text{ and } D = F^2,$$

and conclude that there is a linear $v_n \in K[z]$ such that

$$(F \circ \ell)^2 \circ \ell_n = v_n \circ F^2.$$

Using Lemma 5.4.3.3, if F is obtained from a monomial by composing with linears on both sides, then either $\{\ell_n : n \in \mathcal{M}\}$ is finite or conclusion (2) of Proposition 5.4.3.1 holds.

Thus, whenever (2) of Proposition 5.4.3.1 does not hold, the set $\{\ell_n : n \in \mathcal{M}\}$ is finite, so there exist $n, N \in \mathcal{M}$ such that $\ell_n = \ell_N$ and $n < N$. Then

$$
\begin{aligned}
F^{N-n} \circ F^n = F^N &= (F \circ \ell)^N \circ \ell_n \\
&= (F \circ \ell)^{N-n} \circ (F \circ \ell)^n \circ \ell_n \\
&= (F \circ \ell)^{N-n} \circ F^n,
\end{aligned}
$$

so $F^{N-n} = (F \circ \ell)^{N-n}$, as desired. $\square$

5.4.4. Finishing the proof of Theorem 5.3.0.2. Now we have all ingredients to finish the proof of Theorem 5.3.0.2.

PROOF OF THEOREM 5.3.0.2. Let S be a finite set of places of K such that the ring of S-integers $\mathfrak{o}_{K,S}$ contains α, β, and every coefficient of f and g. Then $\mathfrak{o}_{K,S}$ contains every $f^n(\alpha)$ and $g^n(\beta)$ with $n \in \mathbb{N}$.

Our hypotheses imply that α is not preperiodic for f, and β is not preperiodic for g. Moreover, for every $k \in \mathbb{N}$, the equation $f^k(x) = g^k(y)$ has infinitely many solutions $(x, y) \in \mathfrak{o}_{K,S} \times \mathfrak{o}_{K,S}$.

By Lemma 5.4.2.3, there is some $r \in \mathbb{N}$ such that, for both $n = 1$ and infinitely many $n \in \mathbb{N}$, we have

$$f^{rn} = g^{rn} \circ \ell_n \text{ with } \ell_n \in \overline{K}[z] \text{ linear.}$$

Let $F := f^r$ and $\ell = \ell_1^{-1}$. Then $g^r = F \circ \ell$, and for infinitely many n we have $F^n = (F \circ \ell)^n \circ \ell_n$. If F and $F \circ \ell$ have a common iterate, then so do f and g. By Proposition 5.4.3.1, it remains only to consider the case that

$$F(z) = v^{-1} \circ \epsilon z^d \circ v \text{ and } \ell(z) = v^{-1} \circ \delta z \circ v,$$

where $v \in \overline{K}[z]$ is linear and $\epsilon, \delta \in \overline{K}^*$. Note that $d > 1$.

By hypothesis, the set $\mathcal{M}$ of pairs $(m, n) \in \mathbb{N} \times \mathbb{N}$ satisfying

$$f^m(\alpha) = g^n(\beta)$$

is infinite, and (from non-preperiodicity of both α and β with respect to the actions of f, and respectively of g) its projections onto each coordinate are injective. Thus, for some $s_1, s_2 \in \mathbb{N}$, the set $\mathcal{M}$ contains infinitely many pairs

$$(rm + s_1, rn + s_2) \text{ with } m, n \in \mathbb{N}.$$

Since the projections are injective, $\mathcal{M}$ contains pairs of this form in which $\min(m, n)$ is arbitrarily large. For any $m, n \in \mathbb{N}$ such that $(rm + s_1, rn + s_2) \in \mathcal{M}$, we have

$$F^m(\alpha_1) = (F \circ \ell)^n(\beta_1),$$

where $\alpha_1 := f^{s_1}(\alpha)$ and $\beta_1 := g^{s_2}(\beta)$. Thus

$$v^{-1}\left(\epsilon^{\frac{d^m-1}{d-1}} v(\alpha_1)^{d^m}\right) = F^m(\alpha_1)$$
$$= (F \circ \ell)^n(\beta_1)$$
$$= v^{-1}\left((\epsilon\delta^d)^{\frac{d^n-1}{d-1}} v(\beta_1)^{d^n}\right),$$

so

(5.4.4.1) $$v(\alpha_1)^{d^m} \epsilon^{\frac{d^m-d^n}{d-1}} = \delta^{\frac{d(d^n-1)}{d-1}} v(\beta_1)^{d^n}.$$

We cannot have $v(\alpha_1) = 0$, since otherwise $\alpha_1 = f^{s_1}(x_0)$ is a fixed point of $F = f^r$, contrary to our hypotheses. Likewise $v(\beta_1) \neq 0$. Now let $\epsilon_1, \delta_1 \in \overline{K}$ satisfy

$$\epsilon_1^{d-1} = \epsilon \text{ and } \delta_1^{d-1} = \delta^d,$$

so (5.4.4.1) implies

(5.4.4.2) $$\delta_1 = v(\alpha_1)^{-d^m} \cdot \epsilon_1^{d^n-d^m} \cdot \delta_1^{d^n} \cdot v(\beta_1)^{d^n}.$$

Since (5.4.4.2) holds for pairs (m, n) with $\min(m, n)$ arbitrarily large, there are infinitely many $k \in \mathbb{N}$ for which δ_1 is a d^k-th power in the number field

$$K_0 := \mathbb{Q}(v(\alpha_1), v(\beta_1), \epsilon_1, \delta_1).$$

Letting $\mathfrak{o}_{K_0}$ be the ring of algebraic integers in K_0, it follows that the fractional ideal of $\mathfrak{o}_{K_0}$ generated by δ_1 is a d^k-th power for infinitely many k. Now unique factorization of fractional ideals implies δ_1 is in the unit group U of $\mathfrak{o}_{K_0}$. Moreover, δ_1 is a d^k-th power in U for infinitely many k. Since U is a finitely generated abelian group, δ_1 must be a root of unity whose order N is coprime to d. Thus

$$N \mid (d^q - 1) \text{ for some positive integer } q.$$

Hence

$$(F \circ \ell)^q(z) = v^{-1} \circ (\epsilon \delta^d)^{\frac{d^q-1}{d-1}} z^{d^q} \circ v,$$

and since $\delta^d = \delta_1^{d-1}$ and $\delta_1^{d^q-1} = 1$, we get that $(F \circ \ell)^q = F^q$, as desired. $\qquad\square$

5.5. The general case of Theorem 5.3.0.1

We recall that in Subsection 5.3.1 we proved the special case of Theorem 5.2.0.1 when the coefficients of the polynomials f and g and the starting points are algebraic numbers and also $\deg(f) = \deg(g)$ (see Theorem 5.3.0.1) as a consequence of Theorem 5.3.0.2; then in Section 5.4 we proved Theorem 5.3.0.2.

Now, one can prove a generalization of Theorem 5.3.0.2 when $\deg(f) = \deg(g)$ and the coefficients of the polynomials f and g, and also the starting points α and β are complex numbers using the exact same strategy employed in Section 5.4. Indeed, Theorem 5.4.1.2 (see [**BT00**]) holds even when the corresponding polynomials are defined over a finitely generated extension of $\mathbb{Q}$; even though the statement from [**BT00**] is for polynomials defined over a number field, all the arguments from [**BT00**] are done for polynomials defined over an arbitrary field of characteristic 0. Furthermore, one may use the following generalization due to Lang [**Lan83**, Thm. 8.2.4 and 8.5.1] of Siegel's classical theorem [**Sie29**] (see Theorem 2.3.7.1).

THEOREM 5.5.0.1. *Let K be a finitely generated field of characteristic zero, and let R be a finitely generated subring of K. Let C be a smooth, projective, geometrically irreducible curve over K, and let ϕ be a non-constant function in $K(C)$. Suppose that there are infinitely many points $P \in C(K)$ which are not poles of ϕ and which satisfy $\phi(P) \in R$. Then C has genus zero and ϕ has at most two distinct poles.*

Theorem 5.5.0.1 allows us to extend Theorem 5.4.1.2 to the case of an arbitrary finitely generated field K, thus proving Theorem 5.3.0.2 for polynomials f and g defined over $\mathbb{C}$ of same degree. Then one can employ a similar reduction using the theory of heights as in Subsection 5.3.1 (though one would have to work with heights corresponding to function fields and define properly the minimal field of definition for the polynomials f and g, but essentially the same ideas would work) to prove that the extension of Theorem 5.3.0.2 over $\mathbb{C}$ yields the following result.

THEOREM 5.5.0.2. *Let $f, g \in \mathbb{C}[z]$ such that $\deg(f) = \deg(g) = d > 1$. If there exist $\alpha, \beta \in \mathbb{C}$ such that $\mathcal{O}_f(\alpha) \cap \mathcal{O}_g(\beta)$ is infinite, then there exists $k \in \mathbb{N}$ such that $f^k = g^k$.*

However, we prefer to prove Theorem 5.5.0.2 by inferring it directly from Theorem 5.3.0.1 (without appealing to a generalization of Theorem 5.3.0.2 to $\mathbb{C}$) using specialization techniques. We do this since we find this method of specialization (and its by-product results) useful for future applications in dynamics. We give the proof in the next section (see also [**GTZ08**]).

5.6. The method of specialization and the proof of Theorem 5.5.0.2

The goal of this section is the proof of Theorem 5.5.0.2. We describe first how to obtain Theorem 5.5.0.2 from Theorem 5.3.0.1 using specialization techniques, and then we prove that the hypotheses necessary for our application of a suitable specialization are met. We note that we follow the proof from [**GTZ08**].

PROOF OF THEOREM 5.5.0.2. *(under the additional assumption of the existence of a suitable specialization).*

It suffices to prove Theorem 5.5.0.2 in case $f, g \in K[z]$ and also $\alpha, \beta \in K$, where K is a finitely generated extension of $\mathbb{Q}$. We will prove Theorem 5.5.0.2 by induction on the transcendence degree of $K/\mathbb{Q}$. The base case is Theorem 5.3.0.1. For the inductive step, let E be a subfield of K such that $\operatorname{trdeg}(K/E) = 1$ and $E/\mathbb{Q}$ is finitely generated. Suppose in addition that the diagonal is not periodic under the (f, g) action (i.e., there is no $k \in \mathbb{N}$ for which $f^k = g^k$), and that the set

$$\{(f^m(\alpha), g^n(\beta)) : m, n \in \mathbb{N}\}$$

has infinite intersection with the diagonal. Assume there is a subring R of K, a finite extension E' of E, and a homomorphism

$$\sigma : R \to E',$$

such that denoting by f_σ, g_σ, and α_σ the images of f, g, and α, respectively, under the homomorphism σ, then the following conditions are met:

    (1) R contains α, β, and every coefficient of f and g, but the leading coefficients of f and g have nonzero image under σ;

    (2) $f_\sigma^k \neq g_\sigma^k$ for each $k \in \mathbb{N}$;

    (3) α_σ is not preperiodic for f_α.

Geometrically, we can phrase as follows the conditions (1)—(3). The field K may be viewed as the function field of a smooth, projective, geometrically irreducible curve C defined over E (one can achieve this at the expense of replacing E and K by finite extensions). Then both f and g extend as rational maps on $\mathbb{P}^1$ over an open dense subset of C; specializing at each point $\sigma \in C$ yields rational maps f_σ and g_σ. Similarly, each point $\gamma \in K$ may be viewed as a section of $\mathbb{P}^1_C$ (for the definition of $\mathbb{P}^1_C$, see Subsection 2.1.13), and therefore its intersection with the fiber above $\sigma \in C$ is denoted by γ_σ. Then, in the above purely algebraic setup, the subring R of K is the valuation ring corresponding to the point $\sigma \in C$ (i.e., $R = \mathcal{O}_{C,\sigma}$ with the notation as in Section 2.1), the field E' is the residue field of this valuation, and the homomorphism $R \to E'$ is simply the map $\gamma \mapsto \gamma_\sigma$.

Properties (1) and (3) show that

$$\{(f_\sigma^m(\alpha_\sigma), g_\sigma^n(\beta_\sigma)) : m, n \in \mathbb{N}\} \text{ has infinite intersection with } \Delta,$$

where $\Delta \subset \mathbb{A}^2$ is the usual diagonal line. The inductive hypothesis implies

$$f_\sigma^k = g_\sigma^k$$

for some $k \in \mathbb{N}$, which contradicts property (2). Then Theorem 5.5.0.2 follows immediately. $\qquad\square$

To explain why there exists a σ as in the proof of Theorem 5.5.0.2, we use the geometric setup introduced in the above proof which assumed the existence of a suitable specialization. So, by replacing K by a finite extension, and then replacing E with its algebraic closure in K, we may assume that there exists a smooth, projective, geometrically irreducible curve C over E whose function field is K. Let

$$\pi : \mathbb{P}^1_C \to C$$

be the natural fibration (see Subsection 2.1.13). Any $z \in \mathbb{P}^1_K$ gives rise to a section

$$Z : C \to \mathbb{P}^1$$

of π, and for $\sigma \in C(\overline{E})$, we let $z_\sigma := Z(\sigma)$, and let $E(\sigma)$ be the residue field of K at the valuation corresponding to σ. The polynomial $f \in K[X]$ extends to a rational map (of E-varieties) from $\mathbb{P}^1_C$ to itself, whose generic fiber is f, and whose fiber above any $\sigma \in C$ is f_σ. Note that f_σ is a morphism of degree $\deg(f)$ from the fiber

$$(\mathbb{P}^1_C)_\sigma = \mathbb{P}^1_{E(\sigma)}$$

to itself whenever the coefficients of f have no poles at σ, while the leading coefficient of f does not lie in the maximal ideal of the valuation ring corresponding to σ; hence it is a morphism on $\mathbb{P}^1_{E(\sigma)}$ of degree $\deg(f)$ at all but finitely many σ (we call these σ places of *good reduction* for f; see also Definition 6.1.1.1).

We will show that most choices of σ satisfy conditions (2) and (3) from the proof of Theorem 5.5.0.2; obviously all but finitely many σ satisfy condition (1) for the specialization used in the proof of Theorem 5.5.0.2. First, we prove the following result about specializations of polynomials.

PROPOSITION 5.6.0.3. *For each $r > 0$, there are at most finitely many $\sigma \in C(\overline{E})$ such that $[E(\sigma) : E] \leq r$ and $f_\sigma^k = g_\sigma^k$ for some $k \in \mathbb{N}$.*

PROOF. First, using Theorem 5.2.2.1 in the special case of two polynomials of the same degree we deduce (see [**GTZ08**, Corollary 6.4]) that if f and g are non-linear polynomials of same degree defined over a field K of characteristic 0, if

$$f^k = g^k \text{ for some } k \in \mathbb{N}$$

then

$$f^n = g^n \text{ for some } n \in \mathbb{N} \text{ such that } n \leq N_K,$$

where N_K is the number of roots of unity contained in K.

Pick a point σ on C such that $[E(\sigma) : E] \leq r$ and $f_\sigma^k = g_\sigma^k$ for some $k \in \mathbb{N}$. Let N_σ be the number of roots of unity in $E(\sigma)$. By the observation from the above paragraph, the least $n \in \mathbb{N}$ with $f_\sigma^n = g_\sigma^n$ satisfies $n \leq N_\sigma$.

Now, N_σ is bounded in terms of the degree

$$[E(\sigma) \cap \overline{\mathbb{Q}} : \mathbb{Q}] \leq r \cdot [E \cap \overline{\mathbb{Q}} : \mathbb{Q}].$$

Since E is finitely generated, then

$$[E \cap \overline{\mathbb{Q}} : \mathbb{Q}] \text{ is finite.}$$

So, there is a finite bound on n which depends only on E and r (and not on σ) such that

$$f_\sigma^n = g_\sigma^n.$$

For any fixed $n \in \mathbb{N}$, we have

(5.6.0.4) $$\deg(f_\sigma^n - g_\sigma^n) = \deg(f^n - g^n)$$

for all but finitely many $\sigma \in C$. Since

$$f^n \neq g^n,$$

(5.6.0.4) applied for the finitely many positive integers $n \leq N_\sigma$ yields that there are at most finitely many $\sigma \in C$ such that

$$f_\sigma^n = g_\sigma^n,$$

as desired. $\square$

Next, letting h_C be the Weil height on C associated to a fixed degree-one ample divisor, we prove the following dynamical analogue of Silverman's specialization result for abelian varieties [**Sil83**, Thm. C].

PROPOSITION 5.6.0.5. *There exists $c > 0$ such that, for $\sigma \in C(\overline{E})$ with $h_C(\sigma) > c$, the point α_σ is not preperiodic for f_σ.*

PROOF. First note that E is a global field. The key ingredient in our proof is the following result of Call and Silverman [**CS93**, Thm. 4.1], which relates h_C to the canonical heights

$$\widehat{h}_f : \overline{K} \to \mathbb{R}_{\geq 0} \text{ of } f, \text{ and}$$

$$\widehat{h}_{f_\sigma} : \overline{E} \to \mathbb{R}_{\geq 0} \text{ of } f_\sigma.$$

LEMMA 5.6.0.6 (Call-Silverman [**CS93**]). *For each $z \in K$ we have*

$$(5.6.0.7) \qquad \lim_{h_C(\sigma) \to \infty} \frac{\widehat{h}_{f_\sigma}(z_\sigma)}{h_C(\sigma)} = \widehat{h}_f(z).$$

If $\widehat{h}_f(\alpha) > 0$ then, by Lemma 5.6.0.6, there exists $c > 0$ such that every $\sigma \in C(\overline{E})$ with $h_C(\sigma) > c$ satisfies

$$\frac{\widehat{h}_{f_\sigma}(z_\sigma)}{h_C(\sigma)} > 0.$$

Then $\widehat{h}_{f_\sigma}(z_\sigma) > 0$, so z_σ is not preperiodic for f_σ.

If f is not isotrivial, Lemma 2.6.6.5 implies $\widehat{h}_f(\alpha) > 0$, so the proof is complete. It remains only to consider the case that f is isotrivial and $\widehat{h}_f(\alpha) = 0$.

Then there is a finite extension K' of K and a linear $\ell \in K'[z]$ such that

$$g := \ell^{-1} \circ f \circ \ell$$

has coefficients in E', where E' is the algebraic closure of E in K'. Lemma 2.6.6.6 implies

$$w := \ell^{-1}(\alpha) \in E'.$$

Moreover, since

$$\ell^{-1} \circ f^n(\alpha) = g^n(w)$$

and α is not preperiodic for f, we see that w is not preperiodic for g. Since

$$g \in E'[z] \text{ and } w \in E',$$

we see that for all places σ' of K', the reductions of g and w at σ' equal g, and respectively w (because E' embeds naturally into the residue field at σ'). Hence, for all but finitely many σ' (we only need to exclude the places where ℓ does not have good reduction), if σ is the place of K lying below σ', then z_σ is not preperiodic for f_σ. $\qquad\square$

Propositions 5.6.0.3 and 5.6.0.5 allow us to finish the proof of Theorem 5.5.0.2.

PROOF OF EXISTENCE OF A SUITABLE SPECIALIZATION IN THEOREM 5.5.0.2. Let

$$\phi : C \to \mathbb{P}^1_E$$

be any non-constant rational function, and let $r = \deg(\phi)$. By [**Lan83**, Prop. 4.1.7], there are positive constants c_1 and c_2 such that for all $P \in \mathbb{P}^1(\overline{E})$, the preimage $\sigma = \phi^{-1}(P)$ satisfies

$$h_C(\sigma) \geq c_1 h(P) + c_2.$$

Since there are infinitely many $P \in \mathbb{P}^1(E)$ such that

$$h(P) > \frac{c - c_2}{c_1},$$

we thus obtain infinitely many $\sigma \in C(\overline{E})$ such that

$$h_C(\sigma) > c \text{ and } [E(\sigma) : E] \leq r.$$

Hence, Propositions 5.6.0.3 and 5.6.0.5 yield that there are infinitely many $\sigma \in C(\overline{E})$ satisfying conditions (2) and (3) for a suitable specialization as used in the proof of Theorem 5.5.0.2. Also, clearly all but finitely many of these σ satisfy condition (1) as well. $\qquad\square$

This completes the proof of Theorem 5.5.0.2.

5.7. The case of Theorem 5.2.0.1 when the polynomials have different degrees

So, in Section 5.6 we proved Theorem 5.5.0.2 which is Theorem 5.2.0.1 in the case $\deg(f) = \deg(g)$. In this section we complete the proof of Theorem 5.2.0.1 by reducing the general case to the special case proven in Theorem 5.5.0.2.

The general case of Theorem 5.2.0.1 when $\deg(f) \neq \deg(g)$ is much harder. One needs some delicate results about polynomial decompositions in order to obtain the full Theorem 5.2.0.1. One employs a similar strategy to that used in the special case $\deg(f) = \deg(g)$; the difference is that the polynomial decomposition work is much more difficult. The main reason for this is that, when analyzing functional equations involving f^n and g^n in case $\deg(f) = \deg(g)$, one could use the fact that if $A, B, C, D \in \mathbb{C}[z] \setminus \mathbb{C}$ satisfy

$$A \circ B = C \circ D \text{ and } \deg(A) = \deg(C)$$

then

$$C = A \circ \ell \text{ and } D = \ell^{-1} \circ B$$

for some linear $\ell \in \mathbb{C}[z]$ (see Theorem 5.2.2.1). When f and g have distinct degrees, one must use a different approach. The proof relies on the full strength of the description given in [**MZ**] for the collection of all decompositions of a polynomial; in addition, several new types of polynomial decomposition arguments are used in [**GTZ12**]. The crucial polynomial decomposition result needed is the one proved by Zieve and included in [**MZ**, Theorem 3.1]. In order to state this remarkable result of Zieve regarding polynomial decomposition, we recall the definition of Chebyshev polynomials.

DEFINITION 5.7.0.1. By T_n we mean the (normalized) degree-n *Chebyshev polynomial* (of the first kind), which is defined by the equation

$$T_n(z + z^{-1}) = z^n + z^{-n}.$$

The *classical Chebyshev polynomial* C_n defined by $C_n(\cos\theta) = \cos n\theta$ for all $\theta \in \mathbb{R}$ satisfies

$$2C_n(z/2) = T_n(z).$$

THEOREM 5.7.0.2 (Zieve [**MZ**]). *Pick $f \in \mathbb{C}[z]$ with $\deg(f) = n > 1$, and suppose that there is no linear $\ell \in \mathbb{C}[z]$ such that $\ell \circ f \circ \ell^{-1}$ is either z^n or T_n or $-T_n$. Let $r, s \in \mathbb{C}[z]$ and $d \in \mathbb{N}$ satisfy $r \circ s = f^d$. Then we have*

$$r = f^i \circ R$$
$$s = S \circ f^j$$
$$R \circ S = f^k$$

where $R, S \in \mathbb{C}[z]$ and $i, j, k \in \mathbb{N}_0$ with $k \leq \log_2(n + 2)$.

The proof of Theorem 5.7.0.2 relies on the full strength of the description given in [**MZ**] for the collection of all decompositions of a polynomial; this in turn depends on the classical results of Ritt [**Rit20**] combined with several very clever arguments.

5.7.1. Outline of the proof of Theorem 5.2.0.1. We work under the hypotheses of Theorem 5.2.0.1; hence there exists a finitely generated field K such that $f, g \in K[z]$ and $\alpha, \beta \in K$, and moreover we assume that

$$\mathcal{O}_f(\alpha) \cap \mathcal{O}_g(\beta) \text{ is infinite.}$$

Our goal is to prove that $f^k = g^k$ for some $k \in \mathbb{N}$.

Arguing as in the proof of Theorem 5.3.0.1, for every m, n we find that the polynomial

(5.7.1.1) $f^n(x) - g^m(y)$ has a Siegel factor;

(see Subsection 2.3.7). Indeed, we let S be a finite set of places of K containing all archimedean places of K (if any) such that each coefficient of f and of g is S-integral, and also both α and β are S-integers. Then the plane curve $C_{n,m}$ defined by the equation

$$f^n(x) - g^m(y) = 0$$

contains infinitely many points of the form

$$\left(f^k(\alpha), g^\ell(\beta) \right),$$

for some $k, \ell \in \mathbb{N}$; each such point has both coordinates S-integral. Hence Theorem 5.5.0.1 can be applied to the curve $C_{n,m}$ and derive that it contains an irreducible curve which is a Siegel curve, i.e., it has genus 0 and at most two points at infinity. We recall that the Siegel factor refers to an irreducible factor of the polynomial from (5.7.1.1) for which the corresponding plane curve is a Siegel curve.

In [**GTZ12**] it is shown that the condition expressed by (5.7.1.1) for all $n, m \in \mathbb{N}$ implies that either

$$f^k = g^k \text{ for some } k \in \mathbb{N},$$

or there is a linear $\ell \in \mathbb{C}[z]$ such that

$$(\ell \circ f \circ \ell^{-1}, \ell \circ g \circ \ell^{-1})$$

is either

$$(aX^r, bX^s) \text{ for some } a, b \in \mathbb{C}^*$$

or

$$(\pm T_r, \pm T_s),$$

where T_r is the degree-r Chebyshev polynomial as in Definition 5.7.0.1. Then one uses a consequence of Siegel's theorem to handle these last possibilities.

Siegel's seminal result (see our Theorem 5.5.0.1) was already used for deriving the refined form from [**BT00**] for polynomials F and G such that the curve $F(X) - G(Y) = 0$ has a Siegel factor. The other way Siegel's Theorem is used in the proof from [**GTZ12**] comes in the following consequence due to Lang [**Lan60**], which in turn it is a special case of Laurent's proof [**Lau84**] of the Mordell-Lang conjecture for the multiplicative group $\mathbb{G}_m^N$ (see Theorem 3.4.1.1).

COROLLARY 5.7.1.2 (Lang [**Lan60**]). *Let $a, b \in \mathbb{C}^*$, and let Γ be a finitely generated subgroup of $\mathbb{C}^* \times \mathbb{C}^*$. Then the equation $ax + by = 1$ has at most finitely many solutions $(x, y) \in \Gamma$.*

PROOF. Corollary 5.7.1.2 is proved in [**Lan60**] by applying Theorem 5.5.0.1 to the genus-1 curves $a\alpha X^3 + b\beta Y^3 = 1$, where (α, β) runs through a finite subset of Γ which surjects onto Γ/Γ^3. □

5.7.2. Sketch of the proof of Theorem 5.2.0.1. We conclude Section 5.7 by providing more details to the outline of the proof presented in Subsection 5.7.1.

Now, [**GTZ12**, Proposition 4.1] (see Proposition 5.7.2.1) generalizes [**GTZ08**, Proposition 3.3] (see Proposition 5.4.3.1) as follows.

PROPOSITION 5.7.2.1. *Pick $f, g \in \mathbb{C}[z]$ for which $r := \deg(f)$ and $s := \deg(g)$ satisfy $r, s > 1$. Suppose that for every $m \in \mathbb{N}$, there exists $n \in \mathbb{N}$ and $h \in \mathbb{C}[z]$ such that*

$$g^n = f^m \circ h.$$

Then either f and g have a common iterate, or there is a linear $\ell \in \mathbb{C}[z]$ such that

$$(\ell \circ f \circ \ell^{-1}, \ell \circ g \circ \ell^{-1})$$

is either

$$(az^r, z^s) \text{ for some } a \in \mathbb{C}^*$$

or

$$(T_r \circ \hat{\epsilon}z, T_s \circ \epsilon z) \text{ with } \hat{\epsilon}, \epsilon \in \{1, -1\}.$$

If f and g satisfy the hypothesis of Theorem 5.2.0.1, but they do not satisfy the hypothesis of Proposition 5.7.2.1, one can prove the following result (see [**GTZ12**, Proposition 5.1]).

PROPOSITION 5.7.2.2. *Pick $f, g \in \mathbb{C}[z]$ for which $r := \deg(f)$ and $s := \deg(g)$ satisfy $r, s > 1$. Assume there exists $m \in \mathbb{N}$ with these properties:*

(1) *$g^n \neq f^m \circ h$ for every $h \in \mathbb{C}[z]$ and $n \in \mathbb{N}$; and*
(2) *there are infinitely many $j \in \mathbb{N}$ for which $f^{mj}(X) - g^{mj}(Y)$ has a Siegel factor in $\mathbb{C}[X, Y]$.*

Then there is a linear $\ell \in \mathbb{C}[X]$ for which

$$(\ell \circ f \circ \ell^{-1}, \ell \circ g \circ \ell^{-1})$$

is either

$$(z^r, az^s) \text{ for some } a \in \mathbb{C}^*$$

or

$$(\epsilon_1 T_r, \epsilon_2 T_s) \text{ with } \epsilon_1, \epsilon_2 \in \{1, -1\}.$$

Finally, we combine Propositions 5.7.2.1 and 5.7.2.2 and therefore reduce Theorem 5.2.0.1 to the case that the pair (f, g) has one of the two forms:

(5.7.2.3) $\qquad\qquad (z^r,\ bz^s),\ \text{ with } b \in \mathbb{C}^* \text{ and } r, s \in \mathbb{Z}_{>1};$

(5.7.2.4) $\qquad\qquad (\epsilon_1 T_r,\ \epsilon_2 T_s),\ \text{ with } \epsilon_1, \epsilon_2 \in \{1, -1\} \text{ and } r, s \in \mathbb{Z}_{>1}.$

Proposition 5.7.2.5 finishes the proof of Theorem 5.5.0.2.

PROPOSITION 5.7.2.5. *Pick $f, g \in \mathbb{C}[X]$ such that (f, g) has one of the forms (5.7.2.3) or (5.7.2.4). If there are $\alpha, \beta \in \mathbb{C}$ for which $\mathcal{O}_f(\alpha) \cap \mathcal{O}_g(\beta)$ is infinite, then f and g have a common iterate.*

PROOF. We follow the proof of [**GTZ12**, Proposition 6.5]. Assuming
$$\mathcal{O}_f(\alpha) \cap \mathcal{O}_g(\beta) \text{ is infinite,}$$
let M be the set of pairs $(m, n) \in \mathbb{N} \times \mathbb{N}$ for which $f^m(\alpha) = g^n(\beta)$. Note that any two elements of M have distinct first coordinates, since if M contains (m, n_1) and (m, n_2) with $n_1 \neq n_2$ then
$$g^{n_1}(\beta) = g^{n_2}(\beta)$$
and so, $\mathcal{O}_g(\beta)$ would be finite. Likewise, any two elements of M have distinct second coordinates, so there are elements $(m, n) \in M$ in which $\min(m, n)$ is arbitrarily large.

Suppose (f, g) has the form (5.7.2.3). Since
$$f^m(z) = z^{r^m}$$
and $\mathcal{O}_f(\alpha)$ is infinite, α is neither zero nor a root of unity. We compute
$$g^n(\beta) = b^{\frac{s^n - 1}{s - 1}} \beta^{s^n};$$
putting $\beta_1 := b_1 \beta$ where $b_1 \in \mathbb{C}^*$ satisfies $b_1^{s-1} = b$, it follows that
$$g^n(\beta) = \frac{\beta_1^{s^n}}{b_1},$$
so the infinitude of $\mathcal{O}_g(\beta)$ implies that β_1 is neither zero nor a root of unity. A pair
$$(m, n) \in \mathbb{N} \times \mathbb{N}$$
lies in M if and only if

(5.7.2.6) $\qquad\qquad \alpha^{r^m} = b^{\frac{s^n - 1}{s - 1}} \beta^{s^n},$

or equivalently

(5.7.2.7) $\qquad\qquad b_1 \alpha^{r^m} = \beta_1^{s^n}.$

Since (5.7.2.7) holds for two pairs $(m, n) \in M$ which differ in both coordinates, we have
$$\alpha^k = \beta_1^\ell$$
for some nonzero integers k, ℓ. By choosing k to have minimal absolute value, it follows that the set
$$S := \{(u, v) \in \mathbb{Z}^2 : b_1 \alpha^u = \beta_1^v\}$$
has the form
$$\{(c + kt, d + \ell t) : t \in \mathbb{Z}\}$$

for some $c, d \in \mathbb{Z}$. For $(m, n) \in M$ we have

$$(r^m, s^n) \in S, \text{ so } \frac{r^m - c}{k} = \frac{s^n - d}{\ell}.$$

Since M is infinite, Corollary 5.7.1.2 gives that

$$\frac{c}{k} = \frac{d}{\ell}.$$

In particular, every

$$(m, n) \in M \text{ satisfies } kr^m = \ell s^n.$$

Pick two pairs (m, n) and $(m + m_0, n + n_0)$ in M with $m_0, n_0 \in \mathbb{N}$. Then

$$r^{m_0} = s^{n_0},$$

and S contains both

$$(r^m, s^n) \text{ and } \left(r^{m+m_0}, s^{n+n_0} \right),$$

so

$$\beta_1^{s^n} \alpha^{-r^m} = b_1 = \beta_1^{s^{n+n_0}} \alpha^{-r^{m+m_0}},$$

and thus

$$(\beta_1^{s^n})^{s^{n_0} - 1} = (\alpha^{r^m})^{r^{m_0} - 1}.$$

Since $r^{m_0} = s^{n_0}$, it follows that $b_1^{s^{n_0} - 1} = 1$, and thus

$$f^{m_0} = g^{n_0}.$$

Now suppose (f, g) has the form (5.7.2.4). Then for any $m, n \in \mathbb{N}$ there exist $\epsilon_3, \epsilon_4 \in \{1, -1\}$ such that

$$(f^m, g^n) = (\epsilon_3 T_{r^m}, \epsilon_4 T_{s^n}).$$

Since

$$\mathcal{O}_f(\alpha) \cap \mathcal{O}_g(\beta) \text{ is infinite,}$$

we can choose $\delta \in \{1, -1\}$ such that

$$T_{r^m}(\alpha) = \delta T_{s^n}(\beta)$$

for infinitely many $(m, n) \in \mathbb{N} \times \mathbb{N}$. Pick $\alpha_0, \beta_0 \in \mathbb{C}^*$ such that

$$\alpha_0 + \alpha_0^{-1} = \alpha \text{ and } \beta_0 + \beta_0^{-1} = y.$$

Then there are infinitely many pairs $(m, n) \in \mathbb{N} \times \mathbb{N}$ for which

$$\alpha_0^{r^m} + \alpha_0^{-r^m} = \delta(\beta_0^{s^n} + \beta_0^{-s^n}),$$

so we can choose $\epsilon \in \{1, -1\}$ such that

(5.7.2.8) $$\alpha_0^{r^m} = \delta \beta_0^{\epsilon s^n}$$

for infinitely many $(m, n) \in \mathbb{N} \times \mathbb{N}$. Moreover, since $\mathcal{O}_f(\alpha)$ and $\mathcal{O}_g(\beta)$ are infinite, neither α_0 nor β_0 is a root of unity, so distinct pairs $(m, n) \in \mathbb{N} \times \mathbb{N}$ which satisfy (5.7.2.8) must differ in both coordinates. Now (5.7.2.8) is a reformulation of (5.7.2.7), so we conclude as above that

$$r^{m_0} = s^{n_0}$$

for some $m_0, n_0 \in \mathbb{N}$ such that $\delta^{s^{n_0} - 1} = 1$. If s is odd, it follows that

$$f^{2m_0} = g^{2n_0}.$$

If s is even then we cannot have $\delta = -1$; since

$$f^m = \epsilon_1 T_{r^m} \text{ and } g^n = \epsilon_2 T_{s^n},$$

it follows that $\epsilon_1 = \epsilon_2$, and then again

$$f^{m_0} = g^{n_0},$$

as desired. $\square$

5.8. An alternative proof for the function field case

We can prove a "function field" version of Theorem 5.2.0.1 using only the theory of heights, and also using the fact that we know the special case $\deg(f) = \deg(g)$ in Theorem 5.2.0.1 as stated in Theorem 5.5.0.2. The following result is [**GTZ12**, Theorem 8.1].

THEOREM 5.8.0.1 ([**GTZ12**]). *Let K be a field of characteristic 0, let $f, g \in K[z]$ be polynomials of degree greater than one, and let $\alpha, \beta \in K$. Assume there is no linear $\mu \in \overline{K}[z]$ for which $\mu^{-1}(\alpha), \mu^{-1}(\beta) \in \overline{\mathbb{Q}}$ and both $\mu^{-1} \circ f \circ \mu$ and $\mu^{-1} \circ g \circ \mu$ are in $\overline{\mathbb{Q}}[z]$. If $\mathcal{O}_f(x_0) \cap \mathcal{O}_g(y_0)$ is infinite, then f and g have a common iterate.*

The theory of heights is used to reduce Theorem 5.8.0.1 to the case $\deg(f) = \deg(g)$ and then the result follows from Theorem 5.5.0.2. The advantage in this approach is that one avoids all the complicated polynomial decomposition arguments which were necessary to deal with the general case of Theorem 5.2.0.1. We decided to include Theorem 5.8.0.1 in our book not only since it provides a partial answer to the general case of Theorem 5.2.0.1 avoiding the complicated polynomial decomposition arguments invoked in [**GTZ12**], but also since we believe that the method of proof of Theorem 5.8.0.1 could be useful for other similar questions in arithmetic dynamics. In particular, the proof of Theorem 5.8.0.1 relies heavily on the somewhat surprising property of the canonical height for points under the action of polynomials defined over function fields of taking values only rational numbers (see Lemma 2.6.6.1).

Intuitively, here is the reason for which we are able to prove this "function field" version of Theorem 5.2.0.1. Over a function field, the canonical height with respect to any polynomial is a rational number (see Lemma 2.6.6.1). So, letting H_1 and H_2 be the canonical heights of α respectively β under the action of the polynomials f respectively g (defined over a suitable function field), then for each pair $(m, n) \in \mathbb{N} \times \mathbb{N}$ such that

$$(5.8.0.2) \qquad\qquad f^m(\alpha) = g^n(\beta),$$

we get that

$$(5.8.0.3) \qquad\qquad |\deg(f)^m \cdot H_1 - \deg(g)^n \cdot H_2| < C$$

for some constant C which is independent of (m, n); actually, C depends only on the difference between the usual Weil height and the corresponding canonical heights constructed with respect to f and g (see Proposition 2.6.4.2 (a)—(b)). Because H_1 and H_2 are rational numbers, (5.8.0.3) yields that there exist finitely many possibilities for the quantity

$$\deg(f)^m \cdot H_1 - \deg(g)^n \cdot H_2$$

as we vary m and n. Assuming there exist infinitely many pairs (m, n) satisfying (5.8.0.2) we get that there exists a rational number γ and there exists an infinite set $S \subset \mathbb{N} \times \mathbb{N}$ such that for each $(m, n) \in S$ we have

$$\deg(f)^m \cdot H_1 - \deg(g)^n \cdot H_2 = \gamma.$$

Hence, the plane curve given by the equation

$$H_1 x - H_2 y = \gamma$$

contains infinitely many points in common with the subgroup

$$\Gamma \subset \mathbb{Q}^* \times \mathbb{Q}^*$$

spanned by $(\deg(f), 1)$ and $(1, \deg(g))$. An easy application of an old theorem of Lang [**Lan60**] (see Corollary 5.7.1.2) finishes the proof of Theorem 5.8.0.1.

Now, in order to formalize the argument sketched above, we first prove two easy claims (which are [**GTZ12**, Claims 11.1 and 11.2]).

CLAIM 5.8.0.4. *Let E be any subfield of K, and assume that (f, α) and (g, β) are isotrivial over E (see Definition 2.6.6.7). If $\mathcal{O}_f(\alpha) \cap \mathcal{O}_g(\beta)$ is infinite, then there exists a linear $\mu \in \overline{K}[z]$ such that $\mu \circ f \circ \mu^{-1}, \mu \circ g \circ \mu^{-1} \in \overline{E}[z]$ and $\mu(\alpha), \mu(\beta) \in \overline{E}$.*

PROOF. We reproduce here the proof of [**GTZ12**, Cl. 11.1]. We know that there exist linear $\mu_1, \mu_2 \in \overline{K}[z]$ such that

$$f_1 := \mu_1 \circ f \circ \mu_1^{-1} \in \overline{E}[z] \text{ and } g_1 := \mu_2 \circ g \circ \mu_2^{-1} \in \overline{E}[z],$$

and $\alpha_1 := \mu_1(\alpha) \in \overline{E}$ and $\beta_1 := \mu_2(\beta) \in \overline{E}$. Thus

$$\mathcal{O}_{f_1}(\alpha_1) = \mu_1(\mathcal{O}_f(\alpha)) \text{ and } \mathcal{O}_{g_1}(\beta_1) = \mu_2(\mathcal{O}_g(\beta)).$$

Since $\mathcal{O}_f(\alpha) \cap \mathcal{O}_g(\beta)$ is infinite, there are infinitely many pairs $(x_1, x_2) \in \overline{E} \times \overline{E}$ such that $\mu_1^{-1}(x_1) = \mu_2^{-1}(x_2)$. Thus $\mu := \mu_2 \circ \mu_1^{-1} \in \overline{E}[z]$. Hence

$$\mu_1 \circ g \circ \mu_1^{-1} = \mu^{-1}(\mu_2 \circ g \circ \mu_2^{-1})\mu \in \overline{E}[z],$$

and

$$\mu_1(\beta) = (\mu_1 \circ \mu_2^{-1})(\beta_1) = \mu^{-1}(\beta_1) \in \overline{E},$$

as claimed. $\qquad\square$

CLAIM 5.8.0.5. *Working under the hypotheses of Theorem 5.8.0.1, if*

$$\mathcal{O}_f(\alpha) \cap \mathcal{O}_g(\beta) \text{ is infinite,}$$

then there exist subfields $E \subseteq F \subseteq K$ such that F is a function field of transcendence degree 1 over E, and there exists a linear polynomial $\mu \in \overline{K}[z]$ such that

$$\mu \circ f \circ \mu^{-1}, \mu \circ g \circ \mu^{-1} \in \overline{F}[z], \text{ and } \mu(\alpha), \mu(\beta) \in \overline{F},$$

and either (f, α) or (g, β) is non-isotrivial over E.

PROOF OF CLAIM 5.8.0.5. We reproduce here the proof of [**GTZ12**, Claim 11.2]. Let K_0 be a finitely generated subfield of K such that $f, g \in K_0[z]$ and $\alpha, \beta \in K_0$. Then there exists a finite tower of field subextensions:

$$K_s \subseteq K_{s-1} \subseteq \cdots \subseteq K_1 \subseteq K_0$$

such that K_s is a number field, and for each $i = 0, \ldots, s-1$, the extension K_i/K_{i+1} is finitely generated of transcendence degree 1. Using Claim 5.8.0.4 and the hypotheses of Theorem 5.8.0.1, we conclude that there exists $i = 0, \ldots, s-1$, and there exists a linear $\mu \in \overline{K_0}[z]$ such that

$$\mu \circ f \circ \mu^{-1}, \mu \circ g \circ \mu^{-1} \in \overline{K_i}[z], \text{ and } \mu(\alpha), \mu(\beta) \in \overline{K_i},$$

and either (f, α) or (g, β) is not isotrivial over K_{i+1}. $\qquad\square$

PROOF OF THEOREM 5.8.0.1. We follow the proof of [**GTZ12**, Theorem 8.1]. Let E, F and μ all be as in the conclusion of Claim 5.8.0.5. At the expense of replacing f and g with their respective conjugates by μ, and at the expense of replacing F by a finite extension, we may assume that

$$f, g \in F[z], \text{ and } \alpha, \beta \in F, \text{ and } (f, \alpha) \text{ is not isotrivial over } E.$$

Let $d_1 := \deg(f)$ and $d_2 := \deg(g)$. We construct the canonical heights $\widehat{h}_f$ and $\widehat{h}_g$ associated to the polynomials f and g, with respect to the set of absolute values associated to the function field F/E (see Chapter 2.6). Since (f, α) is non-isotrivial, and since α is not preperiodic for f (note that $\mathcal{O}_f(\alpha) \cap \mathcal{O}_g(\beta)$ is infinite), Lemma 2.6.6.5 yields that

$$H_1 := \widehat{h}_f(\alpha) > 0.$$

Moreover, if

$$H_2 := \widehat{h}_g(y_0),$$

then using Lemma 2.6.6.1, we have that

$$H_1, H_2 \in \mathbb{Q}.$$

Since there exist infinitely many pairs $(m, n) \in \mathbb{N} \times \mathbb{N}$ such that

$$f^m(\alpha) = g^n(\beta),$$

Proposition 2.6.4.2 (a)—(b) yields that

(5.8.0.6) $$|d_1^m \cdot H_1 - d_2^n \cdot H_2| \text{ is bounded}$$

for infinitely many pairs $(m, n) \in \mathbb{N} \times \mathbb{N}$. Since $H_1, H_2 \in \mathbb{Q}$, we conclude that there exist *finitely* many rational numbers $\gamma_1, \ldots, \gamma_s$ such that

$$\gamma_i = d_1^m \cdot H_1 - d_2^n \cdot H_2$$

for each pair (m, n) as in (5.8.0.6); here we are using the fact that there are finitely many rational numbers of bounded denominator, and bounded absolute value. Therefore, there exists a rational number $\gamma := \gamma_i$ (for some $i = 1, \ldots, s$) such that

(5.8.0.7) $$d_1^m H_1 - d_2^n H_2 = \gamma.$$

for infinitely many pairs $(m, n) \in \mathbb{N} \times \mathbb{N}$. Hence, the line $L \subseteq \mathbb{A}^2$ given by the equation

$$H_1 \cdot X - H_2 \cdot Y = \gamma$$

has infinitely many points in common with the rank-2 subgroup

$$\Gamma := \{(d_1^{k_1}, d_2^{k_2}) : k_1, k_2 \in \mathbb{Z}\} \subset \mathbb{G}_m^2(\mathbb{C}).$$

Using Corollary 5.7.1.2 (see also Theorem 3.4.1.1), we obtain that $\gamma = 0$. Since there are infinitely many pairs (m, n) satisfying (5.8.0.7), and since $H_1 \neq 0$, we conclude that there exist positive integers m_0 and n_0 such that

$$d_1^{m_0} = d_2^{n_0}; \text{ thus } \deg(f^{m_0}) = \deg(g^{n_0}).$$

Since $\mathcal{O}_f(\alpha) \cap \mathcal{O}_g(\beta)$ is infinite, we can find $k_0, \ell_0 \in \mathbb{N}$ such that

$$\mathcal{O}_{f^{m_0}}(f^{k_0}(\alpha)) \cap \mathcal{O}_{g^{n_0}}(g^{\ell_0}(\beta)) \text{ is infinite.}$$

Since $\deg(f^{m_0}) = \deg(g^{n_0})$, we can apply Theorem 5.5.0.2 and conclude the proof of Theorem 5.8.0.1. $\qquad\square$

5.9. Possible extensions

In this short section we discuss briefly a somewhat surprising extension to the main results (Theorems 5.1.0.2 and 5.2.0.1) presented in this chapter. For Theorem 5.2.0.1 and its consequence (Theorem 5.1.0.2) we saw that using techniques from the theory of polynomial decomposition coupled with Siegel's theorem (and its subsequent refinements by Bilu and Tichy [**BT00**]) one can prove the Dynamical Mordell-Lang Conjecture for lines V in $\mathbb{A}^N$ under the coordinatewise action of N one-variable polynomials. Using the same approach, coupled also with a topological argument, Wang [**Wan**] proved a Dynamical Mordell-Lang result for endomorphisms of Riemann surfaces:

THEOREM 5.9.0.1 (Wang [**Wan**]). *Let X be a simply connected, open Riemann surface, let $f, g \in \mathrm{End}(X) \setminus \mathrm{Aut}(X)$, and let $\alpha, \beta \in X(\mathbb{C})$. If $\mathcal{O}_f(\alpha) \cap \mathcal{O}_g(\beta)$ is infinite, then f and g have a common iterate.*

The key new ingredient in proving Theorem 5.9.0.1 is a study of the fundamental groups in order to develop a theory of factoring finite maps between Riemann surfaces. Wang [**NW13**] also uses his joint result with Ng [**NW13**] in which they obtained a complete theory of Ritt's factorization for finite maps on the closed unit disk (which correspond to finite Blaschke products).

5.10. The case of plane curves

It is natural to ask how far the method presented in this chapter can be extended. One might hope that the same approach would work for proving the Dynamical Mordell-Lang Conjecture in the case when V is a curve in $(\mathbb{P}^1)^N$ under the action of an endomorphism of $(\mathbb{P}^1)^N$ given by the coordinatewise action of N one-variable rational maps, as we will explain below.

We start by making the following conjecture which we believe holds (see also [**NZa, NZb**] for a proof in the special case of affine lines under the coordinatewise action of rational maps with generic monodromy groups); also, note that Conjecture 5.10.0.1 is a special case of the Dynamical Mordell-Lang Conjecture.

CONJECTURE 5.10.0.1. *Let $N \in \mathbb{N}$, let $\alpha \in (\mathbb{P}^1)^N(\mathbb{C})$, let $C \subseteq (\mathbb{P}^1)^N$ be a curve, and let*

$$\Phi : (\mathbb{P}^1)^N \longrightarrow (\mathbb{P}^1)^N$$

be an endomorphism of the form

$$\Phi(x_1, \ldots, x_N) := (f_1(x_1), \ldots, f_N(x_N))$$

for some rational maps $f_i \in \mathbb{C}(z)$. Then the set of all $n \in \mathbb{N}_0$ such that

$$\Phi^n(\alpha) \in C(\mathbb{C})$$

is a union of at most finitely many arithmetic progressions.

We claim that it suffices to prove Conjecture 5.10.0.1 for plane curves V; i.e., it suffices to prove the following statement.

CONJECTURE 5.10.0.2. *Let $(\alpha_1, \alpha_2) \in \mathbb{P}^1(\mathbb{C}) \times \mathbb{P}^1(\mathbb{C})$, let $C \subset \mathbb{P}^1 \times \mathbb{P}^1$ be a curve, and let*

$$\Phi : \mathbb{P}^1 \times \mathbb{P}^1 \longrightarrow \mathbb{P}^1 \times \mathbb{P}^1$$

be an endomorphism of the form

$$\Phi(x, y) = (f_1(x_1), f_2(x_2))$$

for some rational maps $f_1, f_2 \in \mathbb{C}(z)$. The set of all $n \in \mathbb{N}_0$ such that

$$\Phi^n(\alpha) \in C(\mathbb{C})$$

is a union of at most finitely many arithmetic progressions.

PROOF THAT CONJECTURE 5.10.0.2 YIELDS CONJECTURE 5.10.0.1. The argument we present is similar to the proof of [**BGKT12**, Theorem 1.4]. So, assume Conjecture 5.10.0.2 holds. We prove Theorem 5.10.0.1 by induction on N; the case $N = 2$ is Theorem 5.10.0.2. So, assume Theorem 5.10.0.1 holds for N and next we will prove it for $N + 1$. Let

$$\alpha := (\alpha_1, \ldots, \alpha_{N+1}) \in (\mathbb{P}^1)^{N+1}(\mathbb{C}).$$

Clearly, we can reduce to the case C is irreducible (by intersecting each component of C with the orbit $\mathcal{O} := \mathcal{O}_\Phi(\alpha)$). We may also assume that C projects dominantly onto each of the coordinates of $(\mathbb{P}^1)^{N+1}$; otherwise, we may view C as a curve in $(\mathbb{P}^1)^N$, and apply the inductive hypothesis. We may also assume that no α_i is preperiodic, lest C should fail to project dominantly on the i-th coordinate. Let

$$\pi_1 : (\mathbb{P}^1)^{N+1} \to (\mathbb{P}^1)^N$$

be the projection onto the first N coordinates, let

$$C_1 := \pi_1(C), \text{ and let } \mathcal{O}_1 := \pi_1(\mathcal{O}).$$

By our assumptions, C_1 is an irreducible curve that has an infinite intersection with $\mathcal{O}_1$. By the inductive hypothesis and the reformulation of the Dynamical Mordell-Lang Conjecture as stated in Conjecture 3.1.3.1, C_1 is periodic under the coordinatewise action of Φ restricted on the first N coordinates of $(\mathbb{P}^1)^{N+1}$; we denote by Φ_1 this restriction of Φ on the last N coordinates of $(\mathbb{P}^1)^{N+1}$. We recall from Section 2.2 our definition of periodic (closed) subvarieties:

$$\Phi_1^m(C_1) \subseteq C_1 \text{ for some } m \in \mathbb{N}.$$

Similarly, let C_2 be the projection of C on the last N coordinates of $(\mathbb{P}^1)^{N+1}$. By the same argument, C_2 is periodic under the coordinatewise action of Φ restricted on the last N coordinates of $(\mathbb{P}^1)^{N+1}$.

Thus, C is Φ-preperiodic, because it is an irreducible component of the one-dimensional variety

$$\left(C_1 \times \mathbb{P}^1\right) \cap \left(\mathbb{P}^1 \times C_2\right),$$

and because both $C_1 \times \mathbb{P}^1$ and $\mathbb{P}^1 \times C_2$ are Φ-periodic. The following Claim (proven in [**BGKT12**, Claim 5.2]) will therefore complete the proof of the inductive step in Conjecture 5.10.0.1.

CLAIM 5.10.0.3. *Let X be a variety defined over $\mathbb{C}$, let $\alpha \in X(\mathbb{C})$, let $\Phi : X \longrightarrow X$ be a morphism, and let $C \subseteq X$ be an irreducible curve that has infinite intersection with the orbit $\mathcal{O}_\Phi(\alpha)$. If C is Φ-preperiodic, then C is Φ-periodic.*

PROOF OF CLAIM 5.10.0.3. Assume C is not periodic. Since C is preperiodic, there exist $k_0, n_0 \geq 1$ such that $\Phi^{n_0}(C)$ is periodic of period k_0. Let

$$k := n_0 k_0, \text{ and let } C' := \overline{\Phi^k(C)};$$

so

$$\Phi^k(C') \subseteq C'.$$

(We recall, that $\overline{T}$ denotes the Zariski closure of the set T.) Then $C \neq C'$, since C is not periodic. Since C and C' are irreducible curves (C' is irreducible since it is the Zariski closure of the image of an irreducible variety through a morphism), it follows that

(5.10.0.4) $C \cap C'$ is finite.

On the other hand, there exists $\ell \in \{0, \dots, k-1\}$ such that

$$C(\mathbb{C}) \cap \mathcal{O}_{\Phi^k}(\Phi^\ell(\alpha)) \text{ is infinite,}$$

because we know that $C \cap \mathcal{O}_\Phi(\alpha)$ is infinite. Let $n_1 \in \mathbb{N}$ be the smallest $n \in \mathbb{N}_0$ such that

$$\Phi^{nk+\ell}(\alpha) \in C(\mathbb{C}).$$

Since $C' = \overline{\Phi^k(C)}$ is mapped by Φ^k into itself, we conclude that

$$\Phi^{nk+\ell}(\alpha) \in C'(\mathbb{C})$$

for each $n \geq n_1 + 1$. Therefore

(5.10.0.5) $C(\mathbb{C}) \cap \mathcal{O}_{\Phi^k}(\Phi^\ell(\alpha)) \cap C'$ is infinite.

Statements (5.10.0.4) and (5.10.0.5) are contradictory, proving the claim.   □

Hence it is sufficient that one proves Conjecture 5.10.0.1 for $N = 2$.   □

We note that Conjecture 5.10.0.1 was proven by Xie [**Xieb**, Theorem 0.3] in the special case when the maps f_i are polynomials defined over $\overline{\mathbb{Q}}$.

THEOREM 5.10.0.6 (Xie [**Xieb**]). *Let $f_1, \dots, f_N \in \overline{\mathbb{Q}}[z]$, let $C \subset \mathbb{A}^N$ be a curve defined over $\overline{\mathbb{Q}}$, let $\alpha \in \mathbb{A}^N(\overline{\mathbb{Q}})$, and let*

$$\Phi : \mathbb{A}^N \longrightarrow \mathbb{A}^N$$

be defined by

$$\Phi(x_1, \dots, x_N) := (f_1(x_1), \dots, f_N(x_N)).$$

Then the set

$$S(C, \Phi, \alpha) := \{n \in \mathbb{N}_0 \colon \Phi^n(\alpha) \in C(\overline{\mathbb{Q}})\}$$

is a union of finitely many arithmetic progressions.

We believe it should be possible to extend Theorem 5.10.0.6 to the case polynomials f_i are defined over $\mathbb{C}$ using a specialization argument similar to the one employed in [**BGKT12**] (see also Section 7.2) using the famous classification made by Medvedev and Scanlon [**MS14**] for the invariant plane curves under the action of coordinatewise one-variable complex polynomials.

In order to prove Theorem 5.10.0.6, Xie used the reduction to Conjecture 5.10.0.2 for the case of polynomial endomorphisms. Besides several new results on the dynamics of polynomial endomorphisms of the affine plane (some similar to the ones previously introduced in [**Xie14**]), Xie used a finer theory of valuation rings of $\mathbb{C}[x, y]$ (as proven in [**Xiea**]), but also he used Siegel's theorem similarly to the way we employed it in the proof of Theorem 5.2.0.1.

The general case of rational maps in Conjecture 5.10.0.1 remains open. As proven in this section, it suffices to prove Conjecture 5.10.0.2. So, working under the hypotheses of Conjecture 5.10.0.2, one can assume that α_i is not f_i-preperiodic for $i = 1, 2$. Assuming that C contains infinitely many points in common with

$\mathcal{O}_{(f_1, f_2)}(\alpha_1, \alpha_2)$ we may conclude as in the proof of Theorem 5.2.0.1 (using Theorem 5.5.0.1) that C is a curve of genus 0 with at most 2 points at infinity. In particular this yields that C admits a parametrization of the form

$$(Q_1(t), Q_2(t))$$

where each Q_i is a Laurent polynomial, i.e. there exists $k_i \in \mathbb{N}_0$ such that

$$z^{k_i} Q_i(z) \in \mathbb{C}[z].$$

The reason for this is that C has at most two points at infinity and so the parametrization of C (note that C is a rational curve because it has genus equal to 0) is made by rational functions which map at most two points of $\mathbb{P}^1$ to ∞. Furthermore, for each $n \in \mathbb{N}$, the curve parametrized by

$$(Q_1(f_1^n(t)), Q_2(f_2^n(t)))$$

also has an irreducible component which is a Siegel curve, which in turn leads to analyzing equations of the form

(5.10.0.7) $$a \circ b = c \circ d,$$

where each a and c are Laurent polynomials, while b and d are rational maps. Theoretically this might lead to a solution of Conjecture 5.10.0.2 (similar to the arguments from [**GTZ08, GTZ12**]); Nguyen and Zieve [**NZa, NZb**] announced the proof of this result when C is an affine line under the action of rational maps with generic monodromy groups. The problem with solving completely Question 5.10.0.2 along this line of approach for all rational maps and all curves lies in finding a complete classification of the tuples (a, b, c, d) satisfying (5.10.0.7) (for more details on this difficult problem see [**Zieb**]).

5.11. A Dynamical Mordell-Lang type question for polarizable endomorphisms

We can combine Theorem 4.3.0.1 with the observations regarding heights from this chapter in order to prove the following result, which is another special case of Question 3.6.0.1 in the spirit of Question 5.1.2.1.

THEOREM 5.11.0.1. *Let X be a projective variety defined over a number field K. Let Φ and Ψ be polarizable étale endomorphisms of X of the same degree $d > 1$ with respect to the same very ample line bundle $\mathcal{L}$. Let $\alpha, \beta \in X(K)$. Then the set*

$$S = \{(m, n) \in \mathbb{N}_0 \times \mathbb{N}_0 \colon \Phi^m(\alpha) = \Psi^n(\beta)\}$$

is a union of at most finitely many sets of the form

(5.11.0.2) $$\{(i + kr, j + \ell r) \colon r \in \mathbb{N}_0\}$$

for some $i, j, k, \ell \in \mathbb{N}_0$.

PROOF. If either α or β are preperiodic, then the conclusion is obvious (see Proposition 3.1.2.9); so, from now on, we assume neither α nor β is preperiodic for Φ, respectively for Ψ. Since K is a number field, the points of canonical height equal to 0 are precisely the preperiodic points. This last fact is proven in [**CS93**]; for the case of rational maps, see Proposition 2.6.4.2 (c). Hence we have

$$\gamma := \widehat{h}_\Phi(\alpha) > 0 \text{ and } \delta := \widehat{h}_\Psi(\beta) > 0.$$

Furthermore, the difference between the canonical heights $\widehat{h}_\Phi(\cdot)$ or $\widehat{h}_\Psi(\cdot)$ and the usual Weil height $h_\mathcal{L}(\cdot)$ constructed with respect to the very ample line bundle $\mathcal{L}$ (see Subsection 2.6.8) is uniformly bounded on $X(\overline{K})$. This fact is again proven in [**CS93**]; for the case of rational maps, see Proposition 2.6.4.2 (b). So, we know that there exists a positive constant c such that for all $k \in \mathbb{N}_0$ we have

$$\max\left\{\left|\widehat{h}_\Phi(\Phi^k(\alpha)) - h_\mathcal{L}(\Phi^k(\alpha))\right|, \left|\widehat{h}_\Psi(\Psi^k(\beta)) - h_\mathcal{L}(\Psi^k(\beta))\right|\right\} < c.$$

So, for each (m, n) in the set S from the conclusion of the above theorem, we have

$$\left|\widehat{h}_\Phi(\Phi^m(\alpha)) - \widehat{h}_\Psi(\Psi^n(\beta))\right| < 2c,$$

i.e., $|d^m\gamma - d^n\delta| < 2c$. Because both γ and δ are positive, we conclude that there exists $s \in \mathbb{N}$ independent of (m, n) such that $|m - n| < s$. For each

$$i \in \{-(s-1), \ldots, s-1\},$$

we let

$$S_i := \{(n + i, n) \in \mathbb{N}_0 \times \mathbb{N}_0 \colon \Phi^{n+i}(\alpha) = \Psi^n(\beta)\}.$$

We treat next only the case $i \geq 0$; the case $i < 0$ is similar. Then letting $\alpha_i := \Phi^i(\alpha)$, and letting

$$T_i = \{n \in \mathbb{N}_0 \colon \Phi^n(\alpha_i) = \Psi^n(\beta)\},$$

in order to show that each S_i has the form (5.11.0.2), it suffices to prove T_i is a union of at most finitely many arithmetic progressions. This follows immediately by applying Theorem 4.3.0.1 to the étale endomorphism

$$(\Phi, \Psi) : X \times X \longrightarrow X \times X,$$

the point $(\alpha_i, \beta) \in (X \times X)(K)$, and the diagonal subvariety

$$\Delta := \{(x, x) \colon x \in X\} \subset X \times X.$$

$\square$

It would be interesting to relax the hypotheses in Theorem 5.11.0.1; for example we believe the result should hold even if the two endomorphisms do not have the same degree. On the other hand, the result is not true for arbitrary étale endomorphisms as shown by the following example.

EXAMPLE 5.11.0.3. Let X be an elliptic curve defined over $\overline{\mathbb{Q}}$, let $P \in E(\overline{\mathbb{Q}})$ be a non-torsion point, let

$$\Phi : E \longrightarrow E \text{ be given by } \Phi(Q) := Q + P,$$

and let

$$\Psi : E \longrightarrow E \text{ be given by } \Psi(Q) = 2Q \text{ for each point } Q \in E.$$

Then $\Phi^m(P) = \Psi^n(P)$ if and only if $m = 2^n - 1$.

So, the polarizability condition is essential, but the condition on degrees should not be essential, even though at the moment the known methods do not deliver the result without this condition. Hence, we advance the following question.

QUESTION 5.11.0.4. *Let X be a projective variety defined over a number field K. Let Φ and Ψ be polarizable étale endomorphisms of X with respect to the same very ample line bundle $\mathcal{L}$. Let $\alpha, \beta \in X(K)$. Then the set*

$$S = \{(m, n) \in \mathbb{N}_0 \times \mathbb{N}_0 \colon \Phi^m(\alpha) = \Psi^n(\beta)\}$$

is a union of at most finitely many sets of the form

$$\{(i + kr, j + \ell r) \colon r \in \mathbb{N}_0\}$$

for some $i, j, k, \ell \in \mathbb{N}_0$.

Now, if the canonical heights of the starting points α and β under the action of the endomorphisms Φ (resp. Ψ) were rational numbers (as it is the case for polynomial maps defined over function fields; see Lemma 2.6.6.1), then the exact same argument as in the proof of Theorem 5.8.0.1 may be employed to obtain the conclusion of Theorem 5.11.0.1 even without the hypothesis that $\deg(\Phi) = \deg(\Psi)$. However, it is unknown to us under which hypotheses, the canonical height of a point under a polarizable endomorphism is a rational number. Even in the case of a rational map

$$\Phi \colon \mathbb{P}^1 \longrightarrow \mathbb{P}^1$$

defined over a function field K/k, it is not known to us whether the canonical height of each point in $\mathbb{P}^1(K)$ is a rational number (after a suitable normalization).

The first interesting (yet unknown) case of Question 5.11.0.4 would be for polarizable étale endomorphisms Φ and Ψ of abelian varieties X. We believe in this case, Question 5.11.0.4 has a positive answer and the proof should use the results of [**GS**] for the arithmetic properties of the dynamics of an endomorphism of an abelian variety.

CHAPTER 6

Parametrization of orbits

In Chapter 5 we have seen the extent to which we can use polynomial decomposition techniques in order to prove special cases of the Dynamical Mordell-Lang Conjecture. In this chapter we discuss an alternative approach to Conjecture 1.5.0.1 which uses p-adic parametrizations of the orbit. This approach is reminiscent of the approach described in Chapter 4 for parametrizing orbits under the action of étale endomorphisms. Finding analytic parametrizations for orbits under arbitrary endomorphisms Φ of a projective space $\mathbb{P}^N$ is very difficult; we summarize in Section 6.3 the known results in this direction which can be used to solve Conjecture 1.5.0.1 in some special cases. However, if Φ is a rational map (of one variable), it is well-understood how to parametrize an orbit under the action of Φ. The reason why this is well-understood is that for a rational map Φ defined over $\mathbb{Z}_p$ (which also admits good reduction modulo p; see Subsection 6.1.1 for the definition of good reduction for rational maps), and for a point $\alpha \in \mathbb{P}^1(\mathbb{Z}_p)$ whose reduction x modulo p is fixed by $\overline{\Phi}$ (which is the reduction of Φ modulo p), there are three possibilities for describing the dynamics of Φ within the residue class of α modulo p:

(1) indifferent;
(2) attracting; or
(3) super-attracting.

Informally speaking, in a small neighborhood of an *indifferent* fixed point, the map $\varphi(z)$ is conjugate through a p-adic analytic map to the map

$$z \mapsto z + 1.$$

Similarly, in a small neighborhood of an *attracting* fixed point, the map $\varphi(z)$ is conjugate through a p-adic analytic map to the map

$$z \mapsto \lambda z,$$

where $0 < |\lambda|_p < 1$. Finally, in the case of a *super-attracting* fixed point, the map $\varphi(z)$ is conjugate through a p-adic analytic map to the map

$$z \mapsto z^k,$$

for some integer $k \geq 2$. We discuss in detail these three possibilities in Sections 6.1 and 6.2. We conclude this chapter with a discussion about p-adic analytic parametrizations for orbits of endomorphisms of higher dimensional varieties; see Section 6.3.

Before proceeding to the results of this chapter, we make the following general observation regarding the p-adic parametrizations that we discuss here, in this chapter, and also throughout the book. There are two types of such parametrizations:

- *for a single orbit*; and

- *for an entire p-adic neighborhood.*

For example, the p-adic arc lemma constructed in Theorem 4.4.2.1 is valid for all smooth points of the ambient variety X endowed with an unramified endomorphism Φ; hence the parametrization used in Chapter 4 works for entire p-adic neighborhoods. Similarly, the parametrizations we construct in this chapter for the action of a rational map (see Lemmas 6.2.1.1 and 6.2.2.1) are valid for all orbits of points in a p-adic submanifold of X. Same is true even in some cases of endomorphisms of higher dimensional varieties as presented in Theorem 6.3.0.3. Furthermore, for endomorphisms of semiabelian varieties, the p-adic parametrization of orbits comes from *global* analytic uniformization maps; for more details, see Chapter 9. A similar situation as in the case of endomorphisms of semiabelian varieties is encountered for the action of Drinfeld modules; for more details, see Chapter 12. On the other hand, in order for the p-adic arc lemma to work for the Dynamical Mordedll-Lang Conjecture, one needs simply the p-adic parametrization of that specific orbit; this could be particularly important for the case of endomorphisms of higher dimensional varieties when maybe there is no parametrization valid in an entire p-adic neighborhood.

6.1. Rational maps

First we start with some generalities regarding reduction modulo p of a rational map and we define the concept of good reduction modulo p for a rational map; see Subsection 6.1.1. Then we define the *attracting* and *super-attracting* periodic points; see Definition 6.1.3.1. Later, in Section 6.2, interpreting a rational map with good reduction modulo p as a convergent p-adic analytic series, we will introduce the various analytic parametrization of rational maps in small neighborhoods of fixed points according to the type of that fixed point: attracting or super-attracting.

6.1.1. Good reduction of rational maps.
We follow the exposition from [**BGKT10**] and [**BGKT12**]. If $\varphi : \mathbb{P}^1 \to \mathbb{P}^1$ is a morphism defined over the field K, then (fixing a choice of homogeneous coordinates) there are relatively prime homogeneous polynomials $F, G \in K[X, Y]$ of the same degree $d = \deg \varphi$ such that

$$\varphi([X,Y]) = [F(X,Y) : G(X,Y)].$$

In affine coordinates, $\varphi(z) = F(z,1)/G(z,1) \in K(z)$ is a rational function in one variable. Note that by our choice of coordinates, F and G are uniquely defined up to a nonzero constant multiple. We will need the notion of good reduction of φ, first introduced by Morton and Silverman [**MS94**].

DEFINITION 6.1.1.1. Let K be a field, let v be a non-archimedean valuation on K, let $\mathfrak{o}_v$ be the ring of v-adic integers of K, and let k_v be the residue field at v. Let

$$\varphi : \mathbb{P}^1 \longrightarrow \mathbb{P}^1$$

be a morphism over K given by

$$\varphi([X,Y]) = [F(X,Y) : G(X,Y)],$$

where $F, G \in \mathfrak{o}_v[X, Y]$ are relatively prime homogeneous polynomials of the same degree such that at least one coefficient of F or G is a unit in $\mathfrak{o}_v$. Let

$$\varphi_v := [F_v : G_v],$$

where $F_v, G_v \in k_v[X, Y]$ are the reductions of F and G modulo v. We say that φ has *good reduction* at v if
$$\varphi_v : \mathbb{P}^1(k_v) \longrightarrow \mathbb{P}^1(k_v)$$
is a morphism of the same degree as φ.

Equivalently, φ has good reduction at v if φ extends as a morphism to the fibre of $\mathbb{P}^1_{\mathrm{Spec}(\mathfrak{o}_v)}$ above v. If $\varphi \in K[z]$ is a polynomial, we can give the following elementary criterion for good reduction: φ has good reduction at v if all coefficients of φ are v-adic integers, and its leading coefficient is a v-adic unit. For simplicity, we will always use this criterion when we choose a place v of good reduction for a polynomial φ.

If K is a number field, then for all but finitely many non-archimedean places v of K, the map φ has good reduction (for more details, see [**Sil07**, Sections 2.3-2.5]).

If φ has good reduction at a place v of a number field K, then its reduction modulo v, denoted φ_v, induces a map on the residue classes modulo v, and because there are finitely many such residue classes, we conclude that each residue class modulo v is either periodic, or it maps into a periodic residue class (i.e., it is preperiodic). This prompts us to study the action of φ in a periodic residue class. At the expense of replacing φ by an iterate, it suffices to understand the action of a rational map on a fixed residue class. Here we will see that there are three types of behaviors for φ: attracting, super-attracting, or indifferent.

6.1.2. Conjugate rational maps. First we recall that a *fractional linear transformation* is a map of the form
$$z \mapsto \frac{az + b}{cz + d},$$
for some constants a, b, c, d. Unless otherwise stated, $\mu(z)$ will always denote a linear fractional transformation.

The next definition is a special case of Definition 3.1.2.12 when the ambient variety is $\mathbb{P}^1$.

DEFINITION 6.1.2.1. Two rational functions φ and ψ are *conjugate* if there is a linear fractional transformation μ such that $\varphi = \mu^{-1} \circ \psi \circ \mu$.

In the above definition, if φ and ψ are polynomials, then we may assume that μ is a polynomial of degree one.

6.1.3. Periodic points. Next we define the attracting and super-attracting periodic points.

DEFINITION 6.1.3.1. If K is a field, and $\varphi \in K(z)$ is a rational function, then $x \in \mathbb{P}^1(\overline{K})$ is a *periodic point* for φ if there exists an integer $n \geq 1$ such that
$$\varphi^n(x) = x.$$
The smallest such integer n is the *period* of x, and
$$\lambda = (\varphi^n)'(x)$$
is the *multiplier* of x. If
$$\lambda = 0,$$
then x is called *super-attracting*. If $|\cdot|_v$ is an absolute value on K, and if
$$0 < |\lambda|_v < 1,$$

then x is called *attracting*.

Let x be a periodic point of φ. If $\varphi = \mu^{-1} \circ \psi \circ \mu$, then $\mu(x)$ is a periodic point of ψ, and by the chain rule, it has the same multiplier. In particular, we can define the multiplier of a periodic point at $x = \infty$ by changing coordinates. Also by the chain rule, the multiplier of $\varphi^\ell(x)$ is the same as that of x, because

$$(\varphi^k)'(x) = \prod_{i=0}^{k-1}(\varphi'(\varphi^i(x))) = (\varphi^k)'(\varphi^\ell(x)).$$

6.1.4. Critical points. We say x is a *ramification point* or *critical point* of φ if $\varphi'(x) = 0$. If $\varphi = \mu^{-1} \circ \psi \circ \mu$, then x is a critical point of φ if and only if $\mu(x)$ is a critical point of ψ; in particular, coordinate change can again be used to determine whether $x = \infty$ is a critical point. Note that a periodic point x is super-attracting if and only if at least one of $x, \varphi(x), \varphi^2(x), \ldots, \varphi^{n-1}(x)$ is critical, where n is the period of x.

6.1.5. Exceptional points.

DEFINITION 6.1.5.1. We say that x is an *exceptional point* for the rational map φ if there exist finitely many points $y \in \mathbb{P}^1$ such that for some $n \in \mathbb{N}$ we have $\varphi^n(y) = x$.

A rational map φ of degree $d \geq 2$ can have at most two exceptional points, and if it has two such points, then φ is conjugate either to the map $z \mapsto z^d$, or to the map $z \mapsto z^{-d}$. If φ has precisely one exceptional point, then φ is conjugate to a polynomial map (which is not a single monomial).

6.2. Analytic uniformization

In this section we state formally the three types of analytic uniformization that we encounter near a fixed point of a rational map. Since the same theory applies more generally to p-adic analytic maps, we will state our results in this latter generality. We follow closely the exposition from [**BGKT10**].

Fix a prime p. As usual, let $\mathbb{C}_p$ be the completion of an algebraic closure of $\mathbb{Q}_p$. Given a point $y \in \mathbb{C}_p$ and a real number $r > 0$, write

$$D(y,r) = \{x \in \mathbb{C}_p : |x - y|_p < r\}, \quad \overline{D}(y,r) = \{x \in \mathbb{C}_p : |x - y|_p \leq r\}$$

for the open and closed disks, respectively, of radius r about y in $\mathbb{C}_p$.

We write

$$[y] \subseteq \mathbb{P}^1(\mathbb{C}_p)$$

for the residue class of a point $y \in \mathbb{P}^1(\mathbb{C}_p)$, i.e.,

$$[y] = D(y,1) \text{ if } |y|_p \leq 1,$$

or else

$$[y] = \mathbb{P}^1(\mathbb{C}_p) \setminus \overline{D}(0,1) \text{ if } |y|_p > 1.$$

The action of a p-adic power series $f \in \mathbb{Z}_p[[z]]$ on $D(0,1)$ is either *attracting* (i.e., f contracts distances) or *quasiperiodic* (i.e., f is distance-preserving), depending on its linear coefficient. Rivera-Letelier gives a more precise description of this dichotomy in [**RL03**, Sections 3.1 and 3.2]. The following two Lemmas essentially reproduce his Propositions 3.3 and 3.16 (see also [**BGKT10**, Lemmas 2.1 and 2.2]). Also note that in our Lemmas we verify that the corresponding power series have coefficients in $\mathbb{Q}_p$, not just in $\mathbb{C}_p$.

6.2.1. The attracting and the super-attracting cases.

LEMMA 6.2.1.1. *Let $f(z) = a_0 + a_1 z + a_2 z^2 + \cdots \in \mathbb{Z}_p[[z]]$ be a non-constant power series with $|a_0|_p, |a_1|_p < 1$. Then there is a point $y \in p\mathbb{Z}_p$ such that $f(y) = y$, and $\lim_{n \to \infty} f^n(z) = y$ for all $z \in D(0, 1)$. Write $\lambda = f'(y)$; then $|\lambda|_p < 1$, and:*

(1) *(Attracting). If $\lambda \neq 0$, then there is a radius $0 < r < 1$ and a power series $u \in \mathbb{Q}_p[[z]]$ mapping $\overline{D}(0, r)$ bijectively onto $\overline{D}(y, r)$ with $u(0) = y$, such that for all $z \in D(y, r)$ and $n \geq 0$,*

$$f^n(z) = u(\lambda^n u^{-1}(z)).$$

(2) *(Super-attracting). If $\lambda = 0$, then write f as*

$$f(z) = y + c_m(z - y)^m + c_{m+1}(z - y)^{m+1} + \cdots \in \mathbb{Z}_p[[z - y]]$$

with $m \geq 2$ and $c_m \neq 0$. If c_m has an $(m - 1)$-st root in $\mathbb{Z}_p$, then there are radii $0 < r, s < 1$ and a power series $u \in \mathbb{Q}_p[[z]]$ mapping $\overline{D}(0, s)$ bijectively onto $\overline{D}(y, r)$ with $u(0) = y$, such that for all $z \in D(y, r)$ and $n \geq 0$,

$$f^n(z) = u\left((u^{-1}(z))^{m^n}\right).$$

PROOF. We reproduce here the proof of [**BGKT10**, Lemma 2.1]. Using Hensel's Lemma (see Lemma 2.3.2.1) to the function $f(z) - z$ with $z = 0$, we conclude that f has a $\mathbb{Q}_p$-rational fixed point $y \in D(0, 1)$; that is, $y \in p\mathbb{Z}_p$. Clearly $\lambda = f'(y)$ is also in $p\mathbb{Z}_p$. Replacing $f(z)$ by $f(z + y) - y$ (and, ultimately, replacing $u(z)$ by $u(z) + y$), we may assume hereafter that $y = 0$. By [**RL03**, Proposition 3.2(i)], $\lim_{n \to \infty} f^n(z) = 0$ for all $z \in D(0, 1)$.

If $\lambda \neq 0$, then Rivera-Letelier defines

$$u^{-1}(z) := \lim_{n \to \infty} \lambda^{-n} f^n(z)$$

and proves in [**RL03**, Proposition 3.3(i)] that it has an inverse $u(z)$ under composition that satisfies the desired properties for some radius $0 < r < 1$. Note that $f \in \mathbb{Q}_p[[z]]$, and hence

$$\lambda^{-n} f^n \in \mathbb{Q}_p[[z]]$$

for all $n \geq 1$. Thus, $u^{-1} \in \mathbb{Q}_p[[z]]$, and therefore $u \in \mathbb{Q}_p[[z]]$ as well.

If $\lambda = 0$, then choose $\gamma \in \mathbb{Z}_p \setminus \{0\}$ with $\gamma^{m-1} = c_m$, according to the hypotheses. Define

$$\tilde{f}(z) := \gamma f(\gamma^{-1} z), \text{ so that } \tilde{f}(z) = z^m(1 + g(z)),$$

with $g \in z\mathbb{Q}_p[[z]]$. Rivera-Letelier [**RL03**, Proposition 3.3(ii)] defines

$$h(z) := \sum_{n \geq 0} m^{-n-1} \log\left(1 + g(\tilde{f}^n(z))\right) \in z\mathbb{Q}_p[[z]],$$

where $\log(1 + z) = z - z^2/2 + z^3/3 - \cdots$. Rivera-Letelier then sets

$$\tilde{u}^{-1}(z) := z \exp(h(z)),$$

where $\exp(z) = 1 + z + z^2/2! + \cdots$, and shows that the inverse $\tilde{u}$ of $\tilde{u}^{-1}$ has all the desired properties for $\tilde{f}$; note also that $\tilde{u} \in \mathbb{Q}_p[[z]]$, because

$$\log(1 + \cdot), \exp, g, \tilde{f} \in \mathbb{Q}_p[[z]].$$

Hence, $u(z) = \gamma^{-1} \tilde{u}(z) \in \mathbb{Q}_p[[z]]$ has the desired properties for f, mapping some disk $\overline{D}(0, s)$ bijectively onto some disk $\overline{D}(y, r) \subseteq D(0, 1)$. Finally, the radius s

must be less than 1, or else $u(1) \neq y$ will be fixed by f, contradicting the fact that $\lim_{n\to\infty} f^n(u(1)) = y$.   □

Lemma 6.2.1.1 shows that in the attracting case, the power series $f(z)$ is locally conjugate (through an analytic map $u(z)$) to the much simpler map $z \mapsto \lambda z$. Similarly, in the super-attracting case, Lemma 6.2.1.1 yields that $f(z)$ is analytically conjugate to the map $z \mapsto z^m$. These results are key for us since both the function $z \mapsto \lambda z$ and the function $z \mapsto z^m$ have a very simple form for their n-th iterate and hence we get easily a p-adic analytic parametrization of the orbit under f of any point in the basin of attraction.

COROLLARY 6.2.1.2. *With the notation as in Lemma 6.2.1.1, for each* $x \in D(y,r)$ *and for each* $n \in \mathbb{N}_0$, *if we let* $c := u^{-1}(x)$ *then we have that*

(1) *(Attracting).*
$$f^n(x) = u(\lambda^n \cdot c).$$

(2) *(Super-attracting).*
$$f^n(x) = u\left(c^{m^n}\right).$$

6.2.2. The indifferent case. The next Lemma describes the quasiperiodic domains (or disks) for convergent p-adic power series; we call this case the indifferent case since no point in the quasiperiodic domain gets pulled (under iteration by f) into a periodic point for the map. As noted in [**Sil07**], a point P is called *recurrent* or *quasiperiodic* if it is a limit point of its own forward orbit. In a quasiperiodic domain U, every point in U is quasiperiodic.

LEMMA 6.2.2.1. *Let* $f(z) = a_0 + a_1 z + a_2 z^2 + \cdots \in \mathbb{Z}_p[[z]]$ *be a non-constant power series with* $|a_0|_p < 1$ *but* $|a_1|_p = 1$. *Then for any non-periodic* $x \in p\mathbb{Z}_p$, *there are: an integer* $k \geq 1$, *radii* $0 < r < 1$ *and* $s \geq |k|_p$, *and a power series* $u \in \mathbb{Q}_p[[z]]$ *mapping* $\overline{D}(0,s)$ *bijectively onto* $\overline{D}(x,r)$ *with* $u(0) = x$, *such that for all* $z \in \overline{D}(x,r)$ *and* $n \geq 0$,
$$f^{nk}(z) = u(nk + u^{-1}(z)).$$

The above result is the one-dimensional case of our Theorem 4.4.2.1. We include its proof since it is much simpler in this special case. Again, this result allows us to find p-adic analytic parametrization for any orbit of a point in a quasiperiodic domain for f.

PROOF OF LEMMA 6.2.2.1. We reproduce here the proof of [**BGKT10**, Lemma 2.2].

Since $f \in \mathbb{Z}_p[[z]]$ with $|c_1|_p = 1$ and $|c_0|_p < 1$, f maps $D(0,1)$ bijectively onto itself. Therefore, by [**RL03**, Corollaire 3.12], f is quasiperiodic, which means in particular that for some $r \in (0,1)$ and for some positive integer k, the function
$$f_*(z) := \lim_{|n|_p \to 0} \frac{f^{nk}(z) - z}{nk}$$
converges uniformly on $\overline{D}(x,r)$ to a power series in $\mathbb{C}_p[[z-x]]$. In fact,
$$f_* \in \mathbb{Q}_p[[z-x]]$$
because $(f^{nk}(z) - z)/(nk) \in \mathbb{Q}_p[[z-x]]$ for every n.

Since x is not periodic, $f_*(x) \neq 0$, by [**RL03**, Proposition 3.16(1)]. Define

$$u^{-1} \in \mathbb{Q}_p[[z - x]]$$

to be the antiderivative of $1/f_*$ with $u^{-1}(x) = 0$. Since $(u^{-1})'(x) \neq 0$, we may decrease r so that u^{-1} is one-to-one on $\overline{D}(x, r)$. Also, we replace k by a multiple of itself so that $f^k(x) \in \overline{D}(0, r)$, and we write $\overline{D}(0, s) := u^{-1}(\overline{D}(x, r))$. The proof of [**RL03**, Proposition 3.16(2)] shows that the inverse u of u^{-1}, which must also have coefficients in $\mathbb{Q}_p$, satisfies the desired properties. $\qquad\square$

In fact, the integer k in Lemma 6.2.2.1 is at most $p - 1$, at least in the case that $p > 3$ (see [**BGT10**, Theorem 3.3] and also our Theorem 4.4.2.1 in the one-dimensional case). The next result is an easy consequence of Lemma 6.2.2.1, which yields a p-adic parametrization of the orbit of a point x lying in a quasiperiodic domain for a p-adic analytic map $f(z)$.

COROLLARY 6.2.2.2. *Let* $f(z) = a_0 + a_1 z + a_2 z^2 + \cdots \in \mathbb{Z}_p[[z]]$ *be a non-constant power series with* $|a_0|_p < 1$ *but* $|a_1|_p = 1$. *Then for any* $x \in p\mathbb{Z}_p$, *there are: an integer* $k \geq 1$, *positive radii* $r \in (0, 1)$ *and* s, *and power series* $u_j \in \mathbb{Q}_p[[z]]$ *for* $j = 0, \ldots, k - 1$ *mapping* $\overline{D}(0, s)$ *into* $\overline{D}(x, r)$ *with* $u_j(0) = f^j(x)$, *such that for all* $z \in \overline{D}(x, r)$ *and* $n \geq 0$,

$$f^{nk+j}(x) = u_j(nk).$$

PROOF. If x is non-periodic, the result is precisely Lemma 6.2.2.1. Indeed, we may take $k = (p-1)!$ in Lemma 6.2.2.1 and this value of k would work for each point in $p\mathbb{Z}_p$. Hence for each $j = 0, \ldots, k - 1$ we let u_j be the analytic map constructed in Lemma 6.2.2.1 with respect to $f^j(x)$ and then the conclusion of Corollary 6.2.2.2 follows.

Now, if x is periodic of period, say, k, then simply let $s = r = 1$ and let $u_j(z) = f^j(x)$ be the constant map for each $j = 0, \ldots, k - 1$. $\qquad\square$

6.2.3. Special case of the Dynamical Mordell-Lang Conjecture. Corollary 6.2.2.2 yields the following special case of the Dynamical Mordell-Lang Conjecture (see also [**BGKT12**, Theorem 3.4]).

THEOREM 6.2.3.1. *Let* p *be a prime and* $N \geq 1$. *For each* $i = 1, \ldots, N$, *let* U_i *be an open disk in* $\mathbb{P}^1(\mathbb{C}_p)$, *and let* $f_i : U_i \to U_i$ *be a map for which* U_i *is a quasiperiodicity disk. Let* Φ *denote the action of* $f_1 \times \cdots \times f_N$ *on* $U_1 \times \cdots \times U_N$, *let* $\alpha = (\alpha_1, \ldots, \alpha_N) \in U_1 \times \cdots \times U_N$ *be a point. Let* V *be a subvariety of* $(\mathbb{P}^1)^N$ *defined over* $\mathbb{C}_p$. *Then* $V(\mathbb{C}_p) \cap \mathcal{O}_\Phi(\alpha)$ *is a union of at most finitely many orbits of the form* $\{\Phi^{nk+\ell}(\alpha)\}_{n \geq 0}$ *for nonnegative integers* k *and* ℓ.

PROOF. We let k_i be the natural number corresponding to each pair (α_i, f_i) as in Corollary 6.2.2.2, and then let $k = \mathrm{lcm}[k_1, \ldots, k_N]$. Then for each $\ell = 0, \ldots, k - 1$ we have a p-adic analytic parametrization of $\mathcal{O}_{\Phi^k}(\Phi^\ell(\alpha))$ by some functions $(u_{\ell,1}, \ldots, u_{\ell,N})$. So for each polynomial H in the vanishing ideal of V we form the corresponding p-adic analytic map

$$z \mapsto G_{H,\ell} = H(u_{\ell,1}(z), \ldots, u_{\ell,N}(z))$$

such that if $\Phi^{nk+\ell}(\alpha) \in V(\mathbb{C}_p)$, then $G_{H,\ell}(n) = 0$. Using the discreteness of the zeros of a nonzero p-adic analytic function (Lemma 2.3.6.1), we obtain the desired conclusion. $\qquad\square$

We note that the exact same argument applies when V is replaced by a p-adic analytic variety.

6.3. Higher dimensional parametrizations

In this section we show that under certain technical conditions, there are p-adic analytic parametrizations of orbits under endomorphisms of higher dimensional varieties. Our construction is based on a classical result of Herman and Yoccoz [**HY83**], which was used for the first time in [**GT09**] for the purpose of solving some special cases of the Dynamical Mordell-Lang Conjecture.

We will begin with a theorem of Herman and Yoccoz [**HY83**] on linearization of analytic maps near one of their fixed points. First we set up the notation. Let $\vec{0}$ be the zero vector in $\mathbb{C}_p^g$, and for $\vec{x} := (x_1, \ldots, x_g)$ we let

$$(6.3.0.1) \qquad f(\vec{x}) = \sum_{(i_1,\ldots,i_g) \in \mathbb{N}^g} b_{i_1,\ldots,i_g} x_1^{i_1} \cdots x_g^{i_g}$$

be a power series over $\mathbb{C}_p$ which fixes $\vec{0}$ and has a positive radius of convergence; i.e., there is some $r > 0$ such that (6.3.0.1) converges on $D(\vec{0}, r)$. Furthermore, we assume there exists $A \in \mathrm{GL}_g(\mathbb{C}_p)$ such that

$$f(\vec{x}) = A \cdot \vec{x} + \text{ higher order terms.}$$

In this case, f is a formal diffeomorphism in the terminology of [**HY83**]. More generally, for a formal power series ψ in $\mathbb{C}_p^g$ centered at $\vec{\alpha}$, we define $D\psi_{\vec{\alpha}}$ to be the linear part of the power series. Thus $Df_{\vec{0}} = A$. Note that this coincides with the usual definition of the D-operator from the theory of manifolds (that is, $D\psi_{\vec{\alpha}}$ is the usual Jacobian of ψ at $\vec{\alpha}$—see [**Jos02**, I.1.5]).

Let $\lambda_1, \ldots, \lambda_g$ be the eigenvalues of A. Suppose that there are constants $C, b > 0$ such that

$$(6.3.0.2) \qquad |\lambda_1^{e_1} \cdots \lambda_g^{e_g} - \lambda_i|_p \geq C \left(\sum_{j=1}^g e_j \right)^{-b}$$

for any $1 \leq i \leq g$ and any tuple $(e_1, \ldots, e_g) \in \mathbb{N}^g$ such that $\sum_{j=1}^g e_j \geq 2$ (this is condition (C) from page 413 of [**HY83**]). Note that (6.3.0.2) already implies that no $\lambda_i = 0$. Also, we note that if each λ_i is algebraic, then condition (6.3.0.2) is automatically satisfied as proven in [**Yu90**].

The following result is Theorem 1 of [**HY83**].

THEOREM 6.3.0.3. *Let f and A be as above. There exists $r > 0$, and there exists a bijective, p-adic analytic function $h : D(\vec{0}, r) \longrightarrow D(\vec{0}, r)$ such that*

$$(6.3.0.4) \qquad f(h(\vec{x})) = h(A\vec{x}),$$

for all $\vec{x} \in D(\vec{0}, r)$, where $Dh_{\vec{0}} = \mathrm{Id}$.

In particular, we obtain the following corollary.

COROLLARY 6.3.0.5. *Let $f : \mathbb{C}_p^N \longrightarrow \mathbb{C}_p^N$ be a p-adic analytic map, and let β be a fixed point for f. Assume the Jacobian A of f at β has eigenvalues λ_i satisfying condition (6.3.0.2) above. Then there exists a small p-adic neighborhood $\mathcal{U} \subseteq \mathbb{C}_p^N$*

of β, and there exists a bijective p-adic analytic function $h : \mathcal{U} \longrightarrow \mathcal{U}$ such that for each $\alpha \in \mathcal{U}$ and for each $n \in \mathbb{N}_0$, we have

$$f^n(\alpha) = h\left(A^n h^{-1}(\alpha)\right).$$

PROOF. The proof is immediate once we replace the function h from the conclusion of Theorem 6.3.0.3 with the appropriate conjugate:

$$z \mapsto \beta + h(-\beta + z),$$

defined for each $z \in D(\vec{0}, r)$. $\square$

6.3.1. Periodic points for endomorphisms of higher dimensional varieties. If Φ is an endomorphism of a quasiprojective variety X, and β is a periodic point of Φ, then one can still say (sometimes) whether β is *attracting* or *indifferent*. So, assume X is defined over $\mathbb{C}_p$, and also assume β is fixed by Φ^M; then let

$$J := D\Phi_\beta^M.$$

If each eigenvalue of J is nonzero and it has p-adic absolute value less than 1, then we say that β is *attracting*. On the other hand, if each eigenvalue of J has p-adic absolute value equal to 1, then we say that β is *indifferent*. Often, it happens that some eigenvalues have p-adic absolute value less than 1, while others have absolute value at least equal to 1, in which case the point β has *mixed* behaviour for the dynamics of the endomorphism Φ.

For an arbitrary attracting or indifferent periodic point β (as defined in the above paragraph), we do not always have a p-adic analytic parametrization of an orbit of a point α which is sufficiently close p-adically to β. Even in the case of *super-attracting* fixed points β (i.e., $D\Phi_\beta = 0$), it is not always true that one can find a uniformization for Φ in a sufficiently small neighborhood, as in the case of rational maps (see Lemma 6.2.1.1 (2)). For more details on the delicate issue of uniformizing endomorphisms of $\mathbb{C}^N$ in neighborhoods of super-attracting fixed points, we refer the interested reader to [**BEK12**]. However, in Chapter 9 we are able to present some cases when a p-adic analytic parametrization exists in the vicinity of an attracting or indifferent periodic point (see Theorems 9.1.0.1 and 9.2.0.1). Those parametrizations come from the results of Herman and Yoccoz [**HY83**] (see Theorem 6.3.0.3 and Corollary 6.3.0.5), and they lead naturally to proving various instances of the Dynamical Mordell-Lang Conjecture for endomorphisms of higher dimensional varieties.

CHAPTER 7

The split case in the Dynamical Mordell-Lang Conjecture

Using the parametrization for rational maps introduced in Chapter 6 we can prove certain special cases of the Dynamical Mordell-Lang Conjecture for endomorphisms of $(\mathbb{P}^1)^N$ which are of the form

$$\Phi := (f_1, \ldots, f_N)$$

for some rational maps $f_i \in \mathbb{C}(z)$. So, the results of this chapter refer to the *split case* of the Dynamical Mordell-Lang Conjecture.

The results we present are mainly from two papers [**BGKT12, BGHKST13**] (whose arguments we also follow in our exposition). The results rely on proving that modulo a suitable prime p the orbit of each α_i under f_i (where the starting point $\alpha \in (\mathbb{P}^1)^N$ is $\alpha := (\alpha_1, \ldots, \alpha_N)$) lands in a quasiperiodic domain (see Subsection 6.2.2 and also Theorem 6.2.3.1). Finding such a prime p which would work for *each* pair (f_i, α_i) is *very hard*. Hence the instances when we know this happens are somewhat restrictive either on the maps f_i, or on the subvariety V of $(\mathbb{P}^1)^N$ which is intersected with the orbit $\mathcal{O}_\Phi(\alpha)$, as follows.

- **Result (1):** $\alpha \in (\mathbb{P}^1)^N(\overline{\mathbb{Q}})$, $f_1 = \cdots = f_N = \varphi \in \overline{\mathbb{Q}}(z)$ and φ has no periodic critical points (other than exceptional points; see Subsections 6.1.4 and 6.1.5 for the definition of critical and exceptional points), while V is a curve defined over $\overline{\mathbb{Q}}$. For the precise statement of this result, see Theorem 7.1.0.1 and also [**BGKT12**, Theorem 1.4].
- **Result (2):** $f_1 = \cdots = f_N = \varphi \in \mathbb{C}[z]$ and φ has no periodic critical points (other than exceptional points), while V is a curve defined over $\mathbb{C}$; for the precise statement, see Theorem 7.2.0.1 and also the slightly weaker result from [**BGKT12**, Theorem 1.5].
- **Result (3):** each $f_i \in \overline{\mathbb{Q}}(z)$ is a rational function of degree at least 2 such that there is at most one (distinct) f_i which has at most one critical point that is not f_i-preperiodic (and that all other critical points are preperiodic), while V is any subvariety defined over $\overline{\mathbb{Q}}$. For the precise statement, see Theorem 7.3.0.1 and also [**BGHKST13**, Theorem 4.3].

One sees that there is a balance in the listed results (1)—(3) between the strength of the hypothesis regarding the map Φ and the strength of the hypothesis regarding the subvariety V. In result (1), we have two strong conditions: there is the same rational map φ acting on each coordinate, and also the subvariety V is a curve. However, the actual condition on φ is very weak since *almost all* rational maps (interpreted in any sensible moduli space) have critical points which are *not* preperiodic. More precisely, assume that $\varphi \in K[z]$ is a polynomial of degree $d \geq 2$

for some number field K; i.e.,

$$f(z) = c_d z^d + c_{d-1} z^{d-1} + \cdots + c_1 z + c_0.$$

We let $h : \overline{K} \longrightarrow \mathbb{R}_{\geq 0}$ be the usual Weil height on the algebraic closure of K. For each $T \geq 0$ we denote by

$$S(T) := \{(c_0, \ldots, c_d) \in K^{d+1} : h(c_i) \leq T \text{ for each } i = 0, \ldots, d\}.$$

We also denote by $S_0(T)$ the set of all $(c_0, \ldots, c_d) \in K^{d+1}$ such that each $h(c_i) \leq T$ and moreover, the corresponding polynomial φ has the property that none of its critical points is periodic. Then one can show that

$$\lim_{T \to \infty} \#S_0(T)/\#S(T) = 1.$$

For example, if $d = 2$, using a conjugation by a linear polynomial, one may normalize φ and assume that it is monic and that the coefficient of z is equal to 0; i.e., $\varphi(z) = z^2 + c$. The only critical point of φ (other than the exceptional point from ∞) is $x = 0$ and if 0 is periodic under the action of $z^2 + c$, one can show that $c \in \overline{K}$ has bounded height. This statement follows easily by noting that if 0 is periodic, then c must be integral at all non-archimedean places, while at any archimedean place $|\cdot|_v$, one needs that

$$|c|_v \leq 2.$$

Hence $h(c) \leq 2$ and in particular, this means that there are at most finitely many c's living in any extension of K of bounded degree such that the corresponding polynomial $\varphi(z) = z^2 + c$ does *not* satisfy the hypothesis listed in results (1) and (2) regarding periodic critical points.

In result (2), at the expense of assuming that φ is actually a polynomial, we can allow the map (and also the curve V, and the starting point α) to be defined over $\mathbb{C}$ rather than over $\overline{\mathbb{Q}}$. This extension is possible due to the classification of Medvedev and Scanlon [**MS14**] of periodic plane curves under polynomial actions

$$(x, y) \mapsto (f(x), g(y)).$$

For more details on results (1)—(2), see Sections 7.1 and 7.2, and also [**BGKT12**].

On the other hand, in result (3), we allow for arbitrary subvarieties V, and for possibly different rational maps f_i acting on each coordinate, but we impose a strong hypothesis on the f_i's that they only have at most one critical point which is *not* preperiodic. Result (3) has also two interesting corollaries (for more details, see Theorem 7.3.0.1 and [**BGHKST13**], but also see [**BGKT12**], where a weaker version of result (3) was first proven).

COROLLARY 7.0.0.1. *Let K be a number field, let $V \subseteq (\mathbb{P}^1)^N$ be a subvariety defined over K, let $\alpha = (\alpha_1, \ldots, \alpha_N) \in (\mathbb{P}^1)^N(K)$, let $f \in K[t]$ be a quadratic polynomial, and let $\Phi := (f, \ldots, f)$ act on $(\mathbb{P}^1)^g$ coordinatewise. Then the set of integers $n \in \mathbb{N}_0$ such that $\Phi^n(\alpha) \in V(K)$ is a union of finitely many arithmetic progressions $\{nk + \ell\}_{n \in \mathbb{N}_0}$, where $k, \ell \geq 0$ are nonnegative integers.*

For the next result we need a definition.

DEFINITION 7.0.0.2. A rational map φ is called *post-critically finite* if all of its critical points are preperiodic.

COROLLARY 7.0.0.3. *Let K be a number field, let $V \subseteq \left(\mathbb{P}^1\right)^N$ be a subvariety defined over K, let $\alpha = (\alpha_1, \ldots, \alpha_N) \in (\mathbb{P}^1)^N(K)$, and let $\Phi := (f_1, \ldots, f_N)$ act on $\left(\mathbb{P}^1\right)^N$ coordinatewise, where each $f_i \in K(t)$ is post-critically finite and of degree at least 2. Then the set of integers $n \in \mathbb{N}_0$ such that $\Phi^n(\alpha) \in V(K)$ is a union of finitely many arithmetic progressions $\{nk + \ell\}_{n \in \mathbb{N}_0}$, where $k, \ell \geq 0$ are nonnegative integers.*

The rest of this chapter is devoted to proving the listed results (1)—(3), each result being proven in each of the three remaining Sections of our Chapter.

Before proceeding to the proofs, we note that our results (1)—(3) are not covered by the results of Xie [**Xieb**] who proved the Dynamical Mordell-Lang Conjecture for endomorphisms of $\mathbb{A}^2$ defined over $\overline{\mathbb{Q}}$, even using the reduction of Conjecture 5.10.0.1 to Conjecture 5.10.0.2 proven in Section 5.10. Indeed, result (1) is for coordinatewise action of rational maps which do not necessarily restrict to endomorphisms of $\mathbb{A}^2$, while result (2) is for maps defined over $\mathbb{C}$, not necessarily over $\overline{\mathbb{Q}}$. Finally, result (3) cannot be inferred from a similar statement only for plane curves since it is valid for higher dimensional varieties.

We find it important to include the proof of Theorem 7.1.0.1 since those arguments rely on the p-adic analytic uniformization method presented in Chapter 6, which is in turn connected to the p-adic arc lemma (see Chapter 4), which is the recurrent theme of our book. Furthermore the arguments used in the proof of Theorem 7.1.0.1 ultimately led to the proof of Theorem 7.3.0.1 (see result (3) listed above), which is valid for higher dimensional subvarieties. Finally, we note that the proof of Theorem 7.2.0.1 (see result (2) above) uses a specialization technique which we believe should work more generally to extend Theorem 5.10.0.6 to polynomials with complex coefficients.

7.1. The case of rational maps without periodic critical points

The main result of this section is the following.

THEOREM 7.1.0.1 ([**BGKT12**]). *Let $C \subseteq \left(\mathbb{P}^1\right)^N$ be a curve defined over $\overline{\mathbb{Q}}$, and let*

$$\Phi := (\varphi, \ldots, \varphi)$$

act on $\left(\mathbb{P}^1\right)^N$ coordinatewise, where $\varphi \in \overline{\mathbb{Q}}(z)$ is a rational function with no super-attracting periodic points other than exceptional points. Then for each point $(\alpha_1, \ldots, \alpha_N) \in (\mathbb{P}^1)^N(\overline{\mathbb{Q}})$, the set of integers n such that $\Phi^n(\alpha_1, \ldots, \alpha_N) \in C(\overline{\mathbb{Q}})$ is a union of finitely many arithmetic progressions.

We follow the proof from [**BGKT12**].

7.1.1. Preliminaries. Using the reduction of Conjecture 5.10.0.1 to Conjecture 5.10.0.2 done in Section 5.10, we may assume C is a plane curve. This reduction is essential since the argument used in [**BGKT12**] relies heavily on the assumption that we have two starting points $\alpha, \beta \in \mathbb{P}^1(\overline{\mathbb{Q}})$ and then finding a suitable prime p such that both α and β land in a quasiperiodic domain (modulo p) for the action of φ. Once such a prime p is found, the argument is almost identical with the geometric generalization of the Skolem-Mahler-Lech Theorem, i.e., the p-adic arc lemma (see Chapter 4 and also [**BGT10**]).

Now, if C is replaced by an arbitrary subvariety V, then in order to find a p-adic analytic parametrization of the orbit of $(\alpha_1, \dots, \alpha_N)$ under the action of

$$\Phi = (\varphi, \dots, \varphi)$$

so that the generalized Skolem-Mahler-Lech theorem can be applied, we need to find a prime number p for which *each* α_i lands in a quasiperiodic domain for the action of φ. This is hard in general, and it may not even be possible as some of our heuristics from the Chapter 8 might suggest.

However, when $N = 2$, we can find a suitable prime p; the proof of this fact relies on the following lemma.

LEMMA 7.1.1.1. *Let K be a number field, let $\varphi : \mathbb{P}^1 \longrightarrow \mathbb{P}^1$ be a morphism of degree greater than one defined over K, let $\alpha \in \mathbb{P}^1(K)$ be a point that is not preperiodic for φ, and let $\beta \in \mathbb{P}^1(K)$ be a non-exceptional point for φ. Then there are infinitely many v such that there is some positive integer n for which $\varphi^n(\alpha)$ and β have the same reduction modulo v.*

PROOF. The proof can be found in [**BGKT12**, Lemma 4.1] and it is a consequence of an integrality result of Silverman [**Sil93**]; we reproduce it here since it is a useful tool for various problems in arithmetic dynamics.

Suppose there were only finitely many such v. Let S be the set of all such v, together with all the archimedean places. We may choose coordinates $[x : y]$ for $\mathbb{P}^1$ such that β is the point $[1 : 0]$. Since $[1 : 0]$ is not exceptional for φ, we see that φ^2 is not a polynomial with respect to this coordinate system. Therefore, by [**Sil93**, Theorem 2.2], there are at most finitely many n such that $\varphi^n(\alpha) = [t : 1]$ for $t \in \mathfrak{o}_S$, where $\mathfrak{o}_S$ is the ring of S-integers in K. Hence, for all but finitely many integers $n \geq 0$, there is some $v \notin S$ such that $\varphi^n(\alpha)$ and β have the same reduction modulo v; but this contradicts our original supposition. $\qquad\square$

The next result (see [**BGKT12**, Proposition 4.2]) is key for the proof of Theorem 7.1.0.1.

PROPOSITION 7.1.1.2. *Let K be a number field, let $\varphi : \mathbb{P}^1 \longrightarrow \mathbb{P}^1$ be a morphism of degree greater than one defined over K, and let $\alpha, \beta \in \mathbb{P}^1(K)$ be points that are not preperiodic for φ. Then there are infinitely many finite places v of K such that φ has good reduction at v and such that either:*

(1) *for all $m \geq 0$, $\varphi^m(\alpha)$ and $\varphi^m(\beta)$ do not lie in the residue class of any attracting φ-periodic points; or*

(2) *there are integers $k \geq 1$ and $\ell \geq 0$ and attracting periodic points $\gamma_1, \gamma_2 \in \mathbb{P}^1(K_v)$ of period k such that $\varphi^\ell(\alpha)$ and γ_1 lies in the same residue class modulo v, and also $\varphi^\ell(\beta)$ and γ_2 lies in the same residue class modulo v, and moreover $(\varphi^k)'(\gamma_1) = (\varphi^k)'(\gamma_2)$.*

PROOF. By Lemma 7.1.1.1, there are infinitely many places v of good reduction such that there is some positive integer n for which $\varphi^n(\alpha)$ and β are in the same residue class modulo v. Fix any such v. Then for any periodic point γ, the orbit of α intersects the residue class of γ if and only if the orbit of β does. Thus, if condition (i) of the Proposition fails, we can choose an integer $\ell \geq 0$ and an attracting periodic point γ_1 such that $\varphi^\ell(\alpha)$ and γ_1 are in the same residue class modulo v. By Lemma 6.2.1.1 (see also [**RL03**, Proposition 3.2]), γ_1 lies in the v-adic closure of the orbit of α, and hence $\gamma_1 \in \mathbb{P}^1(K_v)$.

Set $\gamma_2 = \varphi^n(\gamma_1)$. Then $\varphi^\ell(\beta)$ lies in the same residue class modulo v as $\varphi^{n+\ell}(\alpha)$, and thus it is in the same residue class as $\varphi^n(\gamma_1)$ and also as γ_2. Finally, as noted after Definition 6.1.3.1, $(\varphi^k)'(\gamma_1) = (\varphi^k)'(\gamma_2)$. $\qquad\square$

7.1.2. Proof of Theorem 7.1.0.1. As discussed in the previous subsection, we are reduced to the following setting:

- K is a number field;
- $\alpha, \beta \in \mathbb{P}^1(K)$;
- $C \subseteq \mathbb{P}^1 \times \mathbb{P}^1$ is a curve defined over K; and
- $\varphi : \mathbb{P}^1 \longrightarrow \mathbb{P}^1$ is a rational map defined over K.

Our goal is to prove that the set of $n \in \mathbb{N}_0$ such that $(\varphi^n(\alpha), \varphi^n(\beta)) \in C(K)$ is a finite union of arithmetic progressions.

Clearly, we may assume that C is irreducible, since we can prove the statement for each irreducible component of C. If either α or β is preperiodic for φ, then the statement follows easily (see also Proposition 3.1.2.9). So, from now on we assume that neither α nor β is φ-preperiodic.

Suppose that there is a place v of good reduction satisfying condition (i) of Proposition 7.1.1.2. Let $p \in \mathbb{N}$ be the prime number lying in the maximal ideal of the non-archimedean place v, and fix an embedding of K into $\mathbb{C}_p$ respecting v. The desired conclusion is immediate since then the orbits of α and β lie in quasiperiodic domains, and so we can find a p-adic analytic parametrization of the orbit of (α, β) under $\Phi = (\varphi, \varphi)$ (see Theorem 6.2.3.1 and also [**BGKT12**, Theorem 3.4]).

If no such place exists, then by Proposition 7.1.1.2, there must be a place v of good reduction meeting condition (ii) for which neither α nor β lies in the same residue class as an exceptional point (note that since there are at most two exceptional points, and in addition, neither α nor β is an exceptional point for φ because we assumed they are not preperiodic, then for all but finitely many places v, the reductions of both α and β do not coincide with the reduction of an exceptional point for φ).

It follows that the orbits of α and β also avoid the residue classes of exceptional points. In particular, the attracting periodic points γ_1 and γ_2 given in condition (ii) of Proposition 7.1.1.2 cannot be exceptional.

Furthermore, at the expense of replacing both α and β by an iterate $\varphi^\ell(\alpha)$, respectively $\varphi^\ell(\beta)$, we may assume that α and β are already in the residue classes of γ_1, respectively γ_2; note that we are allowed to replace (α, β) by an iterate of it under Φ by Proposition 3.1.2.4. Also, at the expense of replacing φ by an iterate (see Proposition 3.1.2.5), we may assume that both γ_1 and γ_2 are fixed by φ.

By hypothesis, then, γ_1 and γ_2 are attracting but not super-attracting, and therefore the Theorem follows from [**GT09**, Theorem 1.3] (see also our Chapter 9). For the sake of completeness, we sketch here the argument since it is easier than the general case considered in Chapter 9.

So, by Corollary 6.2.1.2 we know that there exist p-adic analytic functions u_1 and u_2 such that for all $n \in \mathbb{N}_0$ we have

$$\varphi^n(\alpha) = u_1(\lambda^n) \text{ and } \varphi^n(\beta) = u_2(\lambda^n),$$

where λ is the common multiplier for the periodic points γ_1 and γ_2 (see Proposition 7.1.1.2). So, if $F(X, Y) = 0$ is the equation of the affine part (inside $\mathbb{A}^1 \times \mathbb{A}^1$)

of the curve $C \subset \mathbb{P}^1 \times \mathbb{P}^1$, then we obtain that

$$(\varphi^n(\alpha), \varphi^n(\beta)) \in C(K) \text{ if and only if } F(u_1(\lambda^n), u_2(\lambda^n)) = 0.$$

Hence considering the p-adic analytic function $H : \mathbb{C}_p \longrightarrow \mathbb{C}_p \times \mathbb{C}_p$ defined by

$$H(z) := F(u_1(z), u_2(z)),$$

we see that if there exist infinitely many $n \in \mathbb{N}_0$ such that

$$(\varphi^n(\alpha), \varphi^n(\beta)) \in C(K),$$

then $H(z) = 0$ has infinitely many solutions $z = \lambda^n$ which converge to 0 (since λ is the uniformizer for γ_i and so, $0 < |\lambda|_p < 1$). By Lemma 2.3.6.1, we conclude that $H = 0$, and so,

$$\Phi^n(\alpha, \beta) \in C(K) \text{ for all } n \in \mathbb{N}_0.$$

7.2. Extension to polynomials with complex coefficients

Using the results of Medvedev and Scanlon [**MS14**] who give a complete classification of all invariant subvarieties of $\mathbb{A}^N$ under the coordinatewise action of N one-variable polynomials, one can extend Theorem 7.1.0.1 to the case when $\varphi \in \mathbb{C}[z]$. The results of this section overlap substantially with [**BGKT12**, Section 7], but strictly speaking, our Theorem 7.2.0.1 is more general than [**BGKT12**, Theorem 1.5] because the latter was stated only for indecomposable polynomials.

THEOREM 7.2.0.1. *Let $\varphi \in \mathbb{C}[z]$ be a non-constant polynomial whose critical points in $\mathbb{C}$ are not periodic, let $\Phi : \mathbb{A}^N \longrightarrow \mathbb{A}^N$ be given by*

$$(x_1, \ldots, x_N) \mapsto (\varphi(x_1), \ldots, \varphi(x_N)),$$

let $C \subseteq \mathbb{A}^N$ be a curve defined over $\mathbb{C}$, and let $\alpha \in \mathbb{A}^N(\mathbb{C})$. Then the set of $n \in \mathbb{N}_0$ such that $\Phi^n(\alpha) \in C(\mathbb{C})$ is a finite union of arithmetic progressions.

As before, using the reduction of Conjecture 5.10.0.1 to Conjecture 5.10.0.2 done in Section 5.10, we can reduce the proof of Theorem 7.2.0.1 to the case $N = 2$. Also, we may assume C is an irreducible curve. In addition, we can easily prove two special cases.

7.2.1. φ is a linear polynomial. If φ is a linear polynomial, then Φ is an automorphism of $\mathbb{A}^N$ and the conclusion follows from the main result of [**Bel06**] (see also our Theorem 4.4.1.1).

7.2.2. φ is conjugate to either $z \mapsto z^d$, or to T_d for some $d \geq 2$. We recall that T_d is itself a conjugate of the Chebyshev polynomial, i.e., T_d is a monic polynomial of degree d with the property that

$$T_d\left(z + \frac{1}{z}\right) = z^d + \frac{1}{z^d}.$$

If φ is conjugate to T_d, then there exists a linear polynomial $\mu \in \mathbb{C}[z]$ such that

$$\varphi = \mu^{-1} \circ T_d \circ \mu.$$

So, using Proposition 3.1.2.13, Theorem 7.2.0.1 for the triple (φ, C, α) follows from the same result being proven for $(T_d, \tilde{\mu}(C), \tilde{\mu}(\alpha))$, where

$$\tilde{\mu} := (\mu, \mu) : \mathbb{A}^2 \longrightarrow \mathbb{A}^2.$$

Hence from now on, we assume $\varphi = T_d$ and thus $\Phi = (T_d, T_d) : \mathbb{A}^2 \longrightarrow \mathbb{A}^2$.

For each $i = 1, 2$, we let $\beta_i \in \mathbb{C}$ such that $\beta_i + 1/\beta_i = \alpha_i$. Therefore, for each $n \in \mathbb{N}_0$ we have that

$$T_d^n(\alpha_i) = \beta_i^{d^n} + \frac{1}{\beta_i^{d^n}}.$$

On the other hand, if $F(x, y) = 0$ is the equation of the plane curve C, then we let $\tilde{C}$ be the plane curve defined by the equation

$$F\left(x + \frac{1}{x}, y + \frac{1}{y}\right) = 0.$$

Clearly, $\Phi^n(\alpha) \in C(\mathbb{C})$ if and only if $\left(\beta_1^{d^n}, \beta_2^{d^n}\right) \in \tilde{C}(\mathbb{C})$. An easy application of Laurent's Theorem [**Lau84**] (see Theorem 3.4.1.1) finishes the proof in this case. Note that we can also apply our Theorem 4.4.1.1 since the map $\tilde{\Phi}(x, y) = (x^d, y^d)$ is étale when restricted to $\mathbb{G}_m^2$.

The case when φ is conjugate with $z \mapsto z^d$ is similar, but only simpler since it reduces immediately to Theorem 4.4.1.1.

7.2.3. The general case of Theorem 7.2.0.1. We are left with proving Theorem 7.2.0.1 for

- $N = 2$;
- C is an irreducible curve; and
- φ is a non-linear polynomial, not conjugate to either a monomial or a Chebyshev polynomial.

Furthermore, using Proposition 3.1.2.9, we may reduce to the case when neither α_1 nor α_2 is φ-preperiodic; we recall that $\alpha = (\alpha_1, \alpha_2)$ is the starting point of the orbit in Theorem 7.2.0.1. Also, using Proposition 3.1.2.13 for replacing φ with a polynomial conjugate with itself, we may assume φ is in *normal form*.

DEFINITION 7.2.3.1. We say that a polynomial $\varphi \in \mathbb{C}[z]$ of degree $d \geq 2$ is in *normal form* if it is monic and the coefficient of z^{d-1} in $\varphi(z)$ equals 0. Furthermore, if we let

$$\varphi(z) = z^d + c_{d-2}z^{d-2} + \cdots + c_0,$$

with $c_i \in \mathbb{C}$, then we say that φ is of *type* (a, b) if a is the smallest nonnegative integer such that $c_a \neq 0$, and b is the largest positive integer such that $\varphi(z) = z^a u(z^b)$ for some polynomial $u \in \mathbb{C}[z]$.

Note that when we normalize φ we may have to enlarge the number field K so that it contains an $(m-1)$-st root of the leading coefficient of φ.

Finally, at the expense of replacing φ by a map φ_1 such that

$$\varphi_1^k = \varphi \text{ (with respect to the composition of functions)},$$

we may assume that φ is not a compositional power of another polynomial. Note that if we prove the result for

$$\Phi_1 := (\varphi_1, \varphi_1),$$

then the result follows for Φ, as proven in Proposition 3.1.2.5. With these reductions in place for φ we can state the pivotal result for our proof, which is a consequence of [**MS14**, Theorem 6.24].

THEOREM 7.2.3.2 (Medvedev-Scanlon [**MS14**]). *Let K be an algebraically closed field of characteristic 0, and let $\varphi \in K[z]$ be a polynomial of degree $d \geq 2$ which is not conjugate to t^d or T_d for any positive integer d. Assume that φ is in normal form, of type (a, b).*

Let Φ denote the action of (φ, φ) on $\mathbb{A}^2$. Let C be a Φ-periodic irreducible plane curve defined over K. Then C is defined by one of the following equations in the variables (x, y) of the affine plane:

    (i) $x = x_0$, *for a φ-periodic point x_0; or*
    (ii) $y = y_0$, *for a φ-periodic point y_0; or*
    (iii) $x = \zeta \varphi^r(y)$, *for some $r \geq 0$; or*
    (iv) $y = \zeta \varphi^r(x)$, *for some $r \geq 0$,*

where ζ is a d-th root of unity, where $d \mid b$ and $\gcd(d, a) = 1$. (Note that if $b = 1$ or $a = 0$, then $d = 1$.)

Now we can prove the remaining case of Theorem 7.2.0.1. We follow closely the arguments provided in the proof of [**BGKT12**, Theorem 7.6].

Let K be a finitely generated field over which C, φ, α_1 and α_2 are defined. Furthermore, at the expense of replacing K by a finite extension, we may assume that C is geometrically irreducible and that K contains all critical points of φ and all $(m - 1)$-st roots of unity.

We prove Theorem 7.2.0.1 by induction on $d := \operatorname{trdeg}_{\mathbb{Q}} K$. If $d = 0$, then K is a number field, and our conclusion follows from Theorem 7.1.0.1.

Assume $d \geq 1$. Then K may be viewed as the function field of a smooth, geometrically irreducible curve Z defined over a finitely generated field E. Thus

$$\operatorname{trdeg}_{\mathbb{Q}} E = d - 1.$$

Moreover, the curve C extends to a one-dimensional scheme, denoted by $\mathfrak{C}$ over Z, all but finitely many of whose fibres $\mathfrak{C}_\gamma$ are irreducible curves.

We claim that there are infinitely many places γ of K for which all of the following statements hold. (By a place of K, we mean a valuation of the function field K/E, see Section 2.6.)

    (a) The fiber $\mathfrak{C}_\gamma$ is an irreducible curve defined over the residue field $E(\gamma)$ of γ, of the same degree as C.
    (b) All nonzero coefficients of φ are units at the place γ; in particular, φ has good reduction at γ, and so we write φ_γ and $\Phi_\gamma := (\varphi_\gamma, \varphi_\gamma)$ for the reductions of φ and respectively Φ at γ.
    (c) The critical points of φ_γ are reductions at γ of the critical points of φ.
    (d) For each critical point z of φ (other than infinity), the reduction z_γ is not a periodic point for φ_γ.
    (e) The map $\mathcal{O} \longrightarrow \mathcal{O}_\gamma$ from the Φ-orbit $\mathcal{O}$ of α under Φ to the Φ_γ-orbit of α_γ, induced by reduction at γ, is injective.
    (f) φ_γ is not conjugate to z^d or T_d (we recall that $d = \deg \varphi$).
    (g) φ_γ is not a compositional power of another polynomial.

Conditions (a)—(c) above are satisfied at all but finitely many places γ of K (see for example [**GW10**, Appendix E] and also the proof of [**Ghi05**, Theorem 4.11] and [**vdDS84**]). We note that a similar result was proven in Proposition 4.4.1.3. The same is true of conditions (f)—(g), by [**BGKT12**, Propositions 7.8 and 7.9]. Condition (d) for preperiodic (but not periodic) critical points also holds at all but finitely many places since for any given finite set of points S for all but finitely many places v, the points in S reduce modulo v to distinct points. Meanwhile, conditions (d)—(e) hold by applying Proposition 5.6.0.5 to α_i and the non-preperiodic critical points, proving the claim.

Let γ be one of the infinitely many places satisfying conditions (a)—(g). From condition (e), we deduce that

$$\mathfrak{C}_\gamma(E(\gamma)) \cap \mathcal{O}_\gamma \text{ is infinite.}$$

Conditions (c)—(d) guarantee that φ_γ has no periodic critical points (other than the exceptional point at infinity). Since $E(\gamma)$ is a finite extension of E, we get

$$\operatorname{trdeg}_{\mathbb{Q}} E(\gamma) = d - 1.$$

By the inductive hypothesis, then, $\mathfrak{C}_\gamma$ is Φ_γ-periodic. By conditions (f)—(g) and Theorem 7.2.3.2, $\mathfrak{C}_\gamma$ is the zero set of an equation from one of the four forms (i)—(iv) in Theorem 7.2.3.2. In fact, if φ has type (a, b), then the degree d in Theorem 7.2.3.2 satisfies

$$d \mid b \text{ and } \gcd(d, a) = 1,$$

because condition (b) implies that φ_γ also has type (a, b). Thus, for one of the four forms (i)—(iv), there are infinitely many places γ satisfying (a)—(g) above such that the equation for $\mathfrak{C}_\gamma$ is of that form. By symmetry, it suffices to consider only forms (i) and (iii) from the conclusion of Theorem 7.2.3.2.

Case 1. Assume there are infinitely many γ satisfying (a)—(g) such that $\mathfrak{C}_\gamma$ is given by an equation

$$x = x(\gamma), \text{ for some } \varphi_\gamma\text{-periodic point } x(\gamma) \in E(\gamma).$$

Then, since the degree of C is preserved by the reduction at γ, we see that the degree of C must be 1. Thus, C is defined by an equation of the form

$$u_1 x + u_2 y + u_3 = 0,$$

for some constants $u_1, u_2, u_3 \in K$. Since there are infinitely many γ such that the above equation reduces at γ to $x = x(\gamma)$, we must have

$$u_2 = 0.$$

Hence, the curve C must be given by an equation

$$x = x_1 \text{ for some } x_1 \in K,$$

contradicting our assumption that C does not project to a point in either of the two coordinates.

Case 2. Assume there are infinitely many γ satisfying (a)—(g) such that $\mathfrak{C}_\gamma$ is given by an equation

$$y = \zeta \varphi_\gamma^r(x),$$

for some $r \geq 0$ and some d-th root of unity ζ, where

$$d \mid b \text{ and } \gcd(d, a) = 1.$$

Since there are only finitely many b-th roots of unity, we may assume ζ is the same for all of the infinitely many γ. Moreover, because $\mathfrak{C}_\gamma$ has the same degree as C, the integer r is the same for all such γ. Thus, there are infinitely many places γ for which the polynomial equation for C reduces modulo γ to

$$y - \zeta \varphi^r(x),$$

and hence the two polynomials are the same. Thus, C is the zero set of the polynomial

$$y - \zeta \varphi^r(x).$$

Since φ is of type (a, b), it follows that C is Φ-periodic, as desired.

7.3. The case of "almost" post-critically finite rational maps

In [**BGHKST13**], the following special case of the Dynamical Mordell-Lang Conjecture was proven.

THEOREM 7.3.0.1 ([**BGHKST13**]). *Let K be a number field, let $V \subseteq \left(\mathbb{P}^1\right)^N$ be a subvariety defined over K, let $\alpha = (\alpha_1, \ldots, \alpha_g) \in (\mathbb{P}^1)^N(K)$, and let*

$$\Phi := (\varphi_1, \ldots, \varphi_N)$$

act on $\left(\mathbb{P}^1\right)^N$ coordinatewise, where each $\varphi_i \in K(t)$ is a rational function of degree at least 2. Suppose that at most one distinct φ_i has a critical point that is not φ_i-preperiodic, and that all other critical points of that φ_i are preperiodic. Then the set of integers $n \in \mathbb{N}$ such that

$$\Phi^n(\alpha) \in V(K)$$

is a union of finitely many arithmetic progressions.

The key to proving the above result is [**BGHKST13**, Theorem 3.1], which we now state.

THEOREM 7.3.0.2 ([**BGHKST13**]). *Let K be a number field, and let*

$$\varphi_1, \ldots, \varphi_N : \mathbb{P}^1 \longrightarrow \mathbb{P}^1$$

be rational maps of degree at least 2 defined over K. Let $\mathcal{A}_1, \ldots, \mathcal{A}_g$ be finite subsets of $\mathbb{P}^1(K)$ such that at most one set $\mathcal{A}_i$ contains a point that is not φ_i-preperiodic, and such that there is at most one such point in that set $\mathcal{A}_i$. Let $\mathcal{T}_1, \ldots, \mathcal{T}_g$ be finite subsets of $\mathbb{P}^1(K)$ such that no $\mathcal{T}_i$ contains any φ_i-preperiodic points. Then there is a positive integer M and a set of primes $\mathcal{P}$ of K having positive density such that for any $i = 1, \ldots, g$, any $\gamma \in \mathcal{T}_i$, any $\alpha \in \mathcal{A}_i$, any $\mathfrak{p} \in \mathcal{P}$, and any $m \geq M$,

$$\varphi_i^m(\gamma) \not\equiv \alpha \pmod{\mathfrak{p}}.$$

We first show how to deduce Theorem 7.3.0.1 from Theorem 7.3.0.2, after which we give the proof of Theorem 7.3.0.2.

PROOF OF THEOREM 7.3.0.1. We reproduce here the proof of [**BGHKST13**, Theorem 4.3].

Arguing by induction (see also Proposition 3.1.2.9), we may assume that no α_i is φ_i-preperiodic.

By Theorem 7.3.0.2, there exist a constant M and a positive proportion of primes $\mathfrak{p}$ of K such that for each $i = 1, \ldots, N$, φ_i has good reduction at $\mathfrak{p}$, and $\varphi_i^m(x_i)$ is not congruent modulo $\mathfrak{p}$ to any critical point of φ_i for all $m \geq M$. Fix any such $\mathfrak{p}$, and note that the derivative of the reduction $(\varphi_{i,\mathfrak{p}})'$ is non-trivial, because φ_i has good reduction and we may assume

$$\mathrm{char}(k_\mathfrak{p}) \nmid \deg \varphi_i.$$

Thus, $\varphi_i'(\varphi_i^m(x_i))$ is a $\mathfrak{p}$-adic unit for all $m \geq M$. It follows that

$$\varphi_i^m(x_i) \not\equiv \gamma \pmod{\mathfrak{p}}$$

for any attracting periodic point γ of φ_i. Therefore each α_i lands in a quasiperiodic domain under the action of φ_i and applying Theorem 6.2.3.1 finishes our proof. $\square$

Before proving Theorem 7.3.0.2 we show how to deduce Corollary 7.0.0.1.

PROOF OF COROLLARY 7.0.0.1. Corollary 7.0.0.1 can be deduced both from Theorem 7.3.0.1, but also directly from Theorem 7.3.0.2. Indeed, as proven in Theorem 7.3.0.2, we can allow in Theorem 7.3.0.1 that there is only one *distinct* φ_i which has only one critical point that is not preperiodic, i.e., we are allowed in Theorem 7.3.0.1 to take

$$\varphi_1 = \cdots = \varphi_N = f,$$

and f has exactly one critical point which is not preperiodic. Now, if f is any quadratic polynomial, then indeed, f has at most one critical point which is not preperiodic because the point at infinity (which is critical for the polynomial f) is clearly fixed by f. Hence Corollary 7.0.0.1 follows from Theorem 7.3.0.1 (or alternatively, directly from Theorem 7.3.0.2). $\square$

On the other hand, Corollary 7.0.0.3 is an immediate consequence of Theorem 7.3.0.1. So, now we are left to complete the proof of Theorem 7.3.0.2, which we do in the following Subsection.

7.3.1. Proof of Theorem 7.3.0.2. We follow the proof of [**BGHKST13**, Theorem 3.1].

We employ in our proof the following standard ramification lemma over $\mathfrak{p}$-adic fields; it says, roughly, that if the field of definition of a point in $\varphi^{-m}(\alpha)$ ramifies at $\tilde{\mathfrak{p}}$, then that point must be a ramification point of φ^m modulo $\tilde{\mathfrak{p}}$, provided the characteristic of the residue field corresponding to $\tilde{\mathfrak{p}}$ does not divide the degree of φ.

LEMMA 7.3.1.1. *Let K be a number field, let $\tilde{\mathfrak{p}}$ be a prime of $\mathfrak{o}_{\overline{K}}$, and let*

$$\varphi : \mathbb{P}^1 \longrightarrow \mathbb{P}^1$$

be a rational function defined over K and of good reduction at $\mathfrak{p} = \tilde{\mathfrak{p}} \cap \mathfrak{o}_K$ such that $\deg \varphi \geq 2$ and that $\deg(\varphi)$ is not divisible by $\mathrm{char}(k_\mathfrak{p})$. Let $\alpha \in \mathbb{P}^1(K)$, let $m \geq 1$ be an integer, let $\beta \in \varphi^{-m}(\alpha) \subseteq \mathbb{P}^1(\overline{K})$, and let $\mathfrak{q} := \tilde{\mathfrak{p}} \cap \mathfrak{o}_{K(\beta)}$. If $\mathfrak{q}$ is ramified over $\mathfrak{p}$, then β is congruent modulo $\tilde{\mathfrak{p}}$ to a ramification point of φ^m.

PROOF. We reproduce here the proof of [**BGHKST13**, Lemma 3.2].

By induction, it suffices to show the lemma in the case $m = 1$; note that the hypothesis about $\deg(\varphi^m)$ not being divisible by $\mathrm{char}(k_{\tilde{\mathfrak{p}}})$ remains valid for all $m \geq 1$.

Let $|\cdot|_{\tilde{\mathfrak{p}}}$ denote the $\tilde{\mathfrak{p}}$-adic absolute value on $\overline{K}$, and let $K_\mathfrak{p}$ be the completion of K with respect to $|\cdot|_{\tilde{\mathfrak{p}}}$. After a change of coordinates, we may assume that $\alpha = 0$ and that $|\beta|_{\tilde{\mathfrak{p}}} \leq 1$.

Writing $\varphi = f/g$, where $f, g \in K[t]$ are relatively prime polynomials, we have $f(\beta) = 0$. Since $\mathfrak{q}$ is ramified over $\mathfrak{p}$, f must have at least one other root congruent to β modulo $\tilde{\mathfrak{p}}$. Thus, the reduction $f_\mathfrak{p}$ of f has a multiple root at β. However, $g_\mathfrak{p}(\beta) \neq 0$, since φ has good reduction. Therefore, the reduction $\varphi_\mathfrak{p}$ has a multiple root at β, and hence $\varphi'_\mathfrak{p}(\beta) = 0$. On the other hand, because $\deg \varphi$ is not divisible by $\mathrm{char}(k_\mathfrak{p})$, there must be some $\gamma \in \mathfrak{o}_{\overline{K}}$ such that $\varphi'_\mathfrak{p}(\gamma) \neq 0$. It follows that there is a root of φ' congruent to β modulo $\tilde{\mathfrak{p}}$. $\square$

We note that Lemma 7.3.1.1 may fail if the characteristic of the residue field for $\tilde{\mathfrak{p}}$ divides the degree of φ. For example, if p is a prime number, $K = \mathbb{Q}$, $\alpha = 1$ and

$$\varphi(z) = z^p,$$

while ζ_p is a primitive p-th root of unity, then $\mathbb{Q}(\zeta_p)/\mathbb{Q}$ is ramified above the prime p even though 1 is not congruent to a ramification point of φ. The part that breaks down from the proof of Lemma 7.3.1.1 in this case is that $\varphi'_p = 0$, where for any polynomial f, we denote by f_p its reduction modulo p.

Next, we use the fact that our residue fields are finite to show that if α is not periodic modulo a large enough prime $\mathfrak{p}$, then for large m, there can be no roots of $\varphi^m(x) - \alpha$ modulo $\mathfrak{p}$. We also obtain some extra information about our fields of definition, which we will need in order to apply the Chebotarev density theorem (see Theorem 2.4.0.1) in our proof of Theorem 7.3.0.2.

LEMMA 7.3.1.2. *Let K be a number field, let $\tilde{\mathfrak{p}}$ be a prime of $\mathfrak{o}_{\overline{K}}$, and let*

$$\varphi : \mathbb{P}^1 \longrightarrow \mathbb{P}^1$$

be a rational function defined over K and of good reduction at $\mathfrak{p} = \tilde{\mathfrak{p}} \cap \mathfrak{o}_K$ such that $\deg \varphi \geq 2$ and that $\deg \varphi$ is not divisible by $\operatorname{char}(k_{\mathfrak{p}})$. Suppose that $\alpha \in \mathbb{P}^1(K)$ is not periodic modulo $\mathfrak{p}$. Then there exists a finite extension E of K with the following property: for any finite extension L of E, there is an integer $M \in \mathbb{N}$ such that for all $m \geq M$ and all $\beta \in \mathbb{P}^1(\overline{K})$ with $\varphi^m(\beta) = \alpha$,

   (i) $\mathfrak{r}$ *does not ramify over $\mathfrak{q}$, and*
   (ii) $[\mathfrak{o}_{L(\beta)}/\mathfrak{r} : \mathfrak{o}_L/\mathfrak{q}] > 1$,

where $\mathfrak{r} := \tilde{\mathfrak{p}} \cap \mathfrak{o}_{L(\beta)}$, and $\mathfrak{q} := \tilde{\mathfrak{p}} \cap \mathfrak{o}_L$.

PROOF. We reproduce here the proof of [**BGHKST13**, Lemma 3.3].
For any $\gamma \in \mathfrak{o}_{\overline{K}}$,

(7.3.1.3)   there is at most one $j \geq 0$ such that $\varphi^j(\gamma) \equiv \alpha \pmod{\tilde{\mathfrak{p}}}$,

since α is not periodic modulo $\mathfrak{p}$. In particular, for each ramification point $\gamma \in \mathbb{P}^1(\overline{K})$ of φ, there are only finitely many integers $n \geq 0$ and points $z \in \mathbb{P}^1(\overline{K})$ such that

(7.3.1.4) $\varphi^n(z) = \alpha$ and $z \equiv \gamma \pmod{\tilde{\mathfrak{p}}}$.

Let E be the finite extension of K formed by adjoining all such points z.

Given any finite extension L of E, let $\mathfrak{q} = \tilde{\mathfrak{p}} \cap \mathfrak{o}_L$. Since $\mathbb{P}^1(\mathfrak{o}_L/\mathfrak{q})$ is finite, (7.3.1.3) implies that for all sufficiently large M, the equation

$$\varphi^M(x) = \alpha \text{ has no solutions in } \mathbb{P}^1(\mathfrak{o}_L/\mathfrak{q}).$$

Fix any such M; note that M must be larger than any of the integers n satisfying (7.3.1.4). Hence, given $m \geq M$ and $\beta \in \mathbb{P}^1(\overline{K})$ such that $\varphi^m(\beta) = \alpha$, we must have

$$[\mathfrak{o}_{L(\beta)}/\mathfrak{r} : \mathfrak{o}_L/\mathfrak{q}] > 1,$$

where $\mathfrak{r} = \tilde{\mathfrak{p}} \cap \mathfrak{o}_{L(\beta)}$, proving conclusion (ii). Furthermore, if β is a root of $\varphi^m(x) - \alpha$, then there are two possibilities: either

   (1) β is not congruent modulo $\tilde{\mathfrak{p}}$ to a ramification point of φ^m, or
   (2) $\varphi^j(\beta) = z$ for some $j \geq 0$ and some point $z \in \mathbb{P}^1(L)$ satisfying equation (7.3.1.4).

In case (1), $\mathfrak{r}$ is unramified over $\mathfrak{q}$ by Lemma 7.3.1.1.

In case (2), choosing a minimal such $j \geq 0$, and applying Lemma 7.3.1.1 with z in the role of α and j in the role of m, $\mathfrak{r}$ is again unramified over $\mathfrak{q}$. Thus, in either case, conclusion (i) of Lemma 7.3.1.2 holds. $\qquad\square$

We now apply Lemma 7.3.1.2 to a set $\mathcal{A}$ of points.

PROPOSITION 7.3.1.5. *Let K be a number field, let $\tilde{\mathfrak{p}}$ be a prime of $\overline{\mathfrak{o}_K}$, and let*

$$\varphi : \mathbb{P}^1 \longrightarrow \mathbb{P}^1$$

be a rational function defined over K and of good reduction at $\mathfrak{p} = \tilde{\mathfrak{p}} \cap \mathfrak{o}_K$ such that $\deg \varphi \geq 2$ and that $\deg \varphi$ is not divisible by $\mathrm{char}(k_{\mathfrak{p}})$. Let

$$\mathcal{A} = \{\alpha_1, \ldots, \alpha_n\}$$

be a finite subset of $\mathbb{P}^1(K)$ such that for each $\alpha_i \subset \mathcal{A}$,

   (i) *if α_i is not periodic, then α_i is not periodic modulo $\mathfrak{p}$; and*
   (ii) *if α_i is periodic, then $\varphi(\alpha_i) = \alpha_i$ (i.e., α_i is fixed by φ) and the ramification index of φ at α_i is the same modulo $\mathfrak{p}$ as over K.*

Then there is a finite extension E of K with the following property: for any finite extension L of E, there is an integer $M \in \mathbb{N}$ such that for all $m \geq M$ and all $\beta \in \mathbb{P}^1(\overline{K})$ with $\varphi^m(\beta) \in \mathcal{A}$ but $\varphi^t(\beta) \notin \mathcal{A}$ for all $t < m$,

   (1) $\mathfrak{r}$ *does not ramify over $\mathfrak{q}$, and*
   (2) $[\mathfrak{o}_{L(\beta)}/\mathfrak{r} : \mathfrak{o}_L/\mathfrak{q}] > 1$,

where $\mathfrak{r} := \tilde{\mathfrak{p}} \cap \mathfrak{o}_{L(\beta)}$ and $\mathfrak{q} := \tilde{\mathfrak{p}} \cap \mathfrak{o}_L$.

PROOF. We reproduce here the proof of [**BGHKST13**, Proposition 3.4].

For each $\alpha_i \in \mathcal{A}$ that is not periodic, we apply Lemma 7.3.1.2 and obtain a field E_i with the property described in that lemma. For each $\alpha_j \in \mathcal{A}$ that is periodic, we apply Lemma 7.3.1.2 to each point

$$\gamma_{jk} \in \varphi^{-1}(\alpha_j) \setminus \{\alpha_j\}$$

and obtain a field E_{jk} with the corresponding property. To do so, of course, we must know that no γ_{jk} is periodic modulo $\tilde{\mathfrak{p}}$. To see that this is true, first note that

$$(7.3.1.6) \qquad\qquad\qquad \gamma_{jk} \not\equiv \alpha_j \pmod{\tilde{\mathfrak{p}}};$$

otherwise the ramification index of φ at α_j would be greater modulo $\mathfrak{p}$ than over K, contradicting our hypotheses. Since α_j is fixed, congruence (7.3.1.6) yields that γ_{jk} is not periodic modulo $\tilde{\mathfrak{p}}$, as desired.

Let E be the compositum of all the fields E_i and E_{jk}. Given any finite extension L of E, then by our choice of E_i and E_{jk}, there are integers $M_i, M_{jk} \in \mathbb{N}$ satisfying the conclusions of Lemma 7.3.1.2. Set

$$M := \max_{i,j,k}(M_i, M_{jk}) + 1.$$

Then for any $m \geq M$ and $\beta \in \mathbb{P}^1(\overline{K})$ such that $\varphi^m(\beta) \in \mathcal{A}$ but $\varphi^t(\beta) \notin \mathcal{A}$ for all $0 \leq t < m$, we have $\varphi^{m-1}(\beta) \notin \mathcal{A}$. Hence, $\varphi^{m-1}(\beta)$ is either some γ_{jk} or is in $\varphi^{-1}(\alpha_i)$ for some non-periodic α_i; that is, β is an element either of some $\varphi^{-(m-1)}(\gamma_{jk})$ or of $\varphi^{-m}(\alpha_i)$ for some non-periodic α_i. Thus, by the conclusions of Lemma 7.3.1.2, β satisfies conditions (i) and (ii), as desired. $\qquad\square$

We will now apply Proposition 7.3.1.5 to several maps $\varphi_1, \ldots, \varphi_g$ at once to obtain a proof of Theorem 7.3.0.2 (again we follow [**BGHKST13**]).

PROOF OF THEOREM 7.3.0.2. We note first that it suffices to prove our result for a finite extension of K. Indeed, if L/K is a finite extension and

$$\mathfrak{q} \cap \mathfrak{o}_K = \mathfrak{r} \text{ for a prime } \mathfrak{q} \subseteq \mathfrak{o}_L,$$

then $\varphi_i^m(\gamma)$ is congruent to α modulo $\mathfrak{q}$ if and only if $\varphi_i^m(\gamma)$ is congruent to α modulo $\mathfrak{r}$. Moreover, given a positive density set of primes $\mathcal{Q}$ of L, the set

$$\mathcal{P} = \{\mathfrak{q} \cap \mathfrak{o}_K : \mathfrak{q} \in \mathcal{Q}\}$$

also has positive density as a set of primes of K.

We may assume, for all $i = 1, \ldots, g$, that every φ_i-preperiodic point $\alpha \in \mathcal{A}_i$ is in fact fixed by φ_i. Indeed, for each such i and α, choose integers $j_\alpha \geq 0$ and $\ell_\alpha \geq 1$ such that

$$\varphi_i^{j_\alpha}(\alpha) = \varphi_i^{j_\alpha + \ell_\alpha}(\alpha).$$

Let $j := \max_\alpha\{j_\alpha\}$, and replace each $\alpha \in \mathcal{A}_i$ by $\varphi_i^j(\alpha)$. Similarly, let

$$\ell := \mathrm{lcm}_\alpha\{\ell_\alpha\},$$

and enlarge each

$$\mathcal{T}_i = \{\gamma_{i1}, \ldots, \gamma_{is_i}\}$$

to include $\varphi_i^b(\gamma_{ic})$ for all $b = 1, \ldots, \ell - 1$ and $c = 1, \ldots, s_i$. Finally, replace each φ_i by φ_i^ℓ, so that for the new data,

(7.3.1.7)   all the φ_i-preperiodic points in $\alpha \in \mathcal{A}_i$ are fixed by φ_i.

If the theorem holds for the new data, then it holds for the original data, since for any $m \geq M$ and any prime $\mathfrak{p}$ at which every φ_i has good reduction,

$$\varphi_i^m(\gamma_{ij}) \equiv \alpha \pmod{\mathfrak{p}} \implies (\varphi_i^\ell)^a(\varphi_i^b(\gamma_{ij})) \equiv \varphi_i^j(\alpha) \pmod{\mathfrak{p}},$$

writing $m + j$ as $a\ell + b$ with $a \geq 0$ and $0 \leq b < \ell$.

We fix the following notation for the remainder of the proof. If there is any index i such that $\mathcal{A}_i$ contains a non-periodic point, we may assume that this happens for $i = 1$, and we denote the non-periodic point by α'. By hypothesis, all points in $\mathcal{A}_1 \setminus \{\alpha'\}$ are φ_1-preperiodic, and we denote them by α_{1j}; similarly, for each $i \geq 2$, all points in $\mathcal{A}_i$ are φ_i-preperiodic, and we denote them by α_{ij}. By (7.3.1.7), we may assume that φ_i fixes α_{ij} for all i, j. Note that there are only

- finitely many primes $\mathfrak{p}$ of bad reduction for any φ_i;
- finitely many primes $\mathfrak{p}$ for which $\mathrm{char}(k_\mathfrak{p}) \mid \deg \varphi_i$ for some i; and
- finitely many primes $\mathfrak{p}$ such that the ramification index of some φ_i at some $\alpha_{ij} \in \mathcal{A}_i$ is greater modulo $\mathfrak{p}$ than over K.

On the other hand, by [**BGKT12**, Lemma 4.3], there are infinitely many primes $\mathfrak{p}$ of K such that α' is not φ_1-periodic modulo $\mathfrak{p}$. Hence, we may choose such a prime $\mathfrak{p}$, and then a prime $\tilde{\mathfrak{p}}$ of $\mathfrak{o}_{\overline{K}}$ for which $\mathfrak{p} = \tilde{\mathfrak{p}} \cap \mathfrak{o}_K$, that simultaneously satisfy, for each $i = 1, \ldots, g$, the hypotheses of Proposition 7.3.1.5 for φ_i and $\mathcal{A}_i$.

Applying Proposition 7.3.1.5, for each $i = 1, \ldots, g$ we obtain finite extensions E_i of K satisfying the conclusions of that result. Let

L be the compositum of the fields $E_1, \ldots, E_g$, and let $\mathfrak{q} = \tilde{\mathfrak{p}} \cap \mathfrak{o}_L$.

Then for all $i = 1, \ldots, g$, all sufficiently large M, and all $\beta \in \overline{K}$ such that

$$\varphi_i^M(\beta) \in \mathcal{A}_i \text{ but } \varphi_i^t(\beta) \notin \mathcal{A}_i \text{ for } 0 \leq t < M,$$

we have

(i) $\mathfrak{r}$ does not ramify over $\mathfrak{q}$, and

(ii) $[\mathfrak{o}_{L(\beta)}/\mathfrak{r} : \mathfrak{o}_L/\mathfrak{q}] > 1$,

where $\mathfrak{r} = \tilde{\mathfrak{p}} \cap \mathfrak{o}_{L(\beta)}$. As noted at the start of this proof, it suffices to prove the Theorem for the field L.

Fix such a sufficiently large integer M, and let F/L be the finite extension obtained by adjoining all points $\beta \in \mathbb{P}^1(\overline{L})$ such that for some $i = 1, \ldots, g$ we have

$$\varphi_i^M(\beta) \in \mathcal{A}_i \text{ but } \varphi_i^t(\beta) \notin \mathcal{A}_i \text{ for all } 0 \leq t < M.$$

Note that F/L is a Galois extension, since each $\mathcal{A}_i$ and each φ_i is defined over L. Moreover, by property (i), F/L is unramified over $\mathfrak{q}$. By property (ii), then, the Frobenius element of $\mathfrak{q}$ belongs to a conjugacy class of $\mathrm{Gal}(\mathrm{F}/\mathrm{L})$ whose members do not fix any of the points β. By the Chebotarev density theorem (Theorem 2.4.0.1), there is a positive density set of primes $\mathcal{S}$ of L whose Frobenius conjugacy classes in $\mathrm{Gal}(\mathrm{F}/\mathrm{L})$ do not fix any of the points β.

Fix any prime $\mathfrak{r} \in \mathcal{S}$. We make the following claim.

CLAIM 7.3.1.8. *Let $m \geq 0$, let $1 \leq i \leq g$, and let $z \in \mathbb{P}^1(L)$ be a point such that $\varphi_i^m(z)$ is congruent modulo $\mathfrak{r}$ to an element of $\mathcal{A}_i$. Then there is some*

$$t \in \{0, \ldots, M\}$$

such that $\varphi_i^t(z)$ is congruent modulo $\mathfrak{r}$ to an element of $\mathcal{A}_i$.

PROOF OF CLAIM 7.3.1.8. Note first that the conclusion is vacuous if $m < M$; thus, we may assume that $m \geq M$. In fact, given any index i and point z as in the claim, we may assume that m is the minimal integer $m \geq M$ satisfying the hypothesis, namely that

$$\varphi_i^m(z) = \varphi_i^M(\varphi_i^{m-M}(z))$$

is congruent modulo $\mathfrak{r}$ to an element of $\mathcal{A}_i$. However, by the defining property of the set of primes $\mathcal{S}$, there cannot be any points $w \in \mathbb{P}^1(L)$ such that $\varphi_i^M(w)$ is congruent modulo $\mathfrak{r}$ to an element of $\mathcal{A}_i$ but $\varphi_i^t(w) \notin \mathcal{A}_i$ for all $0 \leq t < M$. Choosing $w = \varphi_i^{m-M}(z) \in \mathbb{P}^1(L)$, then, there must be some $0 \leq t < M$ such that $\varphi_i^t(\varphi_i^{m-M}(z))$ is congruent modulo $\mathfrak{r}$ to an element of $\mathcal{A}_i$. Thus, $\varphi_i^{m-M+t}(z)$ is congruent modulo $\mathfrak{r}$ to an element of $\mathcal{A}_i$; but $0 \leq m - M + t < m$, contradicting the minimality of m and proving Claim 7.3.1.8. □

Let $\mathcal{U}$ be the subset of $\mathcal{S}$ consisting of primes $\mathfrak{r} \in \mathcal{S}$ such that one or more of the following holds:

(a) $\varphi_i^t(\gamma) \equiv \alpha_{ij} \pmod{\mathfrak{r}}$ for some $i = 1, \ldots, g$, some $\gamma \in \mathcal{T}_i$, some φ_i-periodic $\alpha_{ij} \in \mathcal{A}_i$, and some $0 \leq t < M$; or

(b) $\varphi_1^t(\alpha') \equiv \alpha \pmod{\mathfrak{r}}$ for some $\alpha \in \mathcal{A}_1$ and some $1 \leq t \leq M$.

Note, for each φ_i-periodic $\alpha_{ij} \in \mathcal{A}_i$, there cannot exist a nonnegative integer r and $\gamma \in \mathcal{T}_i$ such that $\varphi_i^r(\gamma) = \alpha_{ij}$, since the elements of $\mathcal{T}_i$ are not φ_i-preperiodic. (However, it *is* possible that $\varphi_1^r(\gamma) = \alpha'$ for some r and some $\gamma \in \mathcal{T}_1$.) Thus $\mathcal{U}$ is a finite subset of $\mathcal{S}$, and hence $\mathcal{S}' := \mathcal{S} \setminus \mathcal{U}$ has positive density. We will now show that Theorem 7.3.0.2 holds for the field L, the integer M, and this set of primes $\mathcal{S}'$.

Suppose there exist a prime $\mathfrak{r} \in \mathcal{S}'$, an index $1 \leq i \leq g$, points $\alpha \in \mathcal{A}_i$ and $\gamma \in \mathcal{T}_i$, and an integer $m \geq M$ such that $\varphi_i^m(\gamma) \equiv \alpha \pmod{\mathfrak{r}}$. By Claim 7.3.1.8, there is an integer $0 \leq t < M$ and a point $\tilde{\alpha} \in \mathcal{A}_i$ such that $\varphi_i^t(\gamma) \equiv \tilde{\alpha} \pmod{\mathfrak{r}}$. By property (b), then we must have $i = 1$ and $\tilde{\alpha} = \alpha'$. Moreover, since

$$\varphi_1^{m-t-1}(\varphi_1(\alpha')) \equiv \alpha \pmod{\mathfrak{r}}, \text{ and since } m - t - 1 \geq 0,$$

Claim 7.3.1.8 tells us that there is some $k \in \{0, \ldots, M\}$ such that $\varphi_1^{k+1}(\alpha')$ is congruent modulo $\mathfrak{r}$ to an element of $\mathcal{A}_1$, contradicting property (b), and hence proving Theorem 7.3.0.2. $\qquad\square$

CHAPTER 8

Heuristics for avoiding ramification

The p-adic parametrization method for treating the Dynamical Mordell-Lang conjecture works for a point $\alpha \in X(K)$ whenever there is a prime $\mathfrak{p}$ of good reduction such that the $\mathcal{O}_\Phi(\alpha)$ does not meet the ramification divisor of Φ modulo $\mathfrak{p}$. In fact, it works under slightly weaker hypotheses, since one only needs that the periodic part of the orbit of α modulo $\mathfrak{p}$ does not meet the ramification divisor of Φ modulo $\mathfrak{p}$. This follows immediately from Theorem 4.4.4.1.

As we have seen in the Chapters 4 and 7, there are some cases where one can prove that such a $\mathfrak{p}$ exists. Unfortunately, as is shown in [**BGHKST13**], there is a heuristic and some numerical evidence that suggests that perhaps no such prime exists for general self-maps of varieties of dimension 5 or more. In this chapter, we explain these ideas in detail.

Most of the material presented in this chapter overlaps with [**BGHKST13**]. The techniques of this chapter are different from the rest of the book, both because they involve speculative heuristics rather than proofs, and because the discussion of the heuristics is probabilistic and analytic rather than arithmetic geometric. We begin with a brief summary of what one should expect if morphisms on special fibers of varieties follow a so-called random model. After that, we provide some proofs about what should happen assuming a random model; here we see that if the random model is accurate, then one should not expect the method of p-adic parametrization to work in dimensions 5 or greater for solving the Dynamical Mordell-Lang Conjecture. Finally, we present more details on how one might hope to analyze the "split case" more generally. At this time, we do not have a strategy for treating the case of dimension 2, 3, and 4 via p-adic parametrization, although the heuristic suggests that it may be possible to do so.

8.1. A random model heuristic

Let X be a variety defined over a number field K, let

$$\Phi : X \longrightarrow X$$

be a morphism, let R_Φ be the ramification locus of Φ, and let $\alpha \in X(K)$. If we have a model $\mathcal{X}$ for X over the ring of integers of K, then Φ extends to a self-map of the special fiber $\mathcal{X}_\mathfrak{p}$ for all but finitely many primes $\mathfrak{p}$. Furthermore, for all but finitely many primes $\mathfrak{p}$ the intersection of the Zariski closure of R_Φ (considered as a subset of the generic fiber of $\mathcal{X}$) with the special fiber $\mathcal{X}_\mathfrak{p}$ is the ramification divisor of $\Phi_\mathfrak{p}$. We will call the primes $\mathfrak{p}$ with these properties "primes of good reduction" for Φ. Let $\alpha_\mathfrak{p}$ be the specialization of α to $\mathcal{X}_\mathfrak{p}$ and let $k_\mathfrak{p}$ denote the residue field of $\mathfrak{p}$. Since $\mathcal{X}_\mathfrak{p}(k_\mathfrak{p})$ is finite, $\alpha_\mathfrak{p}$ must be preperiodic. We let j be the smallest positive integer such that

$$\Phi_\mathfrak{p}^i(\alpha_\mathfrak{p}) = \Phi_\mathfrak{p}^j(\alpha_\mathfrak{p})$$

for some integer $0 \le i < j$. Then $\mathrm{Orb}_{\Phi_{\mathfrak{p}}}(\alpha_{\mathfrak{p}})$ decomposes into a (possibly empty) "tail"

$$(\alpha_{\mathfrak{p}}, \ldots, \Phi_{\mathfrak{p}}^{i-1}(\alpha_{\mathfrak{p}}))$$

and a periodic part

$$(\Phi_{\mathfrak{p}}^{i}(\alpha_{\mathfrak{p}}), \ldots, \Phi_{\mathfrak{p}}^{j-1}(\alpha_{\mathfrak{p}})).$$

We want to come up with a model that predicts the likelihood that the periodic part of $\mathrm{Orb}_{\Phi_{\mathfrak{p}}}(\alpha_{\mathfrak{p}})$ intersects $R_{\Phi_{p}}$.

We think of $\Phi_{\mathfrak{p}}$ as a random map on $\mathcal{X}_{\mathfrak{p}}(k_{\mathfrak{p}})$; that is, to each element $z \in \mathcal{X}_{\mathfrak{p}}(k_{\mathfrak{p}})$, we take $\Phi_{\mathfrak{p}}(z)$ to be a random element of $\mathcal{X}_{\mathfrak{p}}(k_{\mathfrak{p}})$. The birthday paradox (for its application to number theory, see for example [**Pol75**]) shows that on average $|\mathrm{Orb}_{\Phi_{\mathfrak{p}}}(\alpha_{\mathfrak{p}})|$ should be proportional to $\sqrt{\#\mathcal{X}(k_{\mathfrak{p}})}$; a simple analysis also shows the length of the periodic part of $\mathrm{Orb}_{\Phi_{\mathfrak{p}}}(\alpha_{\mathfrak{p}})$ should be about half of $|\mathrm{Orb}_{\Phi_{\mathfrak{p}}}(\alpha_{\mathfrak{p}})|$ on average.

Note however, that the distribution of orbit lengths under the random model is very far from being normal. Suppose that

$$\#\mathcal{X}_{\mathfrak{p}}(k_{\mathfrak{p}}) = M.$$

Then the chance that the length of the orbit of $\alpha_{\mathfrak{p}}$ is equal to i is

$$\frac{M-1}{M} \cdot \frac{M-2}{M} \cdot \ldots \cdot \frac{M-i+1}{M} \cdot \frac{1}{M},$$

so even though the expected value for the orbit length is $\sqrt{\#\mathcal{X}_{\mathfrak{p}}(k_{\mathfrak{p}})}$, the most likely value for the orbit length is 1. In fact, under the random model hypothesis, the orbit length follows a well-known distribution, as we shall soon see.

The number of $k_{\mathfrak{p}}$-points in the ramification locus of $\Phi_{\mathfrak{p}}$ should be about $\#k_{\mathfrak{p}}^{\dim X}$, by the Weil bounds (see [**Del74**]). Typically, the ramification locus R_{Φ} has dimension $\dim X - 1$, so (again by the Weil bounds) there are about $\#k_{\mathfrak{p}}^{\dim X - 1}$ points in $\mathcal{X}_{\mathfrak{p}}(k_{\mathfrak{p}})$ that are in the Zariski closure of R_{Φ}. Thus, the proportion of primes in $\mathcal{X}_{\mathfrak{p}}(k_{\mathfrak{p}})$ that are in the ramification divisor is about $1/\#k_{\mathfrak{p}}$. Hence, the chance of a given iterate being in the ramification divisor is about $1/\#k_{\mathfrak{p}}$. Now, since the lengths of orbits under $\Phi_{\mathfrak{p}}$ increases with dimension (assuming a random model) whereas the proportion of points in $\mathcal{X}_{\mathfrak{p}}(k_{\mathfrak{p}})$ that meet the ramification divisor stays the same, it should not be surprising that that the periodic parts of orbits are less likely to avoid ramification modulo primes in higher dimensions.

Here is a brief summary of what one should expect about orbit lengths and intersection of orbits modulo primes, assuming the random model. We restate briefly the setup for our problem, which is valid for both Sections 8.2 and 8.3.

- $\Phi : X \longrightarrow X$ is an endomorphism of a variety X defined over a nunber field K;
- $\mathfrak{p}$ is a prime of good reduction for Φ with respect to a model $\mathcal{X}$ for X over the ring of algebraic integers of K;
- $\mathcal{X}_{\mathfrak{p}}$ is the special fiber of $\mathcal{X}$ at $\mathfrak{p}$;
- R_{Φ} is the ramification locus of Φ;
- d be the dimension of X, and
- $\alpha \in X(K)$ is a point, and $\alpha_{\mathfrak{p}} \in \mathcal{X}(k_{\mathfrak{p}})$ is its reduction (seen as a point on the special fiber) the special fiber.

Assuming a random model and a large $\#k_{\mathfrak{p}}$, we will find that:

(RM1) The distribution of the length of the periodic part of $\mathrm{Orb}_{\Phi_{\mathfrak{p}}}(\alpha_{\mathfrak{p}})$ follows a suitably normalized version of standard *Gaussian error function*

$$\mathrm{erfc}(s) := \frac{2}{\sqrt{\pi}} \int_s^\infty e^{-t^2} \, dt,$$

which is described in [**AS92**, Chapter 7], for example. We prove a precise form of this in Proposition 8.2.0.5.

(RM2) For $d \geq 3$, the chance that the periodic part of $\mathrm{Orb}_{\Phi_{\mathfrak{p}}}(\alpha_{\mathfrak{p}})$ does not intersect the reduction of R_Φ modulo $\mathfrak{p}$ is proportional to

$$(\#k_{\mathfrak{p}})^{1-\frac{d}{2}} \, .$$

This will be proved in Proposition 8.3.0.2.

To estimate the chance that *each* prime $\mathfrak{p}$ has the property that $\mathrm{Orb}_{\Phi_{\mathfrak{p}}}(\alpha_{\mathfrak{p}})$ intersects the reduction of R_Φ modulo $\mathfrak{p}$, one then takes the infinite product

$$\prod_{\text{finite primes } \mathfrak{p} \text{ of } K} \left(1 - \frac{1}{(\#k_{\mathfrak{p}})^{d/2-1}}\right).$$

Of course, when $K = \mathbb{Q}$, this is simply the reciprocal of the usual Euler product expansion of $\zeta(d/2 - 1)$, where ζ is the usual Riemann ζ-function. Since the Euler product expansion of $\zeta(s)$ converges when $s > 1$, the random model heuristic suggests that when $d \geq 5$, there is a nonzero chance that each prime $\mathfrak{p}$ has the property that $\mathrm{Orb}_{\Phi_{\mathfrak{p}}}(\alpha_{\mathfrak{p}})$ intersects the reduction of R_Φ modulo $\mathfrak{p}$. More generally, for any number field K, the product

$$\prod_{\text{finite primes } \mathfrak{p} \text{ of } K} \left(1 - \frac{1}{(\#k_{\mathfrak{p}})^{d/2-1}}\right)$$

converges to a nonzero number whenever $d \geq 5$. Hence, when $d \geq 5$, for a given point α, there is a chance that at every prime $\mathfrak{p}$, the cyclic part of $\Phi_{\mathfrak{p}}(\alpha_{\mathfrak{p}})$ meets R_Φ modulo $\mathfrak{p}$; thus, there is a chance that there is no prime $\mathfrak{p}$ at which one can treat the dynamical Mordell-Lang problem for $\mathrm{Orb}(\alpha)$ using the method of p-adic analytic parametrization.

The idea of using random maps to model behavior of orbits modulo primes is not new. For example Silverman [**Sil08**] used such a model to formulate questions about the lengths of orbits modulo primes for a point in $\mathbb{P}^d(\mathbb{Q})$ under a morphism

$$\Phi : \mathbb{P}^d \longrightarrow \mathbb{P}^d;$$

he conjectured that the orbit length mod p is greater than $p^{d/2-\epsilon}$ for a full density subset of the primes (see also [**AG09**]). Questions about orbit lengths of polynomials (in particular quadratic polynomials) over finite fields also arise naturally in Pollard's rho method [**Pol75**] for factoring integers (see [**Bac91**]).

Although proofs of the accuracy of the random model are out of reach, numerical data that mirrors RM1 and RM2 can be found in [**BGHKST13**, Section 5]. We also note that data strongly suggests that it should be possible to avoid ramification modulo infinitely many primes in dimension 1 and 2. The case of dimensions 3 and 4 are less clear. A simpler analysis, where one assumes that orbit lengths modulo $\mathfrak{p}$ are normally distributed around the expected value, suggested that it may not be possible to to avoid ramification modulo any prime in some cases; a more detailed analysis, using erfc suggests that is only when $d \geq 5$ that it may not be possible to to avoid ramification modulo any prime. In any case, we currently have no method for producing primes $\mathfrak{p}$ where an orbit avoids ramification modulo $\mathfrak{p}$ when $d \geq 5$.

8.2. Random models and cycle lengths

We begin by proving RM1. Let Z be a (large) finite set, and let

$$f : Z \to Z$$

be a random map in the following sense: for each $z \in Z$, select the image $f(z)$ by randomly selecting an element of Z, with uniform distribution. Given a starting point $z \in Z$, we dneote by $\mathrm{Orb}_f(z)$ the orbit of z under f. Since Z is finite, z is necessarily preperiodic. Let $\pi(z)$ be the number of distinct elements in $\mathrm{Orb}_f(z)$ i.e., $\pi(z)$ is the smallest positive integer such that

$$f^{\pi(z)}(z) = f^s(z)$$

for some $s < \pi(z)$. Then the randomness assumption on f implies that

$$\mathrm{Prob}(\pi(z) > m) = \prod_{j=1}^{m} \left(1 - \frac{j}{|Z|}\right) = \exp\left[\sum_{j=1}^{m} \log\left(1 - \frac{j}{|Z|}\right)\right].$$

From the Taylor series expansion

$$\log(1 - x) = -\left(x + \frac{x^2}{2} + \frac{x^3}{3} + \dots\right),$$

we obtain the inequality

$$(8.2.0.1) \qquad \mathrm{Prob}(\pi(z) > m) \leq \exp\left(-\frac{m(m+1)}{2|Z|}\right),$$

and similarly we find that for $m = o(|Z|^{2/3})$,

$$(8.2.0.2) \qquad \mathrm{Prob}(\pi(z) > m) = \exp\left(-\frac{m^2}{2|Z|}\right) \cdot (1 + o(1)),$$

since

$$(8.2.0.3) \quad \mathrm{Prob}(\pi(z) > m) = \exp\left(-\frac{m(m+1)}{2|Z|} + O(m^3/|Z|^2)\right)$$

$$= \exp\left(-\frac{m^2}{2|Z|} + O(m/|Z| + m^3/|Z|^2)\right).$$

In addition, if we let $\alpha(m) := \mathrm{Prob}(\pi(z) \geq m)$, then

$$\mathrm{Prob}(\pi(z) = m) = \mathrm{Prob}(\pi(z) \geq m) - \mathrm{Prob}(\pi(z) > m)$$

$$= \prod_{j=1}^{m-1} \left(1 - \frac{j}{|Z|}\right) - \prod_{j=1}^{m} \left(1 - \frac{j}{|Z|}\right)$$

$$= \left[1 - \left(1 - \frac{m}{|Z|}\right)\right] \cdot \prod_{j=1}^{m-1} \left(1 - \frac{j}{|Z|}\right) = \frac{m}{|Z|} \cdot \alpha(m).$$

Define $\mathcal{C}$ to be the periodic part of $\mathrm{Orb}(z)$. Conditioning on $\pi(z) = m$, the random map assumption implies that $f^m(z)$ is uniformly selected among

$$\{z, f(z), \dots, f^{m-1}(z)\},$$

and hence
$$\mathrm{Prob}\left(|\mathcal{C}| = \ell \,\Big|\, \pi(z) = m\right) = \frac{1}{m} \quad \text{for any} \quad \ell \leq m.$$
The cycle length probability may thus be written as

$$(8.2.0.4) \quad \mathrm{Prob}(|\mathcal{C}| = \ell) = \sum_{m \geq \ell} \mathrm{Prob}(|\mathcal{C}| = \ell \,|\, \pi(z) = m) \cdot \mathrm{Prob}(\pi(z) = m)$$

$$= \sum_{m \geq \ell} \frac{1}{m} \cdot \mathrm{Prob}(\pi(z) = m) = \sum_{m \geq \ell} \frac{1}{m} \cdot \frac{m}{|Z|} \cdot \alpha(m) = \frac{1}{|Z|} \sum_{m \geq \ell} \alpha(m).$$

PROPOSITION 8.2.0.5. *If* $\ell = o(|Z|^{2/3})$ *then, as* $|Z| \to \infty$,
$$\mathrm{Prob}(|\mathcal{C}| = \ell) = \sqrt{\frac{\pi}{2|Z|}} \cdot \left(\mathrm{erfc}\left(\ell/\sqrt{2|Z|}\right) + o(1)\right).$$

PROOF. By (8.2.0.4), we find that

$$(8.2.0.6) \quad \mathrm{Prob}(|\mathcal{C}| = \ell) = \frac{1}{|Z|} \sum_{m=\ell}^{|Z|} \alpha(m) = \frac{1}{|Z|} \left(\sum_{m=1}^{|Z|} \alpha(m) - \sum_{1 \leq m < \ell} \alpha(m)\right).$$

We begin by evaluating the first sum. Recalling that
$$\alpha(m) = \mathrm{Prob}(\pi(z) > m - 1),$$
if $m = o(|Z|^{2/3})$, then by (8.2.0.2), we have
$$\alpha(m) = \exp\left(-m^2/(2|Z|)\right) \cdot \left(1 + o(1)\right).$$
Moreover, by (8.2.0.1) the inequality
$$\alpha(m) \ll \exp\left(-m^2/(3|Z|)\right)$$
holds for $m \geq 1$. Thus, setting $Q(T) := T^{2/3}/\log T$, we have

$$(8.2.0.7) \quad \sum_{m=1}^{|Z|} \alpha(m) = \sum_{1 \leq m \leq Q(|Z|)} \alpha(m) + \sum_{Q(|Z|) < m \leq |Z|} \alpha(m)$$

$$= \left(1 + o(1)\right) \cdot \sum_{1 \leq m \leq Q(|Z|)} e^{-m^2/(2|Z|)} + O\left(\int_{Q(|Z|)-1}^{\infty} e^{-t^2/(3|Z|)}\, dt\right).$$

To show that the contribution from the integral is negligible, we note the inequality (valid for all $A, B > 0$)

$$\int_A^\infty e^{-t^2/B}\, dt = \sqrt{B} \int_{A/\sqrt{B}}^\infty e^{-s^2}\, ds \leq \frac{B}{A} \int_{A/\sqrt{B}}^\infty s e^{-s^2}\, ds = \frac{B}{2A} e^{-A^2/B}.$$

Thus,

$$(8.2.0.8) \quad \int_{Q(|Z|)-1}^\infty e^{-t^2/(3|Z|)}\, dt \leq \frac{3|Z|}{2Q(|Z|) - 2} \exp\left(-\frac{\left(Q(|Z|) - 1\right)^2}{3|Z|}\right)$$

$$\ll |Z|^{1/3} \log|Z| \exp\left(-\frac{|Z|^{1/3}}{3(\log|Z|)^2}\right) = o(1)$$

as $|Z| \to \infty$.

Meanwhile, note that for any $L \geq 1$,

$$\sum_{1 \leq m \leq L} e^{-m^2/(2|Z|)} = \sqrt{2|Z|} \cdot \left[\int_0^{\lfloor L \rfloor / \sqrt{2|Z|}} e^{-t^2} \, dt + O\left(\frac{\lfloor L \rfloor}{2|Z|} \right) \right]$$

$$= \sqrt{2|Z|} \cdot \int_0^{L/\sqrt{2|Z|}} e^{-t^2} \, dt + O\left(\frac{L}{\sqrt{2|Z|}} + 1 \right).$$

by interpreting the sum as a $1/\sqrt{2|Z|}$-spaced Riemann sum approximation of an integral and noting that

$$|e^{-s^2} - e^{-t^2}| \leq |s - t| \text{ for all } s, t \in \mathbb{R}.$$

Thus, the sum in the right side of (8.2.0.7) is

$$\sum_{1 \leq m \leq Q(|Z|)} e^{-m^2/(2|Z|)} = \sqrt{2|Z|} \cdot \int_0^{Q(|Z|)/\sqrt{2|Z|}} e^{-t^2} \, dt + O\left(\frac{Q(|Z|)}{\sqrt{2|Z|}} + 1 \right)$$

$$= \sqrt{2|Z|} \cdot \left(\int_0^\infty e^{-t^2} \, dt + o(1) \right),$$

and the second sum on the right side of (8.2.0.6) is

$$\sum_{1 \leq m < \ell} \alpha(m) = \left(1 + o(1)\right) \cdot \sum_{m=1}^{\ell-1} e^{-m^2/(2|Z|)}$$

$$= \left(1 + o(1)\right) \cdot \left(\sqrt{2|Z|} \int_0^{\ell/\sqrt{2|Z|}} e^{-t^2} \, dt + O\left(\frac{\ell}{\sqrt{2|Z|}} + 1 \right) \right).$$

Combining equations (8.2.0.6), (8.2.0.7), and (8.2.0.8) with the above Riemann sum estimates, and recalling that $\text{erfc}(s) = \frac{2}{\sqrt{\pi}} \int_s^\infty e^{-t^2} \, dt$, we have

$$\text{Prob}(|\mathcal{C}| = \ell) = \frac{\sqrt{2} \cdot \int_{\ell/\sqrt{2|Z|}}^\infty e^{-t^2} \, dt + o(1)}{|Z|^{1/2}}$$

$$= \sqrt{\frac{\pi}{2|Z|}} \cdot \left(\text{erfc}\left(\ell/\sqrt{2|Z|} \right) + o(1) \right).$$

$$\square$$

8.3. Random models and avoiding ramification

We continue with the notation used in Section 8.1 for X, Φ, R_Φ, $\mathfrak{p}$ and α. We let $\Phi_\mathfrak{p}$ be the induced action of the endomorphism Φ on the special fiber $\mathcal{X}_\mathfrak{p}$ corresponding to the prime $\mathfrak{p}$, and we also let $\alpha_\mathfrak{p}$ be the reduction of α modulo $\mathfrak{p}$. We let $\mathcal{C}_\mathfrak{p}$ denote the periodic part of the forward orbit of $\alpha_\mathfrak{p}$ under the action of $\Phi_\mathfrak{p}$, and let $\mathcal{R}_\mathfrak{p}$ denote the set of points in $\mathcal{X}(\mathfrak{p})$ that are in the Zariski closure of R_Φ (considered as a subset of the generic fiber of $\mathcal{X}$). Finally, we let $d := \dim X$.

We suppose that Φ is *not* unramified, and moreover that R_Φ is nonempty and of codimension 1 in X. By the Weil bounds (see [**Del74**]), we have

$$|\mathcal{X}_\mathfrak{p}(k_\mathfrak{p})| = \#k_\mathfrak{p}^d (1 + o(1)),$$

as $\#k_{\mathfrak{p}}$ goes to infinity. Similarly, if R_{ϕ} is irreducible, then

$$|\mathcal{R}_{\mathfrak{p}}| = \#k_{\mathfrak{p}}^{d-1}(1 + o(1)).$$

When R_{Φ} is not irreducible, $|\mathcal{R}_{\mathfrak{p}}|$ may be larger, but since we are interested in demonstrating the unfeasibility of avoiding $\mathcal{R}_{\mathfrak{p}}$ modulo primes, we may assume that $|\mathcal{R}_{\mathfrak{p}}|$ is as small as possible (but still under the assumption that R_{Φ} has codimension 1 in X). Hence, we may assume that we have

(8.3.0.1)
$$\frac{|R_{\mathfrak{p}}|}{|\mathcal{X}(k_{\mathfrak{p}})|} = \frac{1}{\#k_{\mathfrak{p}}}(1 + o(1)).$$

We say Φ has *random map behavior* modulo $\mathfrak{p}$ if the following two conditions hold:

- $|\mathcal{C}_p|$ has the same probability distribution as the cycle length of a random map on a set of size $|\mathcal{X}(k_{\mathfrak{p}})|$.
- The probability for the event that any given distinct points $y_1, \ldots, y_m \in \mathcal{C}_p$ all belong to $\mathcal{R}_p$ is $\frac{1}{(\#k_{\mathfrak{p}})^m}$.

Note that in light of (8.3.0.1), the second assumption is essentially saying that $\mathcal{C}_p$ and $\mathcal{R}_p$ are suitably independent, i.e., for any given point on the special fiber $\mathcal{X}_{\mathfrak{p}}$ belonging to $\mathcal{C}_p$, respectively to $\mathcal{R}_p$ are two independent events.

PROPOSITION 8.3.0.2. *Assume that the polynomial map* $\Phi : X \longrightarrow X$ *has random map behavior modulo for all but finitely many* $\mathfrak{p}$. *If* $d \geq 3$, *then*

$$\mathrm{Prob}(\mathcal{C}_{fp} \cap \mathcal{R}_{fp} = \emptyset) = \frac{\sqrt{\pi/2}}{(\#k_{\mathfrak{p}})^{d/2-1}} \cdot (1 + o(1)) \quad as \ \#k_{\mathfrak{p}} \to \infty.$$

PROOF. Fix a $\mathfrak{p}$ such that $\#k_{\mathfrak{p}}$ is sufficiently large. We write $q = \#k_{\mathfrak{p}}$. For simplicity of notation, we will write $\mathcal{C}$ and $\mathcal{R}$ instead of $\mathcal{C}_{\mathfrak{p}}$ and $\mathcal{R}_{\mathfrak{p}}$. Conditioning on the cycle length $|\mathcal{C}|$ being equal to ℓ, we find that

$$\mathrm{Prob}\left(\mathcal{C} \cap \mathcal{R} = \emptyset \Big| |\mathcal{C}| = \ell\right) = (1 - 1/q)^{\ell},$$

and hence

$$\mathrm{Prob}(\mathcal{C} \cap \mathcal{R} = \emptyset) = \sum_{\ell=1}^{q^d}(1 - 1/q)^{\ell} \cdot \mathrm{Prob}(|\mathcal{C}| = \ell).$$

We start by bounding the contribution from the large cycles. Since $(1 - 1/q)^{\ell}$ is a decreasing function of ℓ and

$$\sum_{\ell=1}^{q^d} \mathrm{Prob}(|\mathcal{C}| = \ell) = 1,$$

we have

$$\sum_{\ell \geq dq \log q}^{q^d} \left(1 - \frac{1}{q}\right)^{\ell} \cdot \mathrm{Prob}(|\mathcal{C}| = \ell) \leq \left(1 - \frac{1}{q}\right)^{dq \log q} \ll \exp(-d \log q) = q^{-d}.$$

To determine the contribution from the short cycles we argue as follows. By Proposition 8.2.0.5, if

$$\ell \leq dq \log q = o(q^{d/2}),$$

we have

$$\mathrm{Prob}(|\mathcal{C}| = \ell) = \sqrt{\frac{\pi}{2q^d}} \cdot \left(\mathrm{erfc}\left(\ell/\sqrt{2q^d}\right) + o(1)\right) = \sqrt{\frac{\pi}{2q^d}} \cdot (1 + o(1))$$

since $\mathrm{erfc}(0) = 1$. Hence,

$$\sum_{\ell=1}^{dq\log q} \left(1 - \frac{1}{q}\right)^\ell \cdot \mathrm{Prob}(|\mathcal{C}| = \ell) = (1 + o(1)) \cdot \sqrt{\frac{\pi}{2q^d}} \cdot \sum_{\ell=1}^{dq\log q} \sum_{1 \le \ell < dq\log q} \left(1 - \frac{1}{q}\right)^\ell,$$

which, on summing the geometric series, equals

$$(1 + o(1)) \cdot \sqrt{\frac{\pi}{2q^d}} \cdot \frac{1 - O(q^{-d})}{1 - (1 - 1/q)} = (\sqrt{\pi/2} + o(1)) \cdot q^{1-d/2}.$$

$\square$

As noted at the end of Section 8.1, this means if the random model is accurate, then when $d \ge 5$, there may be some $\alpha \in X(K)$ such that one cannot treat the dynamical Mordell-Lang problem for $\mathrm{Orb}_\Phi(\alpha)$ via the method of p-adic analytic parametrization.

8.4. The case of split maps

We believe that the p-adic parametrization method does work for split maps, in the sense that there should always be a prime $\mathfrak{p}$ at which one can avoid ramification. We begin by looking at what the random model heuristic suggests. We work with the following setup:

- $f : \mathbb{P}^1 \longrightarrow \mathbb{P}^1$ is a rational map of degree at least one defined over a number field K;
- $\mathfrak{p}$ is a finite prime of good reduction for f;
- $r_\mathfrak{p}$ is the usual reduction map from $\mathbb{P}^1(K)$ to $\mathbb{P}^1(k_\mathfrak{p})$, where $k_\mathfrak{p}$ the residue field of $\mathfrak{p}$; and
- α and β are points in $\mathbb{P}^1(K)$ such that β is not in the forward orbit of α.

The random model heuristic (see Sections 8.2 and 8.3) suggests that the chance that there is an n such that

$$r_\mathfrak{p}(f^n(\alpha)) = r_\mathfrak{p}(\beta)$$

should be about $1/\sqrt{\#k_\mathfrak{p}}$. Thus, the density of primes $\mathfrak{p}$ such that that there is an n for which

$$r_\mathfrak{p}(f^n(\alpha)) = r_\mathfrak{p}(\beta)$$

should be zero. If this were the case, then one could prove the split case of the dynamical Mordell-Lang conjecture by letting the β_i range over the critical points of the maps involved.

However, the random model heuristic is not accurate in all cases. For example, let

- $f(x) = x^3 + 5$;
- $K = \mathbb{Q}$; and
- $\alpha = 5$ and $\beta = 0$.

Then β is not in the forward orbit of α. On the other hand, for any prime p satisfying

$$p \equiv 2 \pmod 3,$$

the map f reduces to a permutation modulo p, and since

$$f(\beta) = \alpha,$$

this means that there is an n such that

$$r_p(f^n(\alpha)) = \beta.$$

The density of such primes is obviously $\frac{1}{2}$, so the density of primes p such that there is an n for which

$$r_p(f^n(\alpha)) = \beta$$

must be at least $1/2$. In fact, Hamblen, Jones, and Madhu [**HJM15**] proved that the proportion is exactly $1/2$. They proved this by applying the Chebotarev density theorem (see Theorem 2.4.0.1) to the Galois groups of the polynomials $f^m(x)$. It turns out that these Galois groups take on a very specific structure as a subgroup of finite index in a wreath product of the group $\mathbb{Z}_2$ (see [**Odo85, Sto92, Jon08, JKMT**]). More generally, one might hope to use the theory of iterated Galois groups to attack the problem at hand more generally (see also [**Nek05, Jon15, Jon13**]). These considerations, along with the ideas of [**BGHKST13**] lead us to make the following conjecture.

CONJECTURE 8.4.0.3. *Let K be a number field. Let $f_1, \ldots, f_g \in K(x)$ all have degree greater than one and let*

$$\alpha_1, \ldots, \alpha_g, \beta_1, \ldots, \beta_g \in \mathbb{P}^1(K).$$

Suppose that for $i = 1, \ldots, g$ there is no n such that $f_i^n(\alpha_i) = \beta_i$. Let $\mathcal{S}$ be the set of primes $\mathfrak{p}$ such that for $i = 1, \ldots, g$ there is no n such that $r_{\mathfrak{p}}(f_i^n(\alpha_i)) = r_{\mathfrak{p}}(\beta_i)$, where

$$r_{\mathfrak{p}} : \mathbb{P}^1(K) \longrightarrow \mathbb{P}^1(k_{\mathfrak{p}})$$

is the usual reduction map. Then $\mathcal{S}$ has positive density among the primes of K.

This would imply that the dynamical Mordell-Lang conjecture is true for products of rational functions (what we call the "split case"), since one can find a positive density of primes such that a given orbit avoids ramification (see Theorem 4.4.4.1 which proves the Dynamical Mordell-Lang Conjecture under the assumption that such a prime can be found). Conjecture 8.4.0.3 has a similar flavor to Proposition 7.3.0.2. The difference here is that Conjecture 8.4.0.3 applies to any collection of points, not just to collections of points containing at most one wandering point.

In fact, Conjecture 8.4.0.3 is stronger than what is needed to prove the split case of the dynamical Mordell-Lang conjecture, since the points in Conjecture 8.4.0.3 need not be critical points for the maps f_i. Many of the examples of rational functions with pathological Galois groups are post-critically finite rational functions (that is, rational functions all of whose critical points are preperiodic). Moreover, in some cases, one can change the density of primes such that there is an n for which

$$r_p(f^n(\alpha)) = r_p(\beta)$$

by taking a base extension. Take for example, the situation we considered earlier where $f(x) = x^3 + 5$ and $\alpha = 5$ and $\beta = 0$. Then, over $\mathbb{Q}$, the proportion of primes such that there is an n for which

$$r_p(f^n(\alpha)) = r_p(\beta)$$

is $1/2$, but if one extends to $\mathbb{Q}(\xi_3)$ for ξ_3 a primitive third root of unity the proportion drops to zero (see [**HJM15**]). This leads us to pose the following question, which is similar to Conjecture 8.4.0.3.

QUESTION 8.4.0.4. *Let K be a number field. Let $\phi \in K(x)$ have degree greater than one. Let β be critical point of f. Suppose that β is not periodic. Is there a finite extension K' of K such that if S is the set of primes of K' such that $\alpha_\mathfrak{p}$ is periodic under $\phi_\mathfrak{p}$, then S has density zero among the primes of K'?*

A positive answer to Question 8.4.0.4 would also solve the split case of the dynamical Mordell-Lang conjecture (see again Theorem 4.4.4.1).

A first step towards understanding Question 8.4.0.4 via the technique of iterated Galois groups would be an answer to the following question, posed by Sookdeo [**Soo11**].

QUESTION 8.4.0.5. *Let $\varphi : \mathbb{P}^1 \longrightarrow \mathbb{P}^1$ be a rational function defined over a number field K, and let $\beta \in \mathbb{P}^1(K)$ be a wandering point of φ. Is there a constant $C(\varphi, \beta)$ such that for any n the number of $\mathrm{Gal}(\overline{K}/K)$-orbits of points in $\varphi^{-n}(0)$ is bounded above by $C(\varphi, \beta)$?*

A solution to the dynamical Lehmer conjecture on lower bounds for canonical heights of points (see [**Sil07**, Conjecture 3.25]) would yield a positive answer to Question 8.4.0.5.

CHAPTER 9

Higher dimensional results

In this chapter, we use the p-adic parametrization from Section 6.3 to prove several instances of the Dynamical Mordell-Lang Conjecture. We prove Conjecture 1.5.0.1 in two important instances:

   (i) for orbits that are close p-adically to a periodic point β of the endomorphism Φ for which the Jacobian $D\Phi_\beta$ satisfies certain technical hypotehses; and

   (ii) for endomorphisms Φ of semiabelian varieties X.

There are also two different results (see Theorems 9.1.0.1 and 9.2.0.1) that we prove when the orbit of α is *close* p-adically to a suitable periodic point; see the statements of the aforementioned theorems for the precise results. In Theorem 9.1.0.1, the p-adic closure of the orbit of α contains an attracting periodic point, while in Theorem 9.2.0.1, α is in a suitable p-adic neighborhood of an indifferent periodic point. We prove Theorem 9.1.0.1 in Section 9.1, while Theorem 9.2.0.1 is stated in Section 9.2 and then proved in Section 9.5 using also the technical results from linear algebra obtained in Section 9.4. We state the result for endomorphisms of semiabelian varieties in Section 9.3 and then we prove it in Section 9.5 by employing the same strategy used in the proof of Theorem 9.2.0.1.

The method we employ in the proofs of the main results of this chapter is therefore p-adic analytic. Often (but not always) our strategy is a generalization of the Skolem's method discussed first in Section 2.5 and then extended in Chapter 4 to the p-adic arc lemma that we used for proving other instances of the Dynamical Mordell-Lang Conjecture, such as the case of étale endomorphisms (see Theorem 4.3.0.1). After all, the Dynamical Mordell-Lang Conjecture for endomorphisms of semiabelian varieties is a direct consequence of the main result of Chapter 4—see Corollary 4.4.1.2. However, the proofs we present in this chapter are self-contained, and they do not reply on the general form of the p-adic arc lemma developed in Chapter 4 as employed in the proof of Theorem 4.4.1.1.

For more details regarding the results contained in this chapter, we refer the reader to [**GT09**], whose exposition we also follow.

9.1. The Herman-Yoccoz method for periodic attracting points

First, we setup the notation for this section; most of this setup remains valid throughout the entire chapter.

If X is a quasiprojective variety defined over a field L, and

$$\Phi : X \longrightarrow X$$

is a morphism, and both β and $\Phi(\beta)$ are nonsingular points in $X(L)$, then Φ induces an L-linear map

$$D\Phi_\beta : T_\beta \longrightarrow T_{\Phi(\beta)}$$

where T_β is the stalk of the tangent sheaf for X at β (see also Subsection 2.1.6). Since β and $\Phi(\beta)$ are nonsingular, both T_β and $T_{\Phi(\beta)}$ are vector spaces of dimension $\dim X$ over L (see [**Har77**, II.8] and also [**Sha74**]). Note that when $L = \mathbb{C}$ and β and $\Phi(\beta)$ are in a coordinate patch $\mathcal{U}$ on the complex manifold $X^{\text{smooth}}(\mathbb{C})$ (which is the set of smooth points of $X(\mathbb{C})$; see Section 2.1.6), then $D\Phi_\beta$ can be written in coordinates using the partial derivatives of Φ with respect to these coordinates (i.e., $D\Phi_\beta$ is the Jacobian matrix of Φ at β expressed with respect to these coordinates).

Let p be a prime number, and let $\mathbb{C}_p$ be the completion of a fixed algebraic closure of $\mathbb{Q}_p$. We also use the notation $D(\vec{\gamma}, r) \subseteq \mathbb{C}_p^N$ for the open disk centered at $\vec{\gamma} \in \mathbb{C}_p^N$ of radius r; also, sometimes, we drop the vector notation for a point in $\mathbb{C}_p^N$ and simply denote it γ (instead of $\vec{\gamma}$).

Now, using Corollary 6.3.0.5 we can prove the following result (see [**GT09**, Theorem 1.3]). We note that the hypotheses in Theorem 9.1.0.1 are quite restrictive since this result is valid only when the Jacobian of a power Φ^M is a nonzero *homothety*, i.e.

$$D\Phi^M = \lambda \cdot \text{Id},$$

for some nonzero constant λ.

THEOREM 9.1.0.1 ([**GT09**]). *Let p be a prime number, let X be a quasiprojective variety defined over $\mathbb{C}_p$, and let*

$$\Phi : X \longrightarrow X$$

be a morphism defined over $\mathbb{C}_p$. Let $\alpha \in X(\mathbb{C}_p)$, and let V be a closed subvariety of X defined over $\mathbb{C}_p$. Assume the p-adic closure of the orbit $\mathcal{O}_\Phi(\alpha)$ contains a Φ-periodic point β of period dividing M such that β and all of its iterates are nonsingular, and such that the Jacobian of Φ^M at β is a nonzero homothety of p-adic absolute value less than one. Then the set of all $n \in \mathbb{N}_0$ such that $\Phi^n(\alpha) \in V(\mathbb{C}_p)$ is a union of finitely many arithmetic progressions.

PROOF. We follow the arguments from the proof of [**GT09**, Theorem 1.3]. Let $\lambda \in \mathbb{C}_p$ such that

$$D\Phi_\beta^M = \lambda \cdot \text{Id}.$$

According to our hypotheses, we have $0 < |\lambda|_p < 1$. Let $j \in \{0, \ldots, M-1\}$ be fixed. Using the fact that $X(\mathbb{C}_p)$ is a p-adic analytic manifold of dimension g in a neighborhood of each iterate $\Phi^j(\beta)$, we may find an analytic function $\mathcal{F}_j$ defined on a sufficiently small neighborhood $\mathcal{U}_j$ of $\vec{0} \in \mathbb{C}_p^g$ which maps $\mathcal{U}_j$ bijectively onto a neighborhood $\mathcal{V}_j$ of $\Phi^j(\beta)$. Then we write

$$\Psi_j := \mathcal{F}_j^{-1} \circ \Phi^M \circ \mathcal{F}_j$$

as a function of the following form (note that $D(\Psi_j)_{\vec{0}} = \lambda \cdot \text{Id}$):

$$\Psi_j(\vec{x}) := (\mathcal{F}_j^{-1} \circ \Phi^M \circ \mathcal{F}_j)(\vec{x}) = \lambda \cdot \vec{x} + \text{ higher order terms } .$$

Since

$$|\lambda^i - \lambda|_p = |\lambda|_p$$

for $i \geq 2$, we see that (6.3.0.2) is satisfied; so we have a bijective analytic function

$$h_j : D(\vec{0}, r_j) \longrightarrow D(\vec{0}, r_j),$$

for some $r_j > 0$ such that

$$\Psi_j \circ h_j = h_j \circ \lambda \, \text{Id}$$

and furthermore,
$$D\left(h_j\right)_{\vec{0}} = \mathrm{Id},$$
by Theorem 6.3.0.3. Let $r > 0$ such that for each $0 \leq j \leq M - 1$, we have
$$(9.1.0.2) \qquad (\Phi^j \circ \mathcal{F}_0)(D(\vec{0}, r)) \subseteq \mathcal{F}_j(D(\vec{0}, r_j)).$$

Let N_0 be the smallest positive integer such that
$$\Phi^{N_0}(\alpha) \in \mathcal{F}_0(D(\vec{0}, r)).$$
Then
$$\Phi^{N_j}(\alpha) \in \mathcal{F}_j(D(\vec{0}, r_j)),$$
where $N_j := N_0 + j$ for each $j = 1, \ldots, M - 1$. Let $\vec{\alpha_j} \in D(\vec{0}, r_j)$ satisfy
$$h_j(\vec{\alpha_j}) = \mathcal{F}_j^{-1}(\Phi^{N_j}(\alpha)).$$
Since $|\lambda|_p < 1$, we have
$$(9.1.0.3) \qquad \left(\mathcal{F}_j^{-1} \circ \Phi^{kM}\right)\left(\Phi^{N_j}(\alpha)\right) = (\Psi_j^k \circ h_j)(\vec{\alpha_j}) = h_j(\lambda^k \cdot \vec{\alpha_j}).$$

Now, for each polynomial F in the vanishing ideal of V, we construct the function
$$\Theta_{F,j} : \overline{D}(0, 1) \longrightarrow \mathbb{C}_p$$
given by
$$\Theta_{F,j}(z) := F\left((\mathcal{F}_j \circ h_j)\left(z \cdot \vec{\alpha_j}\right)\right).$$
The function $\Theta_{F,j}$ is analytic because each h_j is analytic on $D(\vec{0}, r_j)$, and $\vec{\alpha_j} \in D(\vec{0}, r_j)$.

For each $k \in \mathbb{N}$ such that
$$\Phi^{kM+N_j}(\alpha) \in V(\mathbb{C}_p),$$
we have $\Theta_{F,j}(\lambda^k) = 0$ for each F. Since
$$\lim_{k \to \infty} \lambda^k = 0,$$
we conclude that if there are infinitely many k such that
$$\Phi^{N_0+j+Mk}(\alpha) \in V(\mathbb{C}_p),$$
then $\Theta_{F,j}$ is identically equal to 0 (since the zeros of a p-adic analytic function cannot accumulate, as shown in Lemma 2.3.6.1); hence
$$\Theta_{F,j}(\lambda^k) = 0 \text{ for all } k \in \mathbb{N},$$
which means that F vanishes on all points $\Phi^{N_0+j+Mk}(\alpha)$ for $k \in \mathbb{N}$. Applying this argument for all polynomials F in a finite set of generators for the vanishing ideal of V, we conclude that
$$(9.1.0.4) \qquad \text{either } \Phi^{N_0+j+Mk}(\alpha) \in V(\mathbb{C}_p) \text{ for all } k \in \mathbb{N}$$
$$\text{or } V(\mathbb{C}_p) \cap \mathcal{O}_{\Phi^M}\left(\Phi^{N_0+j}(\alpha)\right) \text{ is finite.}$$

Since
$$\mathcal{O}_\Phi(\alpha) = \{\Phi^i(\alpha) : 0 \leq i \leq N_0 - 1\} \bigcup \left(\bigcup_{j=0}^{M-1} \mathcal{O}_{\Phi^M}(\Phi^{N_0+j}(\alpha))\right),$$
we conclude the proof of Theorem 9.1.0.1. $\qquad \square$

We observe that, just as in Remark 2.5.4.4, we obtain an upper bound for the common difference of the (infinite) arithmetic progressions from the conclusion of the Dynamical Mordell-Lang Conjecture in the setting of Theorem 9.1.0.1; more precisely, the common difference of those arithmetic progressions is bounded above by the order of the period of the point β.

Theorem 9.1.0.1 works on the p-adic analytic side along the same reasoning as the Skolem-Mahler-Lech method and the p-adic arc lemma from Chapter 4: we obtain a p-adic analytic parametrization of the orbit (piecewise) and then for each polynomial F in the vanishing ideal of the subvariety V we construct a one-variable p-adic analytic function $z \mapsto \Theta_{F,j}(z)$ such that

$$\Phi^{Mk+j}(\alpha) \in V(\mathbb{C}_p) \text{ for } k \text{ sufficiently large if and only if } \Theta_{F,j}(\lambda^k) = 0,$$

for each F in a given finite set of generators for the vanishing ideal of V. Using the fact that $\{\lambda^k\}_{k\in\mathbb{N}_0}$ accumulate to 0 since $|\lambda|_p < 1$, we obtain the same dichotomy as in the classical Skolem-Mahler-Lech approach:

(1) either there exist only finitely many k such that $\Phi^{Mk+j}(\alpha) \in V(\mathbb{C}_p)$; or
(2) $\Theta_{F,j} = 0$ for each F, which yields that $\Phi^{Mk+j}(\alpha) \in V(\mathbb{C}_p)$ for all k sufficiently large.

The difference from the p-adic arc lemma constructed in Chapter 4 is the fact that the p-adic analytic functions $\Theta_{F,j}$ are evaluated in this case at points λ^n converging to 0 rather than being evaluated at points $n \in \mathbb{Z} \subset \mathbb{Z}_p$. We also see that it is essential that the Jacobian J of Φ^M at β is a multiple of the identity. Indeed, if J were a diagonal matrix, say, with eigenvalues λ_i which satisfy (6.3.0.2) we could still use Theorem 6.3.0.3 and its Corollary 6.3.0.5 only that this time we would get that for each polynomial F in the vanishing ideal of V there exists a p-adic analytic function $\Theta_{F,j}$ of g variables such that

(9.1.0.5) $\Phi^{Mk+j}(\alpha) \in V(\mathbb{C}_p)$ if and only if $\Theta_{F,j}(\lambda_1^k, \ldots, \lambda_g^k) = 0$ for each F.

Now, if $|\lambda_i|_p = 1$, then we can apply the p-adic arc lemma exactly as in Chapter 4 (see Theorem 9.2.0.1) and conclude the proof of Conjecture 1.5.0.1 in this case.

Assume now that $|\lambda_i|_p < 1$ for each i. Then in this case, (9.1.0.5) is generally insufficient to allow us to apply Lemma 2.3.6.1 since we do not have a *one-variable* p-adic analytic function. In general, the zero set of a multivariable p-adic analytic function forms a positive dimensional p-adic analytic manifold, and thus it has accumulation points. We could still find a one-variable p-adic analytic function if each pair of the eigenvalues λ_i were multiplicatively dependent, and they still satisfied condition (6.3.0.2) (after all, the case of Theorem 9.1.0.1 is when $\lambda_1 = \cdots = \lambda_g$). However, in all other cases, it would generally be impossible to construct a *one-variable* p-adic analytic function $\Theta_{F,j}$ satisfying (9.1.0.5).

On the other hand, assuming each $\lambda_i \in \mathbb{Z}_p$, then arguing as in Chapter 11 (see in particular Lemma 11.8.0.1), the multivariable parametrization from (9.1.0.5) for arbitrary λ_i reduces to finding $k \in \mathbb{N}$ such that

$$G(k, p^k) = 0,$$

where G is a p-adic analytic function of two variables. Indeed, since $\lambda_i \in \mathbb{Z}_p$, then

$$\lambda_i = u_i \cdot p^{e_i}$$

for some nonnegative integers e_i and p-adic units u_i. Then at the expense of replacing M by $(p-1)M$ (and therefore replacing λ_i by λ_i^{p-1}) we may assume

$$|u_i - 1|_p < 1.$$

Hence the existence of a p-adic analytic function G in two variables such that

$$G(k, p^k) = \Theta_{F,j}(\lambda_1^k, \ldots, \lambda_g^k).$$

In Chapter 11 we prove a general statement (see Lemma 11.8.0.1) about the sparseness of the set of positive integers k such that $G(k, p^k) = 0$ (for such a non-trivial p-adic analytic function G; note that if G were identically zero then we would obtain an entire arithmetic progression in the set of iterates n such that $\Phi^n(\alpha) \in V(\mathbb{C}_p)$). For more about the sparseness of the set

$$\{n \in \mathbb{N} \colon \Phi^n(\alpha) \in V\},$$

assuming V does not contain a positive dimensional periodic subvariety, we refer the reader to Chapter 11.

A special case of Theorem 9.1.0.1 is the following result for orbits of points in the vicinity of an attracting periodic point; see Subsection 6.3.1 for the definition of attracting periodic points for endomorphisms of higher dimensional varieties.

THEOREM 9.1.0.6 ([**GT09**]). *Let $N \geq 1$, let p be a prime number, let*

$$\varphi_1, \ldots, \varphi_N \in \mathbb{C}_p(t)$$

be rational functions, and let

$$\Phi := (\varphi_1, \ldots, \varphi_N)$$

act coordinatewise on $\left(\mathbb{P}^1\right)^N$. Let

$$\alpha := (\alpha_1, \ldots, \alpha_N) \in \left(\mathbb{P}^1\right)^N (\mathbb{C}_p),$$

and let $V \subseteq \left(\mathbb{P}^1\right)^N$ be a subvariety defined over $\mathbb{C}_p$. Assume the p-adic closure of the orbit $\mathcal{O}_\Phi(\alpha)$ contains an attracting Φ-periodic point

$$\beta := (\beta_1, \ldots, \beta_N)$$

such that for some positive integer M, we have

$$\Phi^M(\beta) = \beta$$

and

$$(\varphi_1^M)'(\beta_1) = \cdots = (\varphi_N^M)'(\beta_N).$$

Then the set of $n \in \mathbb{N}_0$ such that $\Phi^n(\alpha) \in V(\mathbb{C}_p)$ is a union of finitely many arithmetic progressions.

PROOF. We refer the interested reader to [**GT09**, Theorem 1.5]. □

9.2. The Herman-Yoccoz method for periodic indifferent points

In this section we apply again the Herman-Yoccoz method (see Chapter 6 and Section 6.3) but this time we use an embedding into $\mathbb{Z}_p$ such that each eigenvalue of the Jacobian matrix at the periodic point is a p-adic unit. So, using again the parametrization of an orbit from Corollary 6.3.0.5, one obtains the following result (see [**GT09**, Theorem 1.6]).

THEOREM 9.2.0.1. *Let X be a quasiprojective variety defined over a number field K, let*

$$\Phi : X \longrightarrow X$$

be a morphism defined over K, and let V be a closed subvariety of X defined over K. Let $\beta \in X(K)$ be a periodic point of period dividing M such that β and its iterates are all nonsingular points, and the Jacobian of Φ^M at β is a diagonalizable matrix whose eigenvalues $\lambda_1, \ldots, \lambda_g$ satisfy

$$(9.2.0.2) \qquad \prod_{j=1}^{g} \lambda_j^{e_j} \neq \lambda_i,$$

for each $1 \leq i \leq g$, and any nonnegative integers $e_1, \ldots, e_g$ such that $\sum_{j=1}^{g} e_j \geq 2$.

Then for all but finitely many primes p, there is a p-adic neighborhood $\mathcal{V}_p$ of β (depending only on p and β) such that if

$$\mathcal{O}_\Phi(\alpha) \cap \mathcal{V}_p \text{ is non-empty}$$

for $\alpha \in X(\mathbb{C}_p)$, then the set of $n \in \mathbb{N}_0$ such that $\Phi^n(\alpha) \in V(\mathbb{C}_p)$ is a union of finitely many arithmetic progressions.

Theorem 9.2.0.1 is connected both with Theorem 9.1.0.1 and with Theorem 4.4.1.1. The connection to Theorem 9.1.0.1 lies in the fact that in both the proof of that result and also in the proof of Theorem 9.2.0.1 one starts by choosing an iterate $\Phi^\ell(\alpha)$ that is very close to β. By the work of Herman and Yoccoz [**HY83**] (see our Theorem 6.3.0.3) we obtain the existence of a p-adic function h in a neighborhood of β such that, for a suitable positive integer M, we have

$$\Phi^M \circ h = h \circ A,$$

for some linear function A. When A is a homothety, this means that iterates of $\Phi^\ell(\alpha)$ under Φ^M lie on an analytic line in $\mathbb{C}_p^g$. Under the conditions of Theorem 9.2.0.1, it is necessary to take p-adic logarithms of iterates in order to get a line in $\mathbb{C}_p^g$ but otherwise the proof is the same. Note that existence of the map h in Theorem 9.2.0.1 depends on Yu's [**Yu90**] results on linear forms in p-adic logarithms, which only apply over number fields. Under the conditions of Theorem 9.1.0.1, the map h exists even when the eigenvalues of A are transcendental.

The connection between Theorem 9.2.0.1 and Theorem 4.4.1.1 lies in the fact that in both cases one obtains p-adic analytic functions $z \mapsto \Theta_{F,j}(z)$ (for each $j = 0, \ldots, M-1$, and for each polynomial F in the vanishing ideal of the subvariety V) such that

$$(9.2.0.3) \qquad \Phi^{Mk+j}(\alpha) \in V(\mathbb{C}_p) \text{ if and only if } \Theta_{F,j}(k) = 0 \text{ for each } F.$$

In other words, for Theorem 9.2.0.1 the p-adic arc lemma applies. We postpone the proof of Theorem 9.2.0.1 until Section 9.5. First we discuss in Section 9.3 the

Dynamical Mordell-Lang Conjecture for arbitrary endomorphisms of semiabelian varieties, which motivated the authors of [**GT09**] to prove Theorem 9.2.0.1.

9.3. The case of semiabelian varieties

As we previously wrote, the motivation for obtaining Theorem 9.2.0.1 lay, at the time of writing [**GT09**], in finding a self-contained proof of the Dynamical Mordell-Lang Conjecture for endomorphisms of semiabelian varieties.

THEOREM 9.3.0.1. *Let A be a semiabelian variety defined over $\mathbb{C}$, and let*

$$\Phi : A \longrightarrow A$$

be an arbitrary endomorphism defined over $\mathbb{C}$. Then for every subvariety $V \subseteq A$ defined over $\mathbb{C}$, and for every point $\alpha \in A(\mathbb{C})$, the set of $n \in \mathbb{N}_0$ such that

$$\Phi^n(\alpha) \in V(\mathbb{C})$$

is a union of finitely many arithmetic progressions.

We note that an *arbitrary* endomorphism Φ of a semiabelian variety A is the composition of an algebraic group endomorphism (i.e., an endomorphism of A as an algebraic group) and a translation; see Subsection 2.1.9. Theorem 9.3.0.1 was proven in [**GT09**] in the special case when Φ is an algebraic group endomorphism.

One can prove Theorem 9.3.0.1 by reducing it to deep results of Faltings [**Fal91**] and Vojta [**Voj96**] (see Theorem 3.4.2.1) that settled the classical Mordell-Lang Conjecture; see Section 3.4. Indeed, any endomorphism Φ of a semiabelian variety A is of the form

$$\gamma \mapsto \Phi(\gamma) = \Psi(\gamma) + \beta,$$

where Ψ is an algebraic group endomorphism of A, and $\beta \in A(\mathbb{C})$ is a given point. On the other hand, each algebraic group endomorphism of A is integral over $\mathbb{Z}$. More precisely, if $g := \dim(A)$, then there exists a monic polynomial $P \in \mathbb{Z}[z]$ of degree at most $2g$ such that $P(\Psi) = 0$. Since for any given $\alpha \in A(\mathbb{C})$, we have

$$\Phi^n(\alpha) = \Psi^n(\alpha) + \left(\Psi^{n-1} + \cdots + \Psi + 1\right)(\beta),$$

we conclude that $\mathcal{O}_\Phi(\alpha)$ lies in the finitely generated subgroup Γ of $A(\mathbb{C})$ spanned by $\Psi^i(\alpha)$ and $\Psi^i(\beta)$ for $i = 0, \ldots, 2g - 1$. Then by Theorem 3.4.2.1, we get that

$$\Gamma \cap V(\mathbb{C}) = \bigcup_i \gamma_i + H_i,$$

for some $\gamma_i \in A(\mathbb{C})$ and some subgroups $H_i \subseteq \Gamma$. Thus Theorem 9.3.0.1 reduces to determining the set of all $n \in \mathbb{N}_0$ such that

$$\Phi^n(\alpha) \in (\gamma + H),$$

for a given coset $\gamma + H$ of Γ. Since Φ induces an affine map on the $\mathbb{Z}$-module Γ (i.e., it is a composition of a translation by an element of Γ with a linear map on Γ), a simple combinatorial argument finishes the proof of Theorem 9.3.0.1.

Also, Theorem 9.3.0.1 is an immediate consequence of Theorem 4.4.1.1 since Φ is unramified. However, we show in Section 9.5 how to deduce Theorem 9.3.0.1 without appealing to the more technical (both geometric and p-adic analytic) construction from [**BGT10**] which yields Theorem 4.4.1.1. Also, our method of proof for Theorem 9.3.0.1 does not appeal to the powerful theorem of Vojta [**Voj96**]; instead we apply a strategy common for the proof of both Theorem 9.3.0.1 and of Theorem 9.2.0.1.

9.4. Preliminaries from linear algebra

Before proving Theorems 9.2.0.1 and 9.3.0.1 we need to derive a technical result from linear algebra, which also involves p-adic analytic functions.

Proposition 9.4.0.2 is key for our proofs of both Theorems 9.3.0.1 and 9.2.0.1. We need a definition first (see [**Lan02**, XI.2]).

DEFINITION 9.4.0.1. By a *Jordan matrix* we mean a matrix which is in its Jordan form, i.e. it consists of its Jordan blocks. A *Jordan block* is either a multiple of the identity matrix, or it is an upper-triangular matrix whose entries on the diagonal are all equal, and the only nonzero entries outside the diagonal are the entries on the line above the diagonal which are all equal to one.

Before stating our key linear algebra result, we recall that $M_{n,n}(F)$ always represents the set of all n-by-n matrices with entries in the field F.

PROPOSITION 9.4.0.2. *Let $J \in M_{N,N}(\mathbb{C}_p)$ be a Jordan matrix with the property that each eigenvalue λ_i of J is either equal to 0, or satisfies $|\lambda_i|_p = 1$, and let $b \in \mathbb{C}_p^N$ be an arbitrary vector. We let*

$$\Psi : \mathbb{C}_p^N \longrightarrow \mathbb{C}_p^N \text{ be defined by } \Psi(v) = Jv + b.$$

Then for each $v \in \mathbb{C}_p^N$ there exists a positive integer d (depending only on J) and a p-adic analytic function

$$f := f_{\Psi,v} : \overline{D}(0,1) \longrightarrow \mathbb{C}_p^N$$

such that for each positive integer k, we have

$$f(k) = \Psi^{dk}(v).$$

Proposition 9.4.0.2 is a generalization of [**GT09**, Proposition 3.1].

PROOF OF PROPOSITION 9.4.0.2. Working on subspaces of $\mathbb{C}_p^N$ corresponding the Jordan blocks of J, we see immediately that it suffices to prove the conclusion under the assumption that J is an N-by-N Jordan block; so, from now on, we assume J is an N-by-N Jordan block.

An easy computation shows that for each $d, n \in \mathbb{N}$ we have

$$\Psi^{dn}(v) = J^{dn}v + \left(J^{d(n-1)} + \cdots + J^d + \mathrm{Id} \right) \left(J^{d-1}b + \cdots + Jb + b \right).$$

There are two cases: the (unique) eigenvalue λ of J is either zero or nonzero.

Case 1. $\lambda = 0$.

We let d be the dimension N of J and therefore $J^d = 0$. For each $k \in \mathbb{N}$ we have

$$\Psi^{dk}(v) = J^{d-1}b + \cdots + Jb + b.$$

Hence we may take f be the constant function

$$f(z) := \sum_{i=0}^{d-1} J^i b.$$

Case 2. $|\lambda|_p = 1$.

Using Proposition 2.3.3.3, there exists a positive integer d such that

$$(9.4.0.3) \qquad\qquad |\lambda^d - 1|_p < p^{-\frac{1}{p-1}}.$$

Assume first that

$$\lambda^d \neq 1.$$

Then we see that the matrix

$$J^d - \text{Id} \text{ is invertible,}$$

and so there exists a vector $w \in \mathbb{C}_p^g$ such that

$$(J^d - \text{Id})w = \sum_{i=0}^{d-1} J^i b.$$

Then

$$\Psi^{dk}(v) = J^{dk}v + \left(J^{dk} - \text{Id}\right)w$$

and the result follows easily because all entries of J^{dk} are p-adic analytic functions in the variable k, convergent whenever $|k|_p < 1$. Here we use (9.4.0.3) and also Lemma 2.3.4.2 which yields that

$$z \mapsto (\lambda^d)^z \text{ is analytic if } |z|_p \leq 1.$$

Note that the following N-by-N matrix, defined for each $z \in \overline{D}(0,1)$ by

$$(J^d)^z := (\lambda^d)^z \cdot \begin{pmatrix} 1 & \frac{z}{1!\cdot\lambda} & \frac{z(z-1)}{2!\cdot\lambda^2} & \cdots & \frac{z(z-1)\cdots(z-N+1)}{N!\cdot\lambda^{N-1}} \\ 0 & 1 & \frac{z}{1!\cdot\lambda} & \cdots & \frac{z(z-1)\cdots(z-N+2)}{(N-1)!\cdot\lambda^{N-2}} \\ 0 & 0 & 1 & \cdots & \frac{z(z-1)\cdots(z-N+3)}{(N-2)!\cdot\lambda^{N-3}} \\ \cdots & \cdots & \cdots & \cdots & \cdots \\ 0 & 0 & 0 & \cdots & 1 \end{pmatrix}$$

has the property that $(J^d)^z$ equals the usual power J^{dk} whenever $k \in \mathbb{N}$.

Assume now that

$$\lambda^d = 1.$$

Then

$$J^d = \text{Id} + J_0,$$

where J_0 is a nilpotent matrix, i.e.

$$J_0^N = 0.$$

Then for each z we may define a matrix J^{dz} whose entries are all polynomial functions in z, as follows:

$$J^{dz} = (\text{Id} + J_0)^z := \sum_{n=0}^{N-1} \frac{z(z-1)\cdots(z-n+1)}{n!} \cdot J_0^n.$$

Hence

$$\sum_{i=0}^{n-1} J^{di} = \sum_{k=0}^{N-1} \left(\sum_{i=0}^{n-1} \frac{i(i-1)\cdots(i-k+1)}{k!}\right) \cdot J_0^k = \sum_{k=0}^{N-1} g_k(n) \cdot J_0^k,$$

where each $g_k \in \mathbb{Q}[z]$ is a polynomial of degree $(k+1)$. In conclusion, we can let $f : \overline{D}(0,1) \longrightarrow \mathbb{C}_p^N$ be defined by

$$f(z) = J^{dz}v + \left(\sum_{k=0}^{N-1} g_k(z) \cdot J_0^k\right)\left(\sum_{i=0}^{d-1} J^i b\right),$$

which satisfies the conclusion of Proposition 9.4.0.2. $\square$

9.5. Proofs for Theorems 9.2.0.1 and 9.3.0.1

In this section we conclude the proofs of Theorem 9.2.0.1 and 9.3.0.1.

PROOF OF THEOREM 9.2.0.1. Let $B \in \mathrm{GL}_N(\overline{K})$ such that

$$\Lambda := B(D(\Phi^M)_\beta)B^{-1} \text{ is a diagonal matrix.}$$

At the expense of replacing K by a finite extension, we may assume

$$B \in \mathrm{GL}_N(K).$$

Then for all but finitely many primes p, the entries of both B and B^{-1} have p-adic absolute values at most 1. Let λ_i (for $1 \leq i \leq N$) be the eigenvalues of $D(\Phi^M)_\beta$. According to our hypotheses, each λ_i is nonzero. Thus for all but finitely many primes p, each λ_i is a p-adic unit. Fix a prime p and an embedding of K into $\mathbb{C}_p$ such that

- $|\lambda_i|_p = 1$ for each i; and
- each entry in B and B^{-1} is a p-adic integer.

Let $j \in \{0, \ldots, M-1\}$ be fixed. Clearly, we have

$$D(\Phi^M)_{\Phi^j(\beta)} = D(\Phi^M)_\beta.$$

Since each $\Phi^j(\beta)$ is a nonsingular point, there exists a sufficiently small neighborhood $\mathcal{U}_j \subseteq \mathbb{C}_p^g$ of $\vec{0}$, and an analytic function $\mathcal{F}_j$ that maps $\mathcal{U}_j$ bijectively onto a small neighborhood of $\Phi^j(\beta) \in X(\mathbb{C}_p)$. Let

$$\Psi_j := \mathcal{F}_j^{-1} \circ \Phi^M \circ \mathcal{F}_j.$$

Then

$$\Psi_j(\vec{x}) = (B^{-1}\Lambda B) \cdot \vec{x} + \text{ higher order terms.}$$

Using hypothesis (9.2.0.2) and [**Yu90**, Theorem 1], we conclude that (6.3.0.2) is satisfied by the eigenvalues λ_i. Using Theorem 6.3.0.3, we conclude that there exists a positive number $r_j > 0$ such that

$$D(\vec{0}, r_j) \subseteq \mathcal{U}_j,$$

and there exists a bijective analytic function

$$h_j : D(\vec{0}, r_j) \longrightarrow D(\vec{0}, r_j)$$

such that

$$\Psi_j \circ h_j = h_j \circ (B^{-1}\Lambda B).$$

Let r be a positive number such that for every $j = 0, \ldots, M-1$ we have

$$\Phi^j(\mathcal{F}_0(D(\vec{0}, r))) \subseteq \mathcal{F}_j(D(\vec{0}, r_j)).$$

We let

$$\mathcal{V}_p := \mathcal{F}_0(D(\vec{0}, r))$$

be the corresponding p-adic neighborhood of β in $X(\mathbb{C}_p)$. Suppose that

$$\mathcal{O}_\Phi(\alpha) \cap \mathcal{V}_p \text{ is non-empty.}$$

Then there exists $N_0 \in \mathbb{N}$ such that

$$\Phi^{N_0}(\alpha) \in \mathcal{V}_p,$$

and so,

$$\Phi^{N_j}(\alpha) \in \mathcal{F}_j(D(\vec{0}, r_j))$$

where $N_j := N_0 + j$, for each $j = 0, \ldots, M-1$. Let $\vec{\alpha_j} \in D(\vec{0}, r_j)$ such that

$$h_j(\vec{\alpha_j}) = \mathcal{F}_j^{-1}(\Phi^{N_j}(\alpha)).$$

Then for each $k \in \mathbb{N}$, we have

$$\Phi^{kM+N_j}(\alpha) = (\mathcal{F}_j \circ h_j)\left(B^{-1}\Lambda^k B(\vec{\alpha_j})\right).$$

Note that

$$B^{-1}\Lambda^k B(\vec{\alpha_j}) \in D(\vec{0}, r_j)$$

for each $k \in \mathbb{N}$, since each entry of B, B^{-1}, and Λ is in $\overline{D}(0,1)$,

Let d be a positive integer as in the conclusion of Proposition 9.4.0.2. Then the entries of the matrix $(\Lambda^d)^z$ are p-adic analytic functions of z in the disk $\overline{D}(0,1)$. Therefore, for each fixed $\ell = 0, \ldots, d-1$, the entries of the matrix $\Lambda^\ell \cdot (\Lambda^d)^z$ are p-adic analytic functions of $z \in \overline{D}(0,1)$.

Let F be any polynomial in the vanishing ideal of V. Then, for each $j = 0, \ldots, M-1$ and for each $\ell = 0, \ldots, d-1$, the function

$$\Theta_{F,j,\ell} : \overline{D}(0,1) \longrightarrow \mathbb{C}_p$$

defined by

$$\Theta_{F,j,\ell}(z) = F\left((\mathcal{F}_j \circ h_j)\left(B^{-1}\left(\Lambda^\ell \cdot (\Lambda^d)^z\right)B(\alpha_j)\right)\right)$$

is analytic. Furthermore, for each $k \in \mathbb{N}$ such that

$$\Phi^{N_0+j+M(kd+\ell)}(\alpha) \in V(\mathbb{C}_p)$$

we obtain that $\Theta_{F,j,\ell}(k) = 0$ for each F in the vanishing ideal of V. Since the zeros of a nonzero p-adic analytic function cannot accumulate (see Lemma 2.3.6.1), we conclude that if there are infinitely many $k \in \mathbb{N}$ such that $\Theta_{F,j,\ell}(k) = 0$, then $\Theta_{F,j,\ell}$ is identically equal to zero, and thus F vanishes on all points $\Phi^{N_0+j+M(kd+\ell)}(\alpha)$ for $k \in \mathbb{N}$. Applying this argument to each F in a finite set of generators for the vanishing ideal of V, we conclude that

(9.5.0.4)   either $\Phi^{N_0+j+M(kd+\ell)}(\alpha) \in V(\mathbb{C}_p)$ for all $k \in \mathbb{N}$

$$\text{or } V(\mathbb{C}_p) \cap \mathcal{O}_{\Phi^{Md}}\left(\Phi^{N_0+j+M\ell}(\alpha)\right) \text{ is finite.}$$

Since (9.5.0.4) holds for each $j = 0, \ldots, M-1$ and for each $\ell = 0, \ldots, d-1$, this concludes the proof of Theorem 9.2.0.1. $\square$

We conclude this section by proving Theorem 9.3.0.1.

PROOF OF THEOREM 9.3.0.1. Since Φ is an endomorphism of a semiabelian variety, it is a composition of a an algebraic group endomorphism Ψ with a translation by a point $\beta \in A$, i.e.,

$$\Phi(\gamma) = \Psi(\gamma) + \beta,$$

for each $\gamma \in A$. We let $L := D\Psi_0$ be the Jacobian of Ψ at the identity of A.

We proceed by induction on the dimension of V. The case $\dim V = 0$ follows from Corollary 3.1.2.11. Using the inductive hypothesis, we prove the following reduction.

CLAIM 9.5.0.5. *Let $m \in \mathbb{N}$. It suffices to prove Theorem 9.3.0.1 for the starting point $m\alpha$ and for the endomorphism*

$$\Phi_1 : A \longrightarrow A,$$

defined for all $\gamma \in A$ *by*

$$\Phi_1(\gamma) = \Psi(\gamma) + m\beta.$$

PROOF OF CLAIM 9.5.0.5. As we wrote before, we argue by induction on the dimension of V; the case $\dim(V) = 0$ is the content of Corollary 3.1.2.11 (since Conjecture 1.5.0.1 holds in this case).

Now, assume $\dim(V) \geq 1$. We have that for all $n \in \mathbb{N}$,

$$\Phi_1^n(m\alpha) = m\Phi^n(\alpha).$$

Then, given any subvariety V, we know that the set of n for which

$$\Phi^n(m\alpha) \in mV$$

forms a finite union $\mathcal{P}$ of arithmetic progressions (note that we denote by mV the image of V under the multiplication-by-m-map on A). Thus, if we let W be the inverse image of mV under the multiplication-by-m map (so, W is the finite union of translates of V by torsion points of A of order m) we know that the set of n such that

$$\Phi^n(\alpha) \in W$$

forms the same finite union $\mathcal{P}$ of arithmetic progressions. We let $Z_1, \ldots, Z_s$ be the positive dimensional irreducible components of the Zariski closure of

$$\{\Phi^n(\alpha)\}_{n \in \mathcal{P}};$$

note that if $\{\Phi^n(\alpha)\}_{n \in \mathcal{P}}$ is finite, then also $V(K) \cap \mathcal{O}_\Phi(\alpha)$ is finite, and so, we are done. Hence each Z_i is a Φ-periodic subvariety, i.e.,

$$\Phi(Z_i) \subseteq Z_i; \text{ see Proposition 3.1.2.14.}$$

Also, all but finitely many of the $\Phi^n(\alpha)$ for $n \in \mathcal{P}$ are contained in one of the $Z_i(K)$. Thus, we need only show that for each i, the set of n such that

$$\Phi^n(\alpha) \in V \cap Z_i$$

forms a finite union of arithmetic progressions. Each Z_i is contained in one of irreducible components of W and thus

$$\dim Z_i \leq \dim V.$$

If Z_i is contained in V, then the set of n for which

$$\Phi^n(\alpha) \in Z_i = Z_i \cap V$$

is a finite union of arithmetic progressions, since Z_i is Φ-periodic. If Z_i is not contained in V, then

$$\dim(Z_i \cap V) < \dim(Z_i) \leq \dim V$$

and the set of n such that $\Phi^n(\alpha) \in Z_i \cap V$ is a finite union of arithmetic progressions by the inductive hypothesis. $\square$

At the expense of replacing K by another finitely generated field, we may assume the Jacobian L is defined over K. We choose an embedding over $\mathbb{C}$ of

$$\iota : A \longrightarrow \mathbb{P}^M$$

as an open subset of a projective variety (for some positive integer M). At the expense of enlarging K, we may assume the above embedding ι is defined over K. We write

$$\iota(A) = Z(\mathfrak{a}) \setminus Z(\mathfrak{b})$$

for homogeneous ideals $\mathfrak{a}$ and $\mathfrak{b}$ in $K[x_0, \ldots, x_M]$, where $Z(\mathfrak{c})$ denotes the Zariski closed subset of $\mathbb{P}^M$ on which the ideal $\mathfrak{c}$ vanishes. We choose generators $F_1, \ldots, F_m$ and $G_1, \ldots, G_n$ for $\mathfrak{a}$ and $\mathfrak{b}$ respectively. We let

$$\oplus : A \times A \longrightarrow A \text{ and } \ominus : A \longrightarrow A$$

denote the addition, respectively the inversion map, written with respect to our chosen coordinates on $\mathbb{P}^M$.

The following result is proven by an argument similar to Proposition 4.4.1.4; for more details, see also [**GT09**, Claim 3.3].

CLAIM 9.5.0.6. *There exists a prime number p, and an embedding of K into $\mathbb{Q}_p$ such that:*

(i) *there exists a $\mathrm{Spec}(\mathbb{Z}_p)$-scheme $\mathcal{A}$ whose generic fiber equals A;*
(ii) $\alpha, \beta \in \mathcal{A}(\mathbb{Z}_p)$;
(iii) *L is conjugate over $\mathbb{Z}_p$ to its Jordan canonical form Λ (which is an $N \times N$ matrix), and moreover each of its eigenvalues λ_i is either a p-adic unit or equal to 0;*
(iv) *the maps Φ and $\ominus$ extend as endomorphisms of the $\mathbb{Z}_p$-scheme $\mathcal{A}$, while $\oplus$ extends to a morphism between $\mathcal{A} \times \mathcal{A}$ and $\mathcal{A}$.*

Let p be a prime number for which the conclusion of Claim 9.5.0.6 holds. Then we have a $\mathbb{Z}_p$-scheme $\mathcal{A}$ whose generic fiber equals A such that $\alpha, \beta \in \mathcal{A}(\mathbb{Z}_p)$.

CLAIM 9.5.0.7. *With the above notation, $\mathcal{A}(\mathbb{Z}_p)$ is compact .*

PROOF OF CLAIM 9.5.0.7. Let $\mathcal{V}$ and $\mathcal{W}$ be the Zariski closures of $Z(\mathfrak{a})$ and respectively $Z(\mathfrak{b})$ in $\mathcal{A}$. Then

$$\mathcal{A}(\mathbb{Z}_p) = \mathcal{V}(\mathbb{Z}_p) \cap \left(\mathbb{P}^N \setminus \mathcal{W} \right)(\mathbb{Z}_p)$$

is compact because it is the intersection of two compact subsets of $\mathbb{P}^N(\mathbb{Z}_p)$. Indeed, $\mathcal{V}(\mathbb{Z}_p)$ is compact because it is a closed subset of the compact set $\mathbb{P}^N(\mathbb{Z}_p)$. On the other hand,

$$\left(\mathbb{P}^N \setminus \mathcal{W} \right)(\mathbb{Z}_p)$$

consists of finitely many residue classes of $\mathbb{P}^N(\mathbb{Z}_p)$ and thus, it is compact because $\mathbb{Z}_p$ is compact. The above finitely many residue classes correspond to points in

$$\left(\mathbb{P}^N \setminus Z(\overline{\mathfrak{b}}) \right)(\mathbb{F}_p),$$

where $\overline{\mathfrak{b}}$ is the ideal of $\mathbb{F}_p[x_0, \ldots, x_N]$ generated by the reductions modulo p of each generator G_i of $\mathfrak{b}$. $\qquad\square$

According to [**Bou98**, Proposition 3, p. 216] there exists a p-adic analytic map

$$\exp : \mathbb{C}_p^g \longrightarrow A(\mathbb{C}_p)$$

which maps a sufficiently small neighborhood $D(\vec{0}, r)$ of $\vec{0} \in \mathbb{C}_p^g$ bijectively onto a sufficiently small neighborhood of $0 \in A(\mathbb{C}_p)$. Since $\exp$ is a local isomorphism of the analytic groups $\mathbb{C}_p^g$ and $A(\mathbb{C}_p)$, we conclude that any endomorphism of A corresponds to an affine map of $(\mathbb{C}_p^g, +)$; thus there exists an affine map

$$\varphi : \mathbb{C}_p^g \longrightarrow \mathbb{C}_p^g$$

such that

(9.5.0.8) $$\Phi(\exp(z)) = \exp(\varphi(z)).$$

Computing the Jacobian at $\vec{0}$ in (9.5.0.8), we obtain that

$$D\varphi_{\vec{0}} = L$$

because the embedding into $\mathbb{C}_p$ preserves the Jacobian L of Φ at 0. Therefore

$$\varphi(z) = L \cdot z + b$$

for some given $b \in \mathbb{C}_p^g$.

After replacing α and β by $m \cdot \alpha$ and respectively $m \cdot \beta$, for a positive integer m (as we are allowed to do so, by Claim 9.5.0.5), we may assume

$$\alpha, \beta \in \exp(D(\vec{0}, r)).$$

To see this, we note that $D(\vec{0}, r)$ is an additive subgroup because $|\cdot|_p$ is non-archimedean, so its image in $A(\mathbb{Q}_p)$ under exp is an open subgroup because exp is bijective and analytic on $D(\vec{0}, r)$. The fact that $\mathcal{A}(\mathbb{Z}_p)$ is compact (by Claim 9.5.0.7) means that any open subgroup of $\mathcal{A}(\mathbb{Z}_p)$ has finitely many cosets in $\mathcal{A}(\mathbb{Z}_p)$. Since $\alpha, \beta \in \mathcal{A}(\mathbb{Z}_p)$, there is a positive integer m such that

$$m\alpha, m\beta \in \exp(D(\vec{0}, r)).$$

Let $\vec{v}, \vec{b} \in D(\vec{0}, r)$ such that

$$\exp(\vec{v}) = \alpha \text{ and } \exp(\vec{b}) = \beta.$$

Since the coefficients of L are all p-adic integers, it follows that

$$L^n \vec{v} \in D(\vec{0}, r) \text{ for any } n \in \mathbb{N}.$$

Let d be as in Proposition 9.4.0.2 corresponding to the map φ, and let $f_{\vec{v}} := f_{\varphi, \vec{v}}$ be the p-adic analytic function on $\overline{D}(0, 1)$ such that for every positive integer k we have

$$f_{\vec{v}}(k) = \varphi^{dk}(\vec{v}).$$

Let $\vec{v_j} := \varphi^j(\vec{v})$, for each $j = 0, \ldots, d-1$.

The remaining part of our argument now proceeds as in the proof of Theorem 9.2.0.1. Fix $j \in \{0, \ldots, d-1\}$. Using (9.5.0.8) we obtain that for each $k \in \mathbb{N}$, we have

$$\Phi^{j+kd}(\alpha) = \exp\left(\varphi^{dk}(\vec{v_j})\right).$$

For each polynomial F in the vanishing ideal of V, we define the function $\Theta_{F,j}$ on $\overline{D}(0,1)$ by

$$\Theta_{F,j}(z) = F\left(\exp\left(f_{\vec{v_j}}(z)\right)\right).$$

So, $\Theta_{F,j}$ is analytic, and assuming that there are infinitely many $k \in \mathbb{N}$ such that

$$\Phi^{j+kd}(\alpha) \in V(\mathbb{C}_p),$$

we see that $\Theta_{F,j}$ is identically equal to 0. This means that F vanishes on all points $\Phi^{j+kd}(\alpha)$ for $k \in \mathbb{N}$. We conclude that

(9.5.0.9)   either $\Phi^{j+kd}(\alpha) \in V(\mathbb{C}_p)$ for all $k \in \mathbb{N}$

or $V(\mathbb{C}_p) \cap \mathcal{O}_{\Phi^d}(\Phi^j(\alpha))$ is finite.

This concludes the proof of Theorem 9.3.0.1.   □

CHAPTER 10

Additional results towards the Dynamical Mordell-Lang Conjecture

In this chapter we discuss additional special cases of Conjecture 1.5.0.1. The novelty of these results is that they do not rely either on the p-adic arc lemma (see Chapter 4 and [**BGT10**]), nor on results about polynomial decomposition (see Chapter 5 and [**GTZ08, GTZ12**]). We first describe the results presented in this chapter.

The Dynamical Mordell-Lang Conjecture (and especially its generalization from Question 3.6.0.1) asks for a description of the possible algebraic relations between points in an orbit. In the first two Sections of this chapter we discuss a couple of results due to Thomas Scanlon [**Sca11, Sca**] regarding analytic relations between points in an orbit. Note that in general there is no p-adic analytic version of the Dynamical Mordell-Lang Conjecture (see Proposition 11.10.0.1). However, in Section 10.1, we present the findings of [**Sca**] which can be viewed as an instance of the Dynamical Mordell-Lang Conjecture in the context of v-adic analytic maps (under suitable hypotheses). In Section 10.2 we present a real-analytic instance of the Dynamical Mordell-Lang Conjecture based on the results of [**Sca11**].

In Section 10.3 we present briefly a result of Junyi Xie (see Theorem 10.3.0.1 and also [**Xie14**]) who proved the Dynamical Mordell-Lang Conjecture for birational polynomial self-maps on $\mathbb{A}^2$. We also mention that while writing the last version of this book, we found out that Xie [**Xieb**] proved the Dynamical Mordell-Lang Conjecture for all endomorphisms of $\mathbb{A}^2$ which is based on his deep analysis from [**Xiea**] of valuation subrings of a polynomial ring in 2 variables over a field of characteristic 0. The preprint [**Xieb**] is 95 pages long and it is based on another preprint of Xie [**Xiea**] which is itself 42 pages long. So, while we do not describe the intricate proof of Xie for arbitrary endomorphisms of $\mathbb{A}^2$, we present briefly the special case of birational polynomial self-maps on $\mathbb{A}^2$; we mention that a very detailed description of the strategy employed by Xie for the case of endomorphisms of $\mathbb{A}^2$ can be found in [**Xieb**, pp. 4–7].

10.1. A v-adic analytic instance of the Dynamical Mordell-Lang Conjecture

10.1.1. The philosophy of special points and special subvarieties.
First we recall from Subsection 3.4.3 the interpretation of *special points* and *special subvarieties* in the context of the Dynamical Mordell-Lang Conjecture; this is connected with the classical Mordell-Lang Conjecture and also with other famous conjectures in arithmetic geometry such as the André-Oort, Bogomolov, or Pink-Zilber conjectures. So, for a quasiprojective variety X defined over a field of characteristic 0 endowed with an endomorphism Φ, the *special points* are the ones

contained in a given (infinite) orbit $\mathcal{O}$ under Φ, while the *special* (irreducible) sub-varieties $V \subseteq X$ are periodic (irreducible) subvarieties intersecting $\mathcal{O}$. Then among all irreducible subvarieties V of X, it is precisely the special subvarieties that have the property that

$$V \cap \mathcal{O} \text{ is Zariski dense in } V.$$

More generally, Question 5.1.2.1 (see also Question 3.6.0.1 which generalizes both the classical Mordell-Lang and the Dynamical Mordell-Lang conjectures) asks for a similar description of the *special subvarieties* of $\mathbb{A}^N$ under the action of an endo-morphism Φ of the form

$$(x_1, \ldots, x_N) \mapsto (f_1(x_1), \ldots, f_N(x_N)),$$

for some polynomials $f_i \in K[z]$, where $\mathrm{char}(K) = 0$. This time, the *special points* of $\mathbb{A}^N$ are points belonging to a product of (infinite) orbits:

(10.1.1.1) $$\mathcal{O}_{f_1}(\alpha_1) \times \cdots \times \mathcal{O}_{f_N}(\alpha_N),$$

for some points $\alpha_i \in K$. Then Question 5.1.2.1 (see also Proposition 3.1.2.14) predicts that the special subvarieties of $\mathbb{A}^N$ are periodic subvarieties V under the action of a map of the form

(10.1.1.2) $$(x_1, \ldots, x_N) \mapsto \left(f_1^{k_1}(x_1), \ldots, f_N^{k_N}(x_N) \right)$$

for some $k_1, \ldots, k_N \in \mathbb{N}_0$ not all equal to 0 such that V also intersects the product of orbits from (10.1.1.1).

Scanlon [**Sca**] described the *special* subvarieties in the following related scenario (for the precise statement, see Theorem 10.1.2.4). Let K be a complete discretely valued field and let

$$f : D(0,1) \longrightarrow D(0,1)$$

be a non-constant analytic map from the unit disk to itself. We assume that 0 is an attracting fixed point of f. Let α be an element of K whose orbit $\mathcal{O}_f(\alpha)$ converges to 0. If 0 is a super-attracting fixed point, then Scanlon [**Sca**] proves that every irreducible analytic subvariety of $D(0,1)^N$ meeting $\mathcal{O}_f(\alpha)^N$ in an analytically Zariski dense set is defined by equations of the form

(10.1.1.3) $$x_i = \beta \text{ and } x_j = f(x_k).$$

When 0 is an attracting point but is not super-attracting, then Scanlon [**Sca**] shows that all analytic relations come from algebraic tori.

Hence the results of [**Sca**] describe the possible analytic relations between points in an orbit, as long as the orbit lies in the basin of attraction of a fixed point. In addition, Scanlon's result applies to analytic maps Φ not necessarily algebraic maps (as we considered before in the Dynamical Mordell-Lang Conjecture). So, on one hand, the results of [**Sca**] are more general than the Dynamical Mordell-Lang Conjecture (even in its strong form from Question 3.6.0.1) because Scanlon works with analytic maps and not only that he proves the special subvarieties are indeed the ones periodic under a map of the form given in (10.1.1.2), but also Scanlon [**Sca**] gives the precise form of all special subvarieties (see (10.1.1.3)). On the other hand, the results of [**Sca**] apply only when

$$f := f_1 = \cdots = f_N \text{ and } \alpha := \alpha_1 = \cdots = \alpha_N,$$

and most importantly, the point α is in a suitable neighborhood of an attracting point for the map f. For example, assuming f is a polynomial defined over a number field L, it is usually difficult to find a prime number p and an embedding of L into $\mathbb{Q}_p$ such that this condition on α is verified with respect to the norm $|\cdot|_p$ on $\mathbb{Q}_p$.

10.1.2. Statement of the result. We state in Theorem 10.1.2.4 a special case of [**Sca**] when K is a finite extension of $\mathbb{Q}_p$. We start by defining the analytic Zariski topology on $\mathbb{A}^N$.

DEFINITION 10.1.2.1. We call $V \subseteq \mathbb{A}^N$ an *analytic subvariety* if it is the zero set of finitely many analytic functions in $K[[z_1, \ldots, z_N]]$. A subset S of V is *analytic Zariski dense* in V if V is the smallest analytic subvariety containing S.

Let Φ be a p-adic analytic map which converges on a small open ball $\mathcal{B}$ centered at 0 of radius leass than 1 such that

$$\Phi(z) = \sum_{n \geq M} c_n z^n,$$

where $c_n \in K$ for all n, and $M \geq 1$ such that $0 < |c_M|_p < 1$. In particular, for each $x \in \mathcal{B}$ we assume that

$$|c_n x^n|_p < |c_M x^M|_p \text{ if } n > M.$$

Then 0 is a fixed point for Φ, and if $M > 1$ then 0 is a super-attracting point for Φ; otherwise it is only an attracting fixed point. We consider a point $\alpha \in K$ such that its orbit $\mathcal{O}_\Phi(\alpha)$ intersects the ball $\mathcal{B}$, and thus $\Phi^n(\alpha)$ converges to 0 as $n \to \infty$. To ease the notation, we assume from now on that

$$\alpha \in \mathcal{B} \setminus \{0\}.$$

Note that if $\alpha = 0$, then the orbit consisits of a single point and thus no positive dimensional subvariety of $\mathbb{A}^N$ can have a Zariski dense intersection with a Cartesian power of the orbit; therefore for each nonnegative integers $m < n$ we have

$$|\Phi^m(\alpha)|_p > |\Phi^n(\alpha)|_p.$$

We call a subvariety $V \subseteq \mathbb{A}^N$ *special* if it has an analytically Zariski dense intersection with $\mathcal{O}_\Phi(\alpha)^N$ (the N-th Cartesian power of the orbit of α under Φ). In order to describe the special subvarieties, we start with a definition from [**Sca**].

DEFINITION 10.1.2.2. Let $n \in \mathbb{N}$ and let $\mathcal{I} \subseteq K[z_1, \ldots, z_N]$ be an ideal generated by a set of the form

$$\{z_i - \Phi^{m_i}(\alpha) : i \in I\} \bigcup \{z_j - \Phi^{\ell_{j,k}}(z_k) : (j, k) \in J\}$$

where $I \subseteq \{1, \ldots, N\}$ and $J \subseteq \{1, \ldots, N\}^2$, and each m_i and $\ell_{j,k}$ is a nonnegative integer. We let $V := V_\mathcal{I}$ be the analytic variety defined by $\mathcal{I}$ and we call V an *iterational special variety*.

Unless the equations from Definition 10.1.2.2 are inconsistent, for each $i \in I$ there exists a unique value of m_i appearing in the above equations, and also for each $(j, k) \in J$ there exists a unique value $\ell_{j,k}$ appearing in the defining equations for $V_\mathcal{I}$. In [**Sca**, Proposition 3.3], it is shown that $V = V_\mathcal{I}$ is irreducible, and moreover it has an analytically Zariski dense intersection with $\mathcal{O}_\Phi(\alpha)^N$.

With the above notation for Φ, assume $M = 1$ and let $\lambda := c_1$ (i.e., $\lambda = \Phi'(0)$); so 0 is a fixed attracting, but not super-attracting fixed point for Φ. Then (see

Lemma 6.2.1.1) there exists an analytic function h convergent in a small ball around 0 such that

$$h(\Phi(z)) = \lambda h(z).$$

At the expense of replacing $\mathcal{B}$ with a smaller ball, we may assume h is analytic on $\mathcal{B}$. We extend diagonally the action of h on $\mathbb{A}^N$.

DEFINITION 10.1.2.3. With the above notation, we say that an analytic sub-variety V of $\mathcal{B}^N$ is a *deformed torus*, if there exists a connected algebraic torus T (see Subsection 2.1.9), and there exists a point $\zeta \in \mathcal{B}^N$ such that

$$h(V) = (\zeta \cdot T) \cap \mathcal{B}^N.$$

It is easy to prove (see [**Sca**, Proposition 3.7]) that (with the notation as in Definition 10.1.2.3) if

$$\zeta = \left(\Phi^{t_1}(\alpha), \ldots \Phi^{t_N}(\alpha) \right),$$

and if $A = (a_{i,j})$ is a $k \times N$ matrix with integer coefficients such that the connected algebraic torus T is given by the equations

$$\prod_{i=1}^{N} x_i^{a_{j,i}} = 1 \text{ for } j = 1, \ldots, k,$$

then

$$V \cap \mathcal{O}_\Phi(\alpha)^N = \left\{ (\Phi^{s_1}(\alpha), \ldots, \Phi^{s_N}(\alpha)) : \sum_{i=1}^{N} a_{j,i}(s_i - t_i) = 0 \text{ for } j = 1, \ldots, k \right\}.$$

Moreover, $V \cap \mathcal{O}_\Phi(\alpha)$ is analytically Zariski dense in V. Then the main result of [**Sca**] is the following.

THEOREM 10.1.2.4 (Scanlon [**Sca**]). *With the above notation, we have:*
- *if $M = 1$, then an irreducible analytic variety $V \subseteq \mathcal{B}^N$ meets $\mathcal{O}_\Phi(\alpha)^N$ in an analytically Zariski dense set if and only if V is a deformed torus; and*
- *if $M \geq 2$, then an irreducible analytic variety meets $\mathcal{O}_\Phi(\alpha)$ in an analytically Zariski dense set of and only if it is iterational.*

The proof of Theorem 10.1.2.4 follows by induction on N, the more difficult case being when $M = 1$. In this case, one uses the explicit description of the intersection between a linear subvariety of $\mathbb{G}_m^N$ with a finitely generated subgroup of $\mathbb{G}_m^N(\mathbb{C})$ as proved in [**vdDG06**]. We state below the refinement of the result from [**vdDG06**] for the setting of [**Sca**]

PROPOSITION 10.1.2.5 (Scanlon [**Sca**]). *Let $\Gamma \subseteq \mathbb{C}^*$ be a finitely generated multiplicative subgroup. If $c, c_1, \ldots, c_N \in \mathbb{C}$ then the set*

$$\left\{ (\gamma_1, \ldots, \gamma_N) \in \Gamma^N : c_1 \gamma_1 + \cdots + c_N \gamma_N = c \right\}$$

is a finite union of sets defined by equations of the form

$$x_i = \gamma \text{ and } x_j = \delta x_k$$

for $\gamma, \delta \in \Gamma$.

PROOF. We refer the interested reader to [**Sca**, Lemma 3.11].   □

10.2. A real analytic instance of the Dynamical Mordell-Lang Conjecture

10.2.1. The Dynamical Mordell-Lang principle. To describe the result of [**Sca11**], first we formulate informally the Dynamical Mordell-Lang *principle*. Given a topological set X endowed with a continuous self-map Φ, a closed subset $V \subseteq X$, and a point $\alpha \in X$, we say that the Dynamical Mordell-Lang *principle* holds if the set

$$S(V, \Phi, \alpha) := \{n \in \mathbb{N}_0 : \Phi^n(\alpha) \in V\}$$

is a finite union of arithmetic progressions. Obviously, this *principle* is generally not satisfied; for example, one could consider the discrete topology on X, and therefore $S(V, \Phi, \alpha)$ could be any subset of $\mathbb{N}_0$. Even if X is a metric space, there are examples when $S(V, \Phi, \alpha)$ is not necessarily a finite union of arithmetic progressions; see, for example, Proposition 11.10.0.1 for a counterexample when

$$X = \mathbb{Q}_p \times \mathbb{Q}_p \text{ and } \Phi(x, y) = (x + 1, py).$$

However, if X is Noetherian, then at least one can prove that at the very worst, $S(V, \Phi, \alpha)$ differs from a finite union of arithemtic progressions by a set of Banach density equal to 0 (see Theorem 11.1.0.7); for a comprehensive discussion on this topic, see Chapter 11 (especially Section 11.4).

10.2.2. Description of the result. In [**Sca11**], Scanlon shows that the Dynamical Mordell-Lang principle holds in some cases for real-analytic spaces. More precisely (see Corollary 10.2.4.4), Scanlon proves that if $f_1, \ldots, f_N$ are real analytic functions mapping the interval $(-1, 1)$ into itself for which

$$f_i(0) = 0 \text{ and } |f_i'(0)| \leq 1,$$

and if $\alpha = (\alpha_1, \ldots, \alpha_N)$ is close enough to the origin of $X = \mathbb{A}^N$, and $H(z_1, \ldots, z_N)$ is a real analytic function in N variables, then the set

$$\{n \in \mathbb{N}_0 : H(f_1^n(\alpha_1), \ldots, f_N^n(\alpha_N)) = 0\}$$

is either all of $\mathbb{N}_0$, all of the odd numbers, all of the even numbers, or is finite. In particular the Dynamical Mordell-Lang principle holds for the real analytic map $\Phi := (f_1, \ldots, f_N)$ acting on $\mathbb{A}^N$. Actually, Scanlon [**Sca11**] proves a more general result valid for analytic endomorphisms of $(-1, 1)^N$ which are not necessarily given by the coordinatewise action of N one-variable analytic maps; for more details, see Theorem 10.2.4.3.

The motivation for the proof from [**Sca11**] comes from the Skolem-Mahler-Lech method presented in Chapter 4 only that Scanlon uses real analytic functions instead of p-adic analytic functions. At first glance the translation to the real analytic world seems to be unsuccessful since $\mathbb{Z}$ is not a compact subset of $\mathbb{R}$ and thus the main lemma of zeros (see Lemma 2.3.6.1) for p-adic analytic functions does not have an analogue for real analytic functions. However, in certain instances (as the one outlined above), one can find a function

$$F : \mathbb{R}_+ \longrightarrow X(\mathbb{R})$$

such that for all $n \in \mathbb{N}_0$ we have

$$F(n) = \Phi^n(\alpha),$$

where F is definable in an *o-minimal* expansion of the field of real numbers. Then it would follow from the o-minimality principle that if such an orbit has infinite intersection with an algebraic subvariety of X, then all but finitely many of the points from the orbit lie in the subvariety.

10.2.3. Background on o-minimal theory. We start by briefly introducing the basic notions regarding o-minimal structures (we refer to [**Sca11**] for more details; our presentation follows closely [**Sca11**] but we leave out certain details). First, we recall the definition of *definable sets* in first-order logic.

DEFINITION 10.2.3.1. We give the definition of definable sets with parameters as follows:

- Let $\mathcal{L}$ be a first-order language, let $\mathcal{M}$ be an $\mathcal{L}$-structure with domain M, let $X \subseteq M$, and let $m \in \mathbb{N}$. A set $A \subseteq M^m$ is *definable* in $\mathcal{M}$ with parameters from X if and only if there exists a formula

$$\varphi[x_1, \ldots, x_m, y_1, \ldots, y_n]$$

and elements $b_1, \ldots, b_n \in X$ such that for all $a_1, \ldots, a_m \in M$, we have

$$(a_1, \ldots, a_m) \in A$$

if and only if the formula φ evaluated for $x_i = a_i$ for $i = 1, \ldots, m$ and for $y_j = b_j$ for $j = 1, \ldots, n$ holds in $\mathcal{M}$.
- A set A is definable in $\mathcal{M}$ without parameters if it is definable in $\mathcal{M}$ with parameters from the empty set (that is, with no parameters in the defining formula).
- A function is definable in $\mathcal{M}$ (with parameters) if its graph is definable (with those parameters) in $\mathcal{M}$.

DEFINITION 10.2.3.2. A structure $(M, <, \ldots)$ which is totally ordered by the relation $<$ is *o-minimal* if every definable (with parameters) subset of M is a finite union of singletons and intervals of the form

$$(-\infty, a) := \{x \in M : x < a\},$$
$$(a, b) := \{x \in M : a < x < b\}, \text{ or}$$
$$(b, +\infty) := \{x \in M : b < x\}$$

for some $a, b \in M$.

DEFINITION 10.2.3.3. For each $n \in \mathbb{N}$, let $\mathcal{F}_n$ be a set of real-valued functions

$$f : \mathbb{R}^n \longrightarrow \mathbb{R}$$

and we let $\mathcal{F}$ denote the union of the $\mathcal{F}_n$. By $\mathcal{L}_{\mathcal{F}}$ we mean the first-order language having a binary relation symbol $\leq$, binary function symbols $+$ and $\cdot$, constant symbols r for each $r \in \mathbb{R}$, and n-ary function symbols f for each $f \in \mathcal{F}_n$. By $\mathbb{R}_{\mathcal{F}}$ we mean the $\mathcal{L}_{\mathcal{F}}$-structure having universe $\mathbb{R}$.

In the structure $\mathbb{R}_{\mathcal{F}}$, sets of the form

$$\{(a_1, \ldots, a_n) \in \mathbb{R}^n : f(a_1, \ldots, a_n) \leq g(a_1, \ldots, a_n)\}$$

are definable where $f, g \in \mathcal{F}_n$ or more generally where f and g are obtained from the projection functions, constant functions, functions in $\mathcal{F}$ and addition and multiplication via appropriate compositions. We obtain the class of quantifier-free definable sets by closing off under finite Boolean operations. In general, since first-order logic allows the application of existential quantifiers or equivalently, at the level of the definable sets, images under coordinate projections, then there will be definable sets which cannot be expressed in the simple form of finite Boolean combinations of sets defined by inequalities between the basic functions. Wilkie [**Wil96**] showed that when $\mathcal{F}$ consists of the usual exponential function, then every definable set may be expressed as a projection of a basic set and that

$$\mathbb{R}_{\exp} := (\mathbb{R}, \leq, +, \cdot, \exp)$$

is o-minimal. It is clear that one cannot take $\mathcal{F}$ to consist of arbitrary real analytic functions since, for example, the set

$$\{x \in \mathbb{R} \colon \cos(x) = 0\}$$

is an infinite discrete set and it would be definable in $\mathbb{R}_{\cos}$. However, if we consider only restricted analytic functions on a given finite interval, then the resulting structure is o-minimal. For example we may consider $\mathcal{F}_n$ consisting of all functions

$$f : \mathbb{R}^n \longrightarrow \mathbb{R}$$

for which

(1) there is some open set U containing $[-1, 1]^n$ and a real analytic function $g : U \longrightarrow \mathbb{R}$ for which $f(z) = g(z)$ for each $z \in U$;
(2) $f(z_1, \ldots, z_n) = 0$ if there is some $i = 1, \ldots, n$ such that $|z_i| > 1$.

In this case, the induced structure is denoted $\mathbb{R}_{an}$. Even if we add the exponential function to the previous structure (and thus obtain $\mathbb{R}_{an,\exp}$), the structure remains o-minimal as proven in [**vdDMM94, vdDM94**].

10.2.4. Statement of the results. In order to state the main result of [**Sca11**] we need the following definition.

DEFINITION 10.2.4.1. We say that a real analytic self-map Φ near the origin is *projectively linearizable* if there is a real analytic function h which fixes the origin and is invertible near the origin for which $h \circ \Phi \circ h^{-1}$ is given by a fractional linear transformation near the origin. We say that Φ is *strongly projectively linearizable* if, in addition, every eigenvalue of some matrix representing its projective linearization is real and positive.

We say that Φ is *monomializable* if the differential $D\Phi$ is identically equal to 0 and there are a real analytic h as in the above paragraph, an invertible matrix $A := (a_{i,j})$ all of whose entries are positive integers, but not having any roots of unity among its eigenvalues, and a tuple

$$\lambda := (\lambda_1, \ldots, \lambda_N) \in (\mathbb{R}^*)^N$$

so that near the origin we have

$$h \circ \Phi \circ h^{-1}(z_1, \ldots, z_N)$$
$$= \lambda \cdot (z_1, \ldots, z_N)^A$$
$$= \left(\lambda_1 z_1^{a_{1,1}} z_2^{a_{1,2}} \cdots z_N^{a_{1,N}}, \ldots, \lambda_N z_1^{a_{N,1}} z_2^{a_{N,2}} \cdots z_N^{a_{N,N}} \right).$$

We say that Φ is *strongly monomializable* if, in addition, every eigenvalue of M is real and positive.

As shown in [**Sca11**, Lemma 3.6], if Φ is monomializable then one may choose the vector λ to have all entries in $\{-1, 1\}$. Furthermore, by taking $B := B(N)$ to be the least common multiple of the lengths of the periodic cycles of the maps

$$(z_1, \ldots, z_N) \mapsto (z_1, \ldots, z_N)^A$$

on $(\pm 1)^N$ as A varies over all invertible $N \times N$ matrices with positive integer entries, we obtain the following result (see [**Sca11**, Proposition 3.7]).

PROPOSITION 10.2.4.2 (Scanlon [**Sca11**]). *With the above notation, assuming Φ is strongly monomializable and $\alpha \in \mathbb{R}^N$ is a point sufficiently close to the origin, then there is a function*

$$F : [0, +\infty) \longrightarrow \mathbb{R}^N$$

definable in $\mathbb{R}_{an,\exp}$ satisfying

   (1) $F(0) = \alpha$; and
   (2) $F(z+1) = \Phi^B(F(z))$.

A similar statement is proved in [**Sca11**, Corollary 3.8] if Φ is expressible as a Cartesian product of a strongly projectively linearizable function and a strongly monomializable function. Then we can state the main result of [**Sca11**].

THEOREM 10.2.4.3. *If Φ is a real analytic function for which some positive compositional power is expressible as a product of a strongly projectively linearizable function and a strongly monomializable function and α is sufficiently close to the origin, then for any closed real analytic variety V the set*

$$\{n \in \mathbb{N}_0 \ : \ \Phi^n(\alpha) \in V\}$$

is a finite union of arithmetic progressions.

If Φ is expressible as a product of univariate functions, Scanlon gets a more precise result.

COROLLARY 10.2.4.4 (Scanlon). *Let*

$$f_1, \ldots, f_N : (-1, 1) \longrightarrow (-1, 1)$$

be real analytic functions for which

$$f_i(0) = 0 \ and \ |f_i'(0)| \leq 1,$$

and let $\mathcal{B}^N$ be a small ball centered around the origin such that for each i and for each $z \in \mathcal{B}$ we have

$$|f(z)| \leq |z|.$$

If $\alpha = (\alpha_1, \ldots, \alpha_N) \in \mathcal{B}^N$ and V is a real analytic subvariety of $\mathcal{B}^N$, then

$$\{n \in \mathbb{N}_0 \ : \ (f_1^n(\alpha_1), \ldots, f_N^n(\alpha_N)) \in V\}$$

is either:

   (1) *all of $\mathbb{N}_0$;*
   (2) *all odd positive integers;*
   (3) *all even nonnegative integers;*
   (4) *finite.*

PROOF. Clearly, it suffices to prove Corollary 10.2.4.4 when V is an analytic hypersurface; so assume V is the zero set of some analytic function H in N variables. Corollary 10.2.4.4 follows immediately from Theorem 10.2.4.3 because for each real analytic function f in one variable near the origin which fixes the origin, f^2 is strongly projectively linearizable or strongly monomializable (see the proof of [**Sca11**, Fact 3.9]). $\square$

The methods from [**Sca11**] are quite restrictive since one has to assume α is in the proximity of a fixed attracting point for a real analytic self-map. However, the results obtained are *not* covered by the other methods presented in this book. Furthermore, Scanlon [**Sca11**] speculates that a more refined analysis of his method would lead to proving a gap principle for real analytic maps in the context of the Dynamical Mordell-Lang Conjecture, similar to the one from Chapter 11 (for p-adic analytic maps). Such a result would combine the arguments from [**Sca11**] with the powerful techniques introduced by Pila, Wilkie and Zannier for counting algebraic points of small height on real analytic varieties (for more details, see [**PZ08**]).

10.3. Birational polynomial self-maps on the affine plane

The following result is proven by Xie in [**Xie14**, Theorem A].

THEOREM 10.3.0.1 (Xie [**Xie14**]). *Let K be an algebraically closed field of characteristic* 0, *let $\alpha \in \mathbb{A}^2(K)$, let $C \subset \mathbb{A}^2$ be a curve defined over K, and let*

$$\Phi : \mathbb{A}^2 \longrightarrow \mathbb{A}^2$$

be a birational polynomial morphism. Then the set

$$S := \{n \in \mathbb{N}_0 \colon \Phi^n(\alpha) \in C(K)\}$$

is a union of finitely many arithmetic progressions.

The strategy from [**Xie14**] for proving Theorem 10.3.0.1 is a clever combination of tools from algebraic geometry, number theory and algebraic dynamics. So, assuming α is not preperiodic (otherwise, Theorem 10.3.0.1 holds easily according to Proposition 3.1.2.9), Theorem 10.3.0.1 reduces to proving that if S is infinite, then C is periodic (see also Subsection 3.1.3). The bulk of the argument from [**Xie14**] applies for arbitrary algebraically closed fields K of characteristic 0, with the exception of the last part of the argument which relies on the application of Northcott's Theorem. We sketch the main ingredients of the proof of Theorem 10.3.0.1 when $K = \overline{\mathbb{Q}}$ and refer the reader to [**Xie14**, Section 8] for the general case.

The proof splits into two cases depending on the *dynamical degree* of Φ. Before defining the dynamical degree of a self-map, we recall a few definitions from algebraic geometry.

The *Néron-Severi group* $\mathrm{NS}(X)$ of a variety X is defined as the quotient of the group of divisors modulo algebraic equivalence (for more details on algebraic equivalence of divisors, see [**Har77**, p. 367]). We also recall the definition of the *spectral radius* of a linear operator.

DEFINITION 10.3.0.2. The *spectral radius* $\rho(T, V)$ of a linear operator T acting on a real vector space V is the supremum among the absolute values of its eigenvalues.

As shown in Chapter 2, an endomorphism Φ of a variety X induces a map Φ^* on $\mathrm{NS}(X)$. In fact, this can even be done when one only assumes that Φ is a rational self map of a surface X (see [**Xie14**] for more details). Next we define the *dynamical degree* of a rational self-map.

DEFINITION 10.3.0.3. Let $\Phi : X \longrightarrow X$ be a rational self-map of a variety X and define $\mathrm{NS}(X)_{\mathbb{R}} := \mathrm{NS}(X) \otimes \mathbb{R}$. Then Φ induces a linear operator Φ^* on $\mathrm{NS}(X)_{\mathbb{R}}$

and we define the *dynamical degree* of Φ as

$$(10.3.0.4) \qquad \delta_\Phi := \lim_{n \to \infty} \rho\left((\Phi^n)^*, \mathrm{NS}(X)_\mathbb{R}\right)^{\frac{1}{n}}.$$

We mention that if Φ is an endomorphism (i.e., Φ is regular, or in other words, the indeterminacy locus of Φ is the empty set), then

$$(\Phi^n)^* = (\Phi^*)^n$$

and thus, δ_f is simply the spectral radius of the operator Φ^* acting on $\mathrm{NS}(X)_\mathbb{R}$. For an interesting discussion of the arithmetic properties of the dynamical degree of a morphism (including the explanation of why the limit from (10.3.0.4) exists), we refer the reader to [**KS14**]. Now, if

$$\Phi : \mathbb{A}^2 \longrightarrow \mathbb{A}^2$$

is a birational polynomial morphism of dynamical degree 1, then [**DF01**] and [**FJ11**] yield that (after a suitable conjugation by a linear automorphism of $\mathbb{A}^2$) we may assume that

$$\Phi(x, y) = (ax + b, A(x)y + B(x))$$

for some constants a and b, and some polynomials A and B. Xie [**Xie14**, Section 6] proves directly Theorem 10.3.0.1 for such maps Φ of dynamical degree 1. So, from now on, we assume that

$$\delta_\Phi > 1.$$

Using [**FJ11**, Proposition 2.6 and Theorem 3.1] the following compactification result is obtained (see [**Xie14**, Theorem 2.4]). By a *compactification* of $\mathbb{A}^2$ we mean a smooth projective surface X equipped with a birational map

$$\pi : X \longrightarrow \mathbb{P}^2$$

that is an isomorphism above $\mathbb{A}^2 \subset \mathbb{P}^2$ (for more details, see [**FJ11**]).

THEOREM 10.3.0.5 (Xie [**Xie14**]). *Let $\Phi : \mathbb{A}^2 \longrightarrow \mathbb{A}^2$ be a birational polynomial morphism such that $\delta_\Phi > 1$. Then there exists a compactification X of $\mathbb{A}^2$ satisfying the following properties:*

(i) *the map Φ extends to a self-map on X (also denoted by Φ). In addition, there is no curve $V \subset X$ and no $n \in \mathbb{N}$ such that $f^n(V) \subseteq I(\Phi)$, where $I(\Phi)$ is the indeterminacy locus for the self-map $\Phi : X \longrightarrow X$ (in [**Xie14**], this means that Φ is* algebraically stable*).*

(ii) *there exists a fixed super-attracting point $\beta \in X \setminus \mathbb{A}^2$ for Φ (i.e., $D\Phi^2(\beta) = 0$).*

(iii) *there exists a positive integer n such that $\Phi^n(X \setminus \mathbb{A}^2) = \{\beta\}$.*

We replace C by its Zariski closure inside X. If $f^n(C)$ meets $I(\Phi)$ for all $n \in \mathbb{N}$, then [**Xie14**, Theorem 1.2] yields that C is periodic. So, at the expense of replacing Φ by an iterate of it and replacing C by its image under Φ, we may assume

$$C \setminus \mathbb{A}^2 = \{\beta\}.$$

In [**Xie14**, Theorem 1.1], Xie shows that if there exists an absolute value $|\cdot|_v$ on $\overline{\mathbb{Q}}$ such that $\Phi^n(\alpha)$ converges v-adically to β, and if there are infinitely many points in common for C and for the orbit of α under Φ, then C is fixed by Φ. Now, assuming the point α lies outside the basins of attraction of β with respect to all absolute values on $\overline{\mathbb{Q}}$, one obtains that α has bounded height (see [**Xie14**, Proposition 7.5]). Hence α and all its iterates have bounded height and therefore α is preperiodic (by the Northcott property for algebraic numbers; see Theorem 2.6.3.2), a contradiction. Hence Theorem 10.3.0.1 holds.

It would be very interesting to know how far one can extend the method of Xie [**Xie14**] for proving other cases of the Dynamical Mordell-Lang Conjecture. As mentioned before, Xie [**Xieb**] was able to extend his result from [**Xie14**] to all endomorphisms of $\mathbb{A}^2$, but so far it is unclear whether his method would work for proving Conjecture 1.5.0.1 either for endomorphisms of $\mathbb{A}^N$ when $N > 2$, or for arbitrary endomorphisms of $\mathbb{P}^N$ when $N \geq 2$.

CHAPTER 11

Sparse sets in the Dynamical Mordell-Lang Conjecture

11.1. Overview of the results presented in this chapter

We recall the setting from the Dynamical Mordell-Lang Conjecture:

- X is a quasiprojective variety defined over $\mathbb{C}$,
- $\Phi : X \longrightarrow X$ is an endomorphism,
- $V \subseteq X$ is a subvariety (always closed, unless otherwise mentioned), and
- $\alpha \in X(\mathbb{C})$.

Then Conjecture 3.1.3.2 (which was shown to be equivalent to Conjecture 1.5.0.1 in Subsection 3.1.3) predicts that a subvariety which contains no positive dimensional periodic subvariety must intersect an orbit in finitely many points. We recall that $W \subseteq X$ is periodic (under the action of Φ) if

$$\Phi^m(W) \subseteq W,$$

for some $m \in \mathbb{N}$; for more details on periodic subvarieties, see Subsection 2.2.2.

In this chapter we show that when a subvariety V contains no positive dimensional periodic subvariety intersecting $\mathcal{O}_\Phi(\alpha)$, then the set

$$S = S_V := \{n \in \mathbb{N}_0 \colon \Phi^n(\alpha) \in V(\mathbb{C})\}$$

is very sparse (see Theorem 11.1.0.7, which is [**BGT15b**, Corollary 1.5]). In the special case when $X = (\mathbb{P}^1)^N$ and $\Phi := (\varphi_1, \ldots, \varphi_N)$ is given by the coordinatewise action of N one-variable rational maps, one obtains a more refined result regarding the sparseness of the set S_V (see Theorem 11.1.0.9 and also [**BGKT10**]).

We note that Denis [**Den94**] first raised the question (when X is the projective space defined over a field of arbitrary characteristic) that the above set S_V must be sparse when V does not contain a positive dimensional periodic subvariety. In order to state Denis' question we first define the Banach density of a subset of $\mathbb{N}_0$; this allows us also to give a first interpretation of what *(very) sparse* means.

DEFINITION 11.1.0.1. Let S be a subset of $\mathbb{N}_0$. We define the *Banach density* of S to be

$$\delta(S) := \limsup_{|I| \to \infty} \frac{|S \cap I|}{|I|},$$

where the lim sup is taken over intervals $I \subseteq \mathbb{N}_0$ (also, as always, we denote by $|U|$ the cardinality of a set U). We say that a subset $S \subseteq \mathbb{N}_0$ has *Banach density zero* if $\delta(S) = 0$.

So, a set of Banach density equal to 0 is *sparse*, and actually it is sparser than a set of *natural density* equal to 0. We recall that the *natural density* of a subset

$S \subseteq \mathbb{N}_0$ is defined similar to Definition 11.1.0.1 only that the limit is taken only over intervals I of the form

$$\{0, 1, \ldots, n\}.$$

It is possible to construct (see Example 11.1.0.2) a subset of $\mathbb{N}_0$ of natural density equal to 0, but Banach density equal to 1.

EXAMPLE 11.1.0.2. Let $d \geq 3$ be any integer, and let

$$S := \bigcup_{n=1}^{\infty} \left\{ n^d, n^d + 1, \ldots, n^d + n \right\}.$$

Clearly, the Banach density $\delta(S) = 1$ since S contains arbitrarily long intervals of positive integers. On the other hand, the gaps between two consecutive such intervals grow very fast so that the natural density of S is 0. Indeed, in the interval

$$\{0, 1, \ldots, N\},$$

for N sufficiently large, S contains fewer than $N^{2/d} = o(N)$ integers.

The following is [**Den94**, Question 2].

QUESTION 11.1.0.3 (Denis [**Den94**]). *Let* $\Phi : \mathbb{P}^m \longrightarrow \mathbb{P}^m$ *be an endomorphism defined over a field* K, *let* $V \subseteq \mathbb{P}^m$ *be a subvariety defined over* K, *and let* $\alpha \in \mathbb{P}^m(K)$ *be a non-preperiodic point. Is it true that one of the following two alternatives must hold?*

(1) *The set* $S_V := \{n \in \mathbb{N}_0 : \Phi^n(\alpha) \in V(K)\}$ *has Banach density 0; or*
(2) *There exists a positive dimension subvariety* $W \subseteq V$, *and there exist* $\ell \in \mathbb{N}_0$ *and* $k \in \mathbb{N}$ *such that* $\Phi^\ell(\alpha) \in W(K)$ *and* $\Phi^k(W) \subseteq W$.

Alternatively, one could rephrase Question 11.1.0.3 by asking that whenever S_V has positive Banach density, then condition (2) above holds; so,

$$\Phi^{\ell + nk}(\alpha) \in V(K) \text{ for all } n \in \mathbb{N}_0.$$

Hence, in this case S_V contains an infinite arithmetic progression $\{\ell + nk\}_{n \in \mathbb{N}_0}$. So, one can reformulate Question 11.1.0.3 as follows (and also state the question in the most general context of endomorphisms of quasiprojective varieties).

QUESTION 11.1.0.4. *Let* $\Phi : X \longrightarrow X$ *be an endomorphism of a quasiprojective variety defined over a field* K, *let* $V \subseteq X$ *be a subvariety defined over* K, *and let* $\alpha \in X(K)$ *be a point. If the set*

$$S_V := \{n \in \mathbb{N}_0 : \Phi^n(\alpha) \in V(K)\}$$

has positive Banach density 0, does it follow that S_V *contains an infinite arithmetic progression?*

In Theorem 11.1.0.7 (see [**BGT15b**], where this result was first published) we give a positive answer to Question 11.1.0.4. As a quick observation (see also Proposition 3.1.2.9), Question 11.1.0.4 holds trivially if α is preperiodic.

Denis [**Den94**] showed, for any endomorphism Φ of $\mathbb{P}^m$, if S_V does not contain an infinite arithemetic progression, then S_V cannot be "very dense" (see [**Den94**, Définition 2] and also our next definition).

DEFINITION 11.1.0.5. Let $S \subseteq \mathbb{N}_0$ and let $k \in \mathbb{N}$. We say that S is *very dense* of order k if, for all $\ell \in \mathbb{N}$, there exist positive integers r and b such that

(1) $r \leq \max\{\log^{(k)}(\ell), 1\}$, where $\log^{(k)}$ is the k-th iterate of the natural logarithmic function; and

(2) $ar + b \in S$ for all $0 \leq a \leq \ell$.

Denis (see [**Den94**, Théorème 3]) proves that with the notation from Question 11.1.0.3, if the set of $n \in \mathbb{N}_0$ such that $\Phi^n(\alpha) \in V$ is very dense of order $2\dim(V) - 1$, then V contains a positive dimensional subvariety which is periodic under the action of Φ.

We also note that in both Denis' question from [**Den94**] (see Question 11.1.0.3) and its reformulation (see Question 11.1.0.4) the field K has arbitrary characteristic. When $\mathrm{char}(K) = p > 0$ one cannot expect that alternative (1) in Question 11.1.0.3 could be improved to S_V being a finite set. Indeed, as observed by Denis [**Den94**, Remarque, p. 17] (see also Example 3.4.5.1), the set S_V could be infinite, but very sparse (for example, consisting only of powers of p). It is possible that the set S_V has an even more complicated structure; the following example was first suggested in [**MS04**].

EXAMPLE 11.1.0.6. Let C be a curve of high genus (for example, at least equal to 3) defined over $\mathbb{F}_p$ embedded in its Jacobian J, and let $\alpha \in C(\overline{\mathbb{F}_p}(t)) \setminus C(\overline{\mathbb{F}_p})$. Then the Zariski closure V of $C + C$ (where the addition takes place inside J) *generically* does not contain a translate of a positive dimensional algebraic group (for example, if J is a simple abelian variety, and $C + C$ is not Zariski dense in J). So, the intersection of V with the orbit of $0 \in J$ under the translation-by-α map consists of points of the form $(p^m + p^n) \cdot \alpha$ only.

For more details on a characteristic p version of the Dynamical Mordell-Lang Conjecture, we refer the reader to Chapter 13.

In this chapter we show that Question 11.1.0.3 has a positive answer even in the more general case of rational self-maps on arbitrary quasiprojective varieties.

THEOREM 11.1.0.7 ([**BGT15b**]). *Let X be a quasiprojective variety defined over a field K, let*

$$\Phi : X \longrightarrow X$$

be a rational map defined over K, let $x \in X(K)$ such that $\mathcal{O}_\Phi(x)$ is contained in the domain of definition for Φ, and let Y be a K-subvariety of X. Then the set

$$S := \{n \colon \Phi^n(x) \in Y(K)\}$$

is a union of at most finitely many arithmetic progressions along with a set of Banach density zero.

In the case of a rational self-map Φ on a variety X, and for a point $x \in X$, the orbit $\mathcal{O}_\Phi(x)$ is defined when $\Phi^n(x)$ is not contained in the indeterminacy locus of Φ for each $n \in \mathbb{N}_0$; for more details, see also Question 3.2.0.1 and the discussion from Section 3.2.

Theorem 11.1.0.7 is proven in Section 11.4 using also the technical results from Section 11.2. The results of Sections 11.2, 11.3 and 11.4 are contained in [**BGT15b**]. Actually, we are able to prove a more general result (see Theorem 11.4.2.2, which is [**BGT15b**, Theorem 3.2]) extending Theorem 11.1.0.7 where rational self-maps on algebraic varieties are replaced by continuous maps defined on an open subset of a Noetherian space (for a definition of Noetherian spaces, see Definition 11.4.1.1). Our results for Noetherian spaces show that the Dynamical Mordell-Lang principle (see Subsection 10.2.1) extends quite generally with the caveat that one has to consider sets of Banach density 0 in place of finite sets.

In Section 11.4 we present our most general results towards the Dynamical Mordell-Lang principle for continuous maps on Noetherian spaces (see also [**BGT15b**] where these results were first published). A crucial ingredient for proving Theorem 11.4.2.2 and its consequences is a general result on subsets of $\mathbb{N}_0$ of positive Banach density (see Lemma 11.2.0.1). We use Lemma 11.2.0.1 in Section 11.3 to derive some quantitative results, such as the following statement (see [**BGT15b**, Theorem 1.9]).

THEOREM 11.1.0.8 ([**BGT15b**]). *Let* $C \subseteq \mathbb{P}^\ell$ *be an irreducible curve of degree* d *defined over a field* K, *let*

$$\Phi : \mathbb{P}^\ell \longrightarrow \mathbb{P}^\ell$$

be an endomorphism of degree m *defined over* K, *and let* $\alpha \in \mathbb{P}^\ell(K)$ *be a non-preperiodic point. If* C *is not periodic, then there exists a constant* c_0 *depending only on* d, m, *and* ℓ *such that for all integers* $N \geq 2$, *we have*

$$\#\{1 \leq n \leq N : \Phi^n(\alpha) \in C(K)\} \leq \frac{c_0 N}{\log(N)}.$$

In Section 11.5 (which contains mainly the results from [**BGKT10**]) we state a more precise result than Theorem 11.1.0.7 in the special case $X = (\mathbb{P}^1)^N$ and Φ is given by the coordinatewise action of N (one-variable) complex rational maps.

THEOREM 11.1.0.9 ([**BGKT12**]). *Let* $f_1, \ldots, f_N \in \mathbb{C}(z)$ *be rational functions, and let*

$$\Phi = (f_1, \ldots, f_N)$$

denote their coordinatewise action on $(\mathbb{P}^1)^N$. *Let*

$$\alpha = (\alpha_1, \ldots, \alpha_N) \in (\mathbb{P}^1)^N(\mathbb{C})$$

be a non-preperiodic point, and let $V \subseteq (\mathbb{P}^1)^N$ *be a subvariety such that* $\Phi^n(\alpha) \in V$ *for infinitely many* $n \in \mathbb{N}$. *If* V *contains no positive dimensional periodic subvariety intersecting* $\mathcal{O}_\Phi(\alpha)$, *then*

$$\#\{0 \leq k \leq n : \Phi^k(\alpha) \in V(\mathbb{C})\} = o\left(\log^{(m)}(n)\right),$$

for each $m \in \mathbb{N}$, *where* $\log^{(m)}$ *is the* m*-th iterated logarithm.*

In Section 11.5 we prove Theorem 11.1.0.9 as a corollary of Theorem 11.5.0.2 (which is stated in Section 11.5 and proven in Section 11.8). Sections 11.6 and 11.7 contain the technical setup and the necessary lemmas used in Section 11.8 for proving Theorem 11.5.0.2.

In Section 11.11 we prove a general approximating result (see Theorem 11.11.1.1) for parametrizing p-adically an orbit. Our result can be viewed as an approximating version of the p-adic arc lemma. We use Theorem 11.11.1.1 to deduce some partial results towards the Dynamical Mordell-Lang Conjecture (see Theorem 11.11.3.1).

11.2. Sets of positive Banach density

Denis [**Den94**] uses in his proof the famous theorem of Szemerédi [**Sze75**] regarding the existence of (arbitrarily long, but finite) arithmetic progressions contained in subsets of $\mathbb{N}_0$ of positive Banach density. More precisely, Szemerédi

[**Sze75**] showed that for all $\epsilon > 0$ and $k \in \mathbb{N}$, there exists a positive number $N(k, \epsilon)$ such that for all integers $n > N(k, \epsilon)$, if

$$R \subseteq \{1, \ldots, n\} \text{ such that } |R| > \epsilon n,$$

then R contains an arithmetic progression with (at least) k elements. In other words, if $S \subseteq \mathbb{N}_0$ is a set of positive natural density, then it contains arbitrarily long (but finite) arithmetic progressions. Assuming that the set S_V from Question 11.1.0.4 is *very dense* of order $2 \dim(V) - 1$ (see Definition 11.1.0.5) allows Denis [**Den94**] to use Szmerédi's Theorem to infer a positive answer to Question 11.1.0.3 under these hypotheses. Essentially, the argument from [**Den94**] relies on the following observation: if an irreducible subvariety $V \subseteq \mathbb{P}^m$ contains the first ℓ iterates of a point $\alpha \in \mathbb{P}^m(K)$ under an endomorphism Ψ, and ℓ is sufficiently large (compared to the degrees of both Ψ and of V), then V is periodic under Ψ.

Theorem 11.1.0.7 is based on a result (see Lemma 11.2.0.1) of the same flavour as Szemerédi's theorem [**Sze75**], but going in a slightly different direction. Szemerédi's theorem yields the existence of long arithmetic progressions contained in a subset $S \subseteq \mathbb{N}_0$ of positive density; our Lemma 11.2.0.1 yields the existence of a positive integer k and of a positive density subset $Q \subset S$ such that its translation by k is also a subset of S.

LEMMA 11.2.0.1 ([**BGT15b**]). *Let S be a set of positive integers having positive Banach density. Let $N \geq \lfloor 1/\delta(S) \rfloor + 1$, where $\lfloor x \rfloor$ as usual denotes the greatest integer less than or equal to x. Then there is a positive integer k and a subset $Q \subseteq S$ such that*

(1) *$k \leq N - 1$;*
(2) *$\delta(Q) \geq \frac{N\delta(S) - 1}{2N^2(N-1)} > 0$; and*
(3) *for all $a \in Q$, we have $a + k \in S$.*

PROOF. We reproduce here the proof from [**BGT15b**, Lemma 2.1]. By assumption, $\delta(S) > \frac{1}{N}$, and so there exist intervals I_n with $|I_n| \to \infty$ such that

$$\frac{|S \cap I_n|}{|I_n|} > \frac{\delta(S) + \frac{1}{N}}{2}.$$

Let

$$P = \{i \colon |\{iN + 1, \ldots, (i+1)N\} \cap S| \geq 2\}.$$

We claim that P has positive Banach density. To see this, let

$$J_n = \{i \colon \{iN + 1, \ldots, (i+1)N\} \subseteq I_n\}.$$

Then

$$|J_n| \leq \frac{|I_n|}{N} \text{ and } |J_n| \to \infty, \text{ as } n \to \infty.$$

For $i \in J_n \setminus P$ we have

$$S \cap \{iN + 1, \ldots, (i+1)N\}$$

has size at most 1 and for $i \in P \cap J_n$ we have

$$S \cap \{iN + 1, \ldots, (i+1)N\}$$

has size at most N. Since there are at most $2N$ elements of I_n that are not accounted for by taking the union of the

$$\{iN + 1, \ldots, (i+1)N\}$$

with $i \in J_n$, we see that

$$\frac{\left(\delta(S) + \frac{1}{N}\right) \cdot |I_n|}{2} \leq |I_n \cap S| \leq |J_n \setminus P| + N|P \cap J_n| + 2N.$$

Using the fact that $|J_n| \leq \frac{|I_n|}{N}$, we see

$$\frac{\left(\delta(S) + \frac{1}{N}\right) \cdot N|J_n|}{2} \leq |J_n \setminus P| + N|P \cap J_n| + 2N.$$

Dividing by $N|J_n|$ now gives

$$\frac{\delta(S) + \frac{1}{N}}{2} \leq \frac{|J_n \setminus P|}{N|J_n|} + \frac{|P \cap J_n|}{|J_n|} + \frac{2}{|J_n|}.$$

Since $|J_n \setminus P| \leq |J_n|$, we get

$$\frac{\delta(S) + \frac{1}{N}}{2} \leq \frac{1}{N} + \frac{|P \cap J_n|}{|J_n|} + \frac{2}{|J_n|},$$

which gives

$$\frac{\delta(S) - \frac{1}{N}}{2} \leq \frac{|P \cap J_n|}{|J_n|} + \frac{2}{|J_n|}.$$

Since $|J_n| \to \infty$, we see that $\delta(P) \geq \frac{\delta(S) - \frac{1}{N}}{2}$.

For each $i \in P$, we pick $a_i, b_i \in \{iN + 1, \ldots, (i+1)N\} \cap S$ with $0 < b_i - a_i < N$. For $j \in \{1, \ldots, N-1\}$, we let

$$P_j := \{i \in P : b_i - a_i = j\}.$$

Then

$$P = \bigcup_{j=1}^{N-1} P_j$$

and since Banach density is subadditive, we have

$$\delta(P) \leq \sum_{j=1}^{N-1} \delta(P_j).$$

Thus there is some $k \in \{1, \ldots, N-1\}$ such that $\delta(P_k) \geq \frac{\delta(P)}{N-1}$. Let

$$Q := \{a_i : i \in P_k\} \subseteq S.$$

Then $a + k \in S$ for all $a \in Q$ and a simple computation yields

$$\delta(Q) \geq \frac{\delta(P_k)}{N} \geq \frac{N\delta(S) - 1}{2N^2(N-1)} > 0,$$

as desired. $\square$

The following Corollary of Lemma 11.2.0.1 is used in the proof of Theorem 11.3.0.4, and we also consider that it is of independent interest for possible other applications of Lemma 11.2.0.1. We state here [**BGT15b**, Corollary 2.2].

COROLLARY 11.2.0.2 ([**BGT15b**]). *Let S be a set of positive integers having positive Banach density. Then there is a positive integer $k < \frac{2}{\delta(S)}$ and a subset $Q \subseteq S$ such that*

(a) *$\delta(Q) \geq \frac{\delta(S)^3}{24}$; and*
(b) *for all $a \in Q$, we have $a + k \in S$.*

PROOF. The proof is from [**BGT15b**]. We let $\delta := \delta(S)$, and we apply Lemma 11.2.0.1 with $N = \lfloor \frac{2}{\delta} \rfloor$ (which is at least equal to $\lfloor \frac{1}{\delta} \rfloor + 1$ since $\delta \le 1$). This shows the existence of a set $Q \subseteq S$ satisfying property (b) above; in addition

$$\delta(Q) \ge \frac{N\delta - 1}{2N^2(N-1)}.$$

So, in order to show that condition (a) holds, it suffices to prove that

$$\frac{N\delta - 1}{2N^2(N-1)} \ge \frac{\delta^3}{24},$$

which is equivalent to proving that

$$\frac{N\delta - 1}{N - 1} \ge \frac{N^2\delta^3}{12} = \frac{2}{3N} \cdot \left(\frac{N\delta}{2} \right)^3.$$

Since $N = \lfloor \frac{2}{\delta} \rfloor \le \frac{2}{\delta}$, then it suffices to show that $\frac{N\delta-1}{N-1} \ge \frac{2}{3N}$, which is equivalent with showing that

$$(11.2.0.3) \qquad\qquad \delta \ge \frac{5}{3N} - \frac{2}{3N^2}.$$

Since $N = \lfloor \frac{2}{\delta} \rfloor$, we have $\frac{2}{\delta} < N + 1$ and so,

$$\delta > \frac{2}{N + 1}.$$

Then inequality (11.2.0.3) follows since

$$(11.2.0.4) \qquad \frac{2}{N+1} - \left(\frac{5}{3N} - \frac{2}{3N^2} \right) = \frac{N-5}{3N(N+1)} + \frac{2}{3N^2} \ge 0.$$

Inequality 11.2.0.4 is obvious for all $N \ge 5$, while for $N \in \{2, 3, 4\}$ the inequality can be checked directly (note that $N = \lfloor \frac{2}{\delta} \rfloor \ge 2$ because $\delta \le 1$). $\square$

11.3. General quantitative results

The following result is [**BGT15b**, Theorem 4.1] and it is an easy application of Lemma 11.2.0.1.

THEOREM 11.3.0.1 ([**BGT15b**]). *Let X be a quasiprojective variety defined over a field K, let*

$$\Phi : X \longrightarrow X$$

be an endomorphism defined over K, let $C \subseteq X$ be an irreducible curve, and let $\alpha \in X(K)$ be a point that is not preperiodic under Φ. If the set

$$S := \{n \in \mathbb{N} \colon \Phi^n(\alpha) \in C(K)\}$$

has Banach density $\delta > 0$, then S contains an infinite arithmetic progression of common difference

$$k = \frac{1}{\delta}, \text{ and } \Phi^k(C) \subseteq C.$$

In particular, C is periodic under the action of Φ.

PROOF. The proof is from [**BGT15b**]. It follows from Lemma 11.2.0.1 applied with $N = \lfloor \frac{1}{\delta} \rfloor + 1$ that there exists a positive integer $k \leq \lfloor \frac{1}{\delta} \rfloor$ and a subset $Q \subset S$ of positive density such that for each $n \in Q$, also $\Phi^{n+k}(\alpha) \in C(K)$. So

$$\Phi^n(\alpha) \in C \cap \Phi^{-k}(C).$$

Hence, $C \cap \Phi^{-k}(C)$ contains an infinite set of points (because α is not preperiodic under the action of Φ). Since C is an irreducible curve, we see then that

(11.3.0.2) $C \subseteq \Phi^{-k}(C)$; thus $\Phi^k(C) \subseteq C$.

This yields the desired infinite arithmetic progression of common difference $k \leq \frac{1}{\delta}$. If $k < 1/\delta$, then the existence of this arithmetic progression would imply that

$$\delta \geq 1/k > \delta,$$

a contradiction. Thus, $k = \frac{1}{\delta}$; also (11.3.0.2) yields that C is periodic. $\square$

Theorem 11.3.0.1 has the following interesting consequence.

COROLLARY 11.3.0.3. *Let X be a quasiprojective variety defined over a field K, let*

$$\Phi : X \longrightarrow X$$

be an endomorphism defined over K, let $C \subseteq X$ be an irreducible curve, and let $\alpha \in X(K)$ be a point that is not preperiodic under Φ. If the set

$$S := \{n \in \mathbb{N} \colon \Phi^n(\alpha) \in C(K)\}$$

has positive Banach density δ, then

$$\delta = \frac{1}{k},$$

for some $k \in \mathbb{N}$, and moreover,

$$S = \{\ell + nk \colon n \in \mathbb{N}_0\},$$

for some $\ell \in \mathbb{N}_0$.

In other words, under the hypotheses of Corollary 11.3.0.3, not only that the Dynamical Mordell-Lang Conjecture holds, but we also know that there is *precisely* one arithmetic progression appearing in the conclusion of Conjecture 1.5.0.1.

PROOF OF COROLLARY 11.3.0.3. The conclusion of Theorem 11.3.0.1 yields immediately that

$$\delta = \frac{1}{k} \text{ for some } k \in \mathbb{N}.$$

Furthermore, $\Phi^k(C) \subseteq C$, which yields that given $\ell \in \mathbb{N}_0$ be the smallest integer such that

$$\Phi^\ell(\alpha) \in C(K),$$

then $\{\ell + nk \colon n \in \mathbb{N}_0\} \subseteq S$. Now, if there exists some

$$\ell_1 \in \mathbb{N} \setminus \{\ell + nk \colon n \in \mathbb{N}_0\}$$

such that $\Phi^{\ell_1}(\alpha) \in C(K)$, then we would get again that

$$\{\ell_1 + nk \colon n \in \mathbb{N}_0\} \subset S,$$

and since the two arithmetic progressions $\{\ell + nk\}_{n \in \mathbb{N}_0}$ and $\{\ell_1 + nk\}_{n \in \mathbb{N}_0}$ are disjoint, we would get that

$$\delta(S) \geq \frac{2}{k} > \delta,$$

a contradiction. In conclusion,

$$S = \{\ell + nk \colon n \in \mathbb{N}_0\},$$

as claimed. $\square$

Applying the technique of the proof of Theorem 11.3.0.1 recursively in the case of endomorphisms of $\mathbb{P}^\ell$ one can obtain a similar result for all projective subvarieties of $\mathbb{P}^\ell$. This may be viewed as a weak dynamical analogue of a result of Evertse-Schlickewei-Schmidt [**ESS02**, Theorem 1.2] on effective bounds for the common differences of arithmetic progressions that arise from linear recurrence sequences.

THEOREM 11.3.0.4 ([**BGT15b**]). *For each real number $\delta > 0$, and each $D, \ell, e,$ $m \in \mathbb{N}$, there exists a positive real number*

$$M := M(\delta, m, \ell, D, e)$$

with the following property: for any endomorphism $\Phi : \mathbb{P}^\ell \longrightarrow \mathbb{P}^\ell$ given by homogenous polynomials of degree at most m defined over any algebraically closed field K, any irreducible subvariety $V \subseteq \mathbb{P}^\ell$ of dimension at most e and degree at most D, and any $\alpha \in \mathbb{P}^\ell(K)$, if the set

$$S := \{n \in \mathbb{N} \colon \Phi^n(\alpha) \in V(K)\}$$

has Banach density at least equal to δ, then S contains an infinite arithmetic progression of common difference at most M.

PROOF. The proof is from [**BGT15b**, Theorem 4.2]. One uses induction on the dimension e of V, with the base case being Theorem 11.3.0.1; so

$$M(\delta, m, \ell, D, 1) := \frac{1}{\delta}.$$

Assume now $e \geq 2$. If $V \subseteq \Phi^{-k}(V)$ for some integer

$$k < N := \left\lfloor \frac{2}{\delta} \right\rfloor,$$

then V is periodic under Φ and moreover it contains an infinite arithmetic progression of common difference at most $\lfloor \frac{2}{\delta} \rfloor$. So, assume $V \cap \Phi^{-k}(V)$ has dimension at most $e - 1$ for each $k < N$. By Corollary 11.2.0.2 applied with $N = \lfloor \frac{2}{\delta} \rfloor$, there is a set $Q \subset S$ of density at least equal to $\frac{\delta^3}{24}$ and there is an integer $k < N$ such that

$$\Phi^n(\alpha) \in V \cap \Phi^{-k}(V)$$

for each $n \in Q$. Then with the use of Bézout-type inequalities (see [**Hei83**, Theorem 1] or [**Lan83**, Lemma 3.5, p. 65], for example), one sees that the degree of $V \cap \Phi^{-k}(V)$ is at most

$$m^{k(\ell-e)} D^2$$

because $\Phi^{-k}(V)$ has degree at most $m^{k(\ell-e)} D$. Hence, by the pigeonhole principle, there exists an irreducible subvariety

$$W \subseteq V \cap \Phi^{-k}(V)$$

of dimension at most $e - 1$ such that the set

$$S_W := \{n \in \mathbb{N} \colon \Phi^n(\alpha) \in W(K)\}$$

has density at least equal to

$$\frac{\delta^3}{24m^{k(\ell-e)}D^2}.$$

Then the induction hypothesis yields the following recursive formula

$$M(\delta, m, \ell, D, e) = M\left(\frac{\delta^3}{24m^{\frac{2(\ell-e)}{\delta}}D^2}, m, \ell, \lfloor m^{\frac{2(\ell-e)}{\delta}}D^2\rfloor + 1, e - 1\right).$$

$\square$

Theorem 11.1.0.8 is proven in [**BGT15b**, Theorem 1.9], whose arguments we follow.

PROOF OF THEOREM 11.1.0.8. The proof relies on the following quantitative version of the argument from the proof of Theorem 11.3.0.1.

LEMMA 11.3.0.5. *Let* $k \in \mathbb{N}$, *and let* K, C, Φ, α, m, d *be as in Theorem* 11.1.0.8. *Let* $c_1 = 2m^\ell d^2$. *Then we have*

$$\#\{n \geq 1\colon \Phi^n(\alpha) \in C(K) \text{ and } \Phi^{n+k}(\alpha) \in C(K)\} \leq c_1^k.$$

PROOF OF LEMMA 11.3.0.5. Let n be a positive integer such that

$$\Phi^n(\alpha) \in C(K) \text{ and also } \Phi^{n+k}(\alpha) \in C(K);$$

then $\Phi^n(\alpha) \in C \cap \Phi^{-k}(C)$. We know that C is an irreducible non-periodic curve. Then $C \cap \Phi^{-k}(C)$ is a proper intersection of projective curves, and therefore, by Bézout's Theorem (see [**Hei83**, Theorem 1], for example), the number of points in the intersection is bounded above by

$$\deg(C) \cdot \deg(\Phi^{-k}(C)) \leq \left(2m^\ell d^2\right)^k = c_1^k.$$

$\square$

Based on Lemma 11.3.0.5, the rest of the proof is a simple counting argument. We let $\{n_k\}_{k\geq 1}$ be the increasing sequence of positive integers n for which

$$\Phi^n(x) \in C(K).$$

The proof of Theorem 11.1.0.8 is a consequence of the following result.

LEMMA 11.3.0.6. *With notation as in Lemma* 11.3.0.5, *let*

$$c_2 = (4c_1 \log(c_1) \log_{c_1}(2c_1))^{-1}.$$

Then $n_k \geq c_2 k \log(k)$ *for each* $k \geq 1$.

PROOF OF LEMMA 11.3.0.6. We first estimate the size of n_k for $k > (2c_1)^{4c_1}$. Note that $c_1 \geq 2$. Then there exists a unique positive integer j such that

$$\sum_{i=1}^{j} c_1^i < k \leq \sum_{i=1}^{j+1} c_1^i,$$

which, since $c_1 \geq 2$, gives

(11.3.0.7) $$k \leq 2c_1^{j+1}, \qquad \text{or equivalently} \qquad c_1^j \geq k/2c_1.$$

Using Lemma 11.3.0.5, we obtain

$$n_k = n_1 + \sum_{i=1}^{k-1}(n_{i+1} - n_i) \geq \sum_{i=1}^{j} i \cdot c_1^i \geq jc_1^j.$$

Then, using equation (11.3.0.7), we find

(11.3.0.8)
$$n_k \geq jc_1^j \geq \frac{k}{2c_1}\log_{c_1}(k/2c_1) = k(\log_{c_1} k)\frac{\left(1 - \log_{c_1}(2c_1)/\log_{c_1}(k)\right)}{2c_1}$$

$$\geq \frac{1}{4c_1}k\log_{c_1}(k),$$

where the last inequality follows from the fact that we are assuming for now that $k > (2c_1)^{4c_1}$.

Suppose now that $k \leq (2c_1)^{4c_1}$. Since $n_i \geq i$ for every i, we see then that

$$n_k \geq k \geq \frac{1}{4c_1\log_{c_1}(2c_1)}k\log_{c_1}(k).$$

Hence we have $n_k \geq c_2 k \log(k)$ for every $k \geq 1$.   □

Now, for each $N \geq 2$, if $k := \#\{1 \leq n \leq N\colon \Phi^n(x) \in C(K)\}$ then $n_k \leq N$. Using Lemma 11.3.0.6 we obtain

$$N \geq n_k \geq c_2 k \log(k),$$

and therefore, $k \leq \frac{c_0 N}{\log(N)}$ with $c_0 := \frac{2}{c_2}$ (which depends on d, m and ℓ only).   □

11.4. The Dynamical Mordell-Lang problem for Noetherian spaces

11.4.1. Noetherian spaces. Before stating the main result of this section (which follows closely [**BGT15b**]), we define *Noetherian spaces*.

DEFINITION 11.4.1.1. Let X be a topological space. We say that X is *Noetherian* if it satisfies the descending chain condition for its closed subsets, i.e., there exists no infinite descending chain of proper closed subsets.

Any quasiprojective variety endowed with the Zariski topology is a Noetherian space. Also, a compact p-adic manifold endowed with the *rigid analytic topology* is a Noetherian space. For example, on $D(0,1)^m$ (where $D(0,1)$ is the p-adic unit ball) endowed with the rigid analytic topology, the closed sets are the zero sets of p-adic analytic functions f from the *Tate algebra*, i.e. power series

$$f(z_1, \ldots, z_m) := \sum_{i_1,\ldots,i_m \in \mathbb{N}_0} c_{i_1,\ldots,i_m} z_1^{i_1} \cdots z_m^{i_m} \in \mathbb{Z}_p[[z_1, \ldots, z_m]]$$

such that $|c_{i_1,\ldots,i_m}|_p \to 0$ as $i_1 + \cdots + i_m \to \infty$.

11.4.2. The main results of this section. The following result is [**BGT15b**, Theorem 1.4].

THEOREM 11.4.2.1. *Let X be a Noetherian topological space, and let $\Phi \colon X \longrightarrow X$ be a continuous function. Then for each $x \in X$ and for each closed subset Y of X, the set $S := \{n \in \mathbb{N}\colon \Phi^n(x) \in Y\}$ is a union of at most finitely many arithmetic progressions along with a set of Banach density zero.*

We note that both in the context of algebraic varieties over fields of positive characteristic (see Examples 3.4.5.1 and 11.1.0.6) and in the context of rigid analytic spaces (see Proposition 11.10.0.1 and [**BGKT10**]), there are examples of Noetherian spaces X having a closed subset $V \subset X$, a continuous self-map Φ on X, and a point $\alpha \in X$ such that the set S from the conclusion of Theorem 11.4.2.1 is infinite but has Banach density 0.

Using different techniques, Petsche [**Pet**] proved Theorem 11.1.0.7 when Φ is an endomorphism of an affine variety X. Petsche uses methods from topological dynamics and ergodic theory; in particular, he uses Berkovich spaces (for a comprehensive introduction to Berkovich spaces, see [**BR10**]) and a strong topological version of Furstenberg's Poincaré Recurrence Theorem (see [**Fur81**]). Theorem 11.4.2.1 can also be derived using arguments that come from a deep result of ergodic theory on Noetherian spaces proved by Charles Favre (this is Théorème 2.5.8 in Favre's PhD thesis [**Fav**]); see also [**Gig**, Example A.3.2] and [**Gig14**, Theorem 1.6] for an alternative proof of Favre's result using measure-theoretic methods.

Actually, a more general result holds (see [**BGT15b**, Theorem 3.2]), which is a generalization of Theorem 11.1.0.7.

THEOREM 11.4.2.2 ([**BGT15b**]). *Let X be a Noetherian space, let $U \subseteq X$ be an open subset, and let $\Phi : U \longrightarrow X$ be a continuous map. Let $x \in X$ such that $\Phi^n(x) \in U$ for each nonnegative integer n. Then for each closed set $Y \subseteq X$, the set*

$$S := \{n \in \mathbb{N} \colon \Phi^n(x) \in Y\}$$

is a union of at most finitely many arithmetic progressions along with a set of Banach density zero.

PROOF. Let

$$Z := \bigcap_{n \geq 0} \Phi^{-n}(U).$$

We know that Z is non-empty since x (and therefore $\mathcal{O}_\Phi(x)$) is contained in Z. We endow Z with the inherited topology from X. Then Z is also a Noetherian space. Furthermore, by its definition, we obtain that Φ restricts to a self-map

$$\Phi_Z : Z \longrightarrow Z.$$

Next we show that Φ_Z is continuous. Indeed, let $V \subseteq X$ be an open set. We need to show that $\Phi_Z^{-1}(V \cap Z)$ is open in Z. This follows immediately once we prove that

$$\Phi_Z^{-1}(V \cap Z) = \Phi^{-1}(V) \cap Z$$

because $\Phi : U \longrightarrow X$ is continuous and so $\Phi^{-1}(V)$ is open in U and (because U is an open subset of X) it is also open in X which yields that $\Phi^{-1}(V) \cap Z$ is open in Z. To see that

$$\Phi_Z^{-1}(V \cap Z) = \Phi^{-1}(V) \cap Z$$

we note that for each $y \in \Phi_Z^{-1}(V \cap Z) \subseteq Z$ we have $\Phi_Z(y) \in V$. So, $\Phi(y) \in V$ and thus $y \in \Phi^{-1}(V) \cap Z$. Conversely, if

$$y \in \Phi^{-1}(V) \cap Z,$$

then $y \in Z$ and so $\Phi_Z(y) \in V \cap Z$ as claimed. Therefore $\Phi_Z : Z \longrightarrow Z$ is a continuous map on a Noetherian space. Hence by Theorem 11.4.2.1, the set of all $n \in \mathbb{N}$ such that

$$\Phi_Z^n(x) \in Y \cap Z$$

is a union of at most finitely many arithmetic progressions along with a set of Banach density zero. Since $\Phi = \Phi_Z$ on Z, we see

$$\Phi^n(x) \in Y \text{ if and only if } \Phi_Z^n(x) \in Y \cap Z,$$

which concludes our proof. $\qquad\square$

So, even though at a first glance, Theorem 11.4.2.2 is more general than Theorem 11.4.2.1, its proof reduces to applying the latter theorem in a proper setting. The rest of this section is devoted to proving Theorem 11.4.2.1, but first we start by noting some useful reductions.

11.4.3. A reformulation of Theorem 11.4.2.1. Theorems 11.4.2.1 and 11.4.2.2 yield that once removing finitely many arithmetic progressions contained in S, we obtain a very sparse set. The key for the proof of Theorem 11.4.2.1 is the following result (see [**BGT15b**, Proposition 3.1]).

PROPOSITION 11.4.3.1 ([**BGT15b**]). *Let X be a Noetherian topological space, let $\Phi : X \longrightarrow X$ be a continuous map, let $x \in X$, let Y be a closed subset of X, and let*

$$S := S_Y := \{n\colon \Phi^n(x) \in Y\}.$$

If S has positive Banach density, then it contains an infinite arithmetic progression.

PROOF. Consider the set $\mathcal{V}$ of all closed subsets V of X with the property that

$$S_V := \{n\colon \Phi^n(x) \in V\}$$

has positive Banach density but does not contain an infinite arithmetic progression. If $\mathcal{V}$ is empty, then there is nothing to prove. Thus we may assume, towards a contradiction, that $\mathcal{V}$ is non-empty. We let W be a minimal element of $\mathcal{V}$ with respect to the inclusion of sets (note that such an element exists since X is Noetherian). By Lemma 11.2.0.1, we have a positive integer k and a subset $Q \subseteq S_W$ with $\delta(Q) > 0$ such that $a + k \in S_W$ for all $a \in Q$.

If $n \in Q$, then $\Phi^n(x) \in W$ and $\Phi^{n+k}(x) \in W$. Thus

$$\Phi^n(x) \in W \cap \Phi^{-k}(W)$$

whenever $n \in Q$. If $\Phi^{-k}(W) \supseteq W$ then S_W has the property that $n + k \in S_W$ whenever $n \in S_W$ and since S_W is non-empty, it contains an infinite arithmetic progression, which contradicts the fact that $W \in \mathcal{V}$. Thus

$$Z := W \cap \Phi^{-k}(W)$$

is a proper closed subset of W (since Φ is continuous and W is closed) and so we have $\Phi^n(x) \in Z$ for all $n \in Q$. Since Q has positive Banach density, we obtain that $S_Z \supseteq Q$ also has positive Banach density and therefore S_Z contains an infinite arithmetic progression. Since $S_Z \subseteq S_W$, we see that S_W contains an infinite arithmetic progression, a contradiction. This concludes our proof. $\quad\square$

We finish the proof of Theorem 11.4.2.1 in the following subsection.

11.4.4. Proof of Theorem 11.4.2.1. The proof is given in [**BGT15b**, Section 3], and it follows from Proposition 11.4.3.1 similarly to the equivalence of the Conjectures 1.5.0.1 and 3.1.3.2 (which was shown in Chapter 3). Before proceeding to our proof, we recall that for any set $U \subseteq \mathbb{N}_0$ and any $c \in \mathbb{N}$, we let

$$c \cdot U := \{cj \colon j \in U\},$$

and we also let

$$c + U := U + c := \{c + j \colon j \in U\}.$$

We assume towards a contradiction that the conclusion of Theorem 11.4.2.1 does not hold. Then let $\mathcal{V}$ be the collection of all closed subsets V of X for which there exists:

- a continuous map $g : V \to V$,
- a closed subset W of V, and
- a point $y \in V$ such that $\{n \colon g^n(y) \in W\}$ cannot be expressed as a finite union of arithmetic progressions along with a set of Banach density zero.

By assumption, $X \in \mathcal{V}$ and so we may choose a minimal element $V \in \mathcal{V}$. Then there is some continuous map $g : V \longrightarrow V$, some closed subset W of V, and some point $y \in V$ such that

$$S := \{n \colon g^n(y) \in W\}$$

cannot be expressed as a finite union of arithmetic progressions along with a set of Banach density zero. We necessarily have that $W_i := g^{-i}(W)$ is a proper closed subset of V (note that g is continuous), since otherwise S would contain every integer greater than or equal to i (and thus it would be the union of an arithmetic progression with a finite set). Moreover, by our choice of V, W and y, it follows that $\delta(T) > 0$ and thus by Proposition 11.4.3.1, there exist $a, b \in \mathbb{N}$ such that

$$S \supseteq \{an + b \colon n \geq 0\}.$$

Let C_i denote the closure of

$$T_i := \{g^{(an+b)}(y) \colon n \geq i\}.$$

Then

$$C_0 \supseteq C_1 \supseteq \cdots$$

is a descending chain of closed sets and hence there is some m such that

$$C_m = C_{m+1} = \cdots.$$

We take $V_0 = C_m$. Then

$$g^{-a}(V_0) \supseteq g^{-a}(T_{m+1}) \supseteq T_m$$

and since $g^{-a}(V_0)$ is closed we thus see it contains the closure of T_m, which is V_0. Then $V_0 \subseteq W$ is closed and we have

$$g^{-a}(V_0) \supseteq V_0.$$

We let V_j denote the closed set

$$g^{-j}(V_0) \text{ for } j \in \{1, \ldots, a - 1\}.$$

Since $V_j \subseteq W_{a+j} \subsetneq V$, we see that each V_j is a proper subset for $0 \leq j \leq a - 1$. Then

$$g^{-a}(V_j) = g^{-a}(g^{-j}(V_0)) = g^{-j}(g^{-a}(V_0)) \supseteq g^{-j}(V_0) = V_j,$$

and so for $j \in \{0, \ldots, a-1\}$, we have $g^{-j+na+b}(y) \in V_j$ for every $n > m$. Moreover, since

$$g^{-a}(V_j) \supseteq V_j,$$

we have that $h := g^a$ restricts to continuous maps

$$h : V_j \longrightarrow V_j \text{ for each } j \in \{0, \ldots, a-1\}.$$

We let $y_j := g^{-j+a+b}(y)$. It follows from the minimality of V that

$$S_j := \{n \geq m \colon h^n(y_j) \in W \cap V_j\}$$

is a finite union of arithmetic progressions along with a set of Banach density zero. On the other hand,

$$S_j = \{n \geq m \colon g^{-j+a(n+1)+b}(y) \in W\},$$

for each $j = 0, \ldots, a-1$. Then, up to a finite set, we have

$$S = \bigcup_{j=0}^{a-1} (aS_j + b + a - j).$$

Hence S is a finite union of arithmetic progressions along with a set of Banach density zero.

11.5. Very sparse sets in the Dynamical Mordell-Lang problem for endomorphisms of $(\mathbb{P}^1)^N$

In this section we present a sharpening of Theorem 11.1.0.7 in the case the ambient variety is $X = (\mathbb{P}^1)^N$ and Φ is an endomorphism of X, i.e.,

$$\Phi(x_1, \ldots, x_N) = (f_1(x_1), \ldots, f_N(x_N)),$$

where each $f_i \in \mathbb{C}(z)$. The results of this section are mainly from [**BGKT10**]. For any point $\alpha \in (\mathbb{P}^1)^N(\mathbb{C})$ we show that if

$$S_V = \{n \geq 0 : \Phi^n(\alpha) \in V(\mathbb{C})\}$$

does not contain any infinite arithmetic progressions, then S_V must be a *very sparse* set of integers (improving on the conclusion of both Proposition 11.4.3.1 and of Theorem 11.1.0.8). In particular, we show that for any k and any sufficiently large N, the number of $n \leq N$ such that $\Phi^n(\alpha) \in V(\mathbb{C})$ is less than $\log^{(k)} N$, where $\log^{(k)}$ denotes the k-th iterate of the log function (see Theorem 11.1.0.9). This result can be interpreted as an analogue of the gap principle of Davenport-Roth [**DR55**] and Mumford [**Mum65**] in the context of the classical Mordell's Conjecture. We state Mumford's theorem (for more details, see [**Lan83**, Theorem 8.1, p. 135] and [**HS00**, Theorem B.6.6, p. 218]).

THEOREM 11.5.0.1 (Mumford [**Mum65**]). *Let C be a curve of genus greater than 1 defined over a number field K, and we embed C into its Jacobian J. We write the set $C(K)$ as the sequence $\{x_n\}$ ordered increasingly with respect to the size of their canonical height $\widehat{\mathrm{h}}$ computed inside J. Then there exists a positive integer M and a real number $c > 1$ such for all n, we have $\widehat{\mathrm{h}}(x_{n+M}) \geq c \cdot \widehat{\mathrm{h}}(x_n)$.*

We recall that Faltings [**Fal83**] proved Mordell's Conjecture, i.e., that $C(K)$ is a finite set if the genus of C is greater than 1. However, Theorem 11.5.0.1 was proven by Mumford several years prior to Faltings and for many years it was believed that Mumford's result would be the key to a solution to Mordell's Conjecture. This belief was proven right by Vojta [**Voj89**] who provided an alternative proof of Mordell's Conjecture building on Mumford's original result and also adding several deep ideas of his own regarding Diophantine approximation. Vojta's proof [**Voj89**] inspired Faltings [**Fal91**] to prove the Mordell-Lang Conjecture for abelian varieties, and later, Vojta [**Voj96**] obtained the most general result in the Mordell-Lang Conjecture by proving it for all semiabelian varieties (see Section 3.4 for our discussion of the classical Mordell-Lang Conjecture).

The precise result in the context of arithmetic dynamics is the following (see [**BGKT10**, Theorem 1.4]).

THEOREM 11.5.0.2 ([**BGKT10**]). *Let* $f_1, \ldots, f_N \in \mathbb{C}(z)$ *be rational functions, and let* $\Phi = (f_1, \ldots, f_N)$ *denote their coordinatewise action on* $(\mathbb{P}^1)^N$. *Let*

$$\alpha = (\alpha_1, \ldots, \alpha_N) \in (\mathbb{P}^1)^N(\mathbb{C}),$$

and let $V \subseteq (\mathbb{P}^1)^N$ *be a proper subvariety such that* $\Phi^n(\alpha) \in V$ *for infinitely many* $n \in \mathbb{N}$. *Then there exist positive integers* $M, \ell \geq 1$ *and a real number* $c > 1$ *such that one of the following two statements holds:*

(i) $\Phi^{\ell+nM}(\alpha) \in V$ *for all nonnegative integers* n.
(ii) *For any sufficiently large integers* $n > m \geq 0$ *such that* $n \equiv m \pmod{M}$ *and* $\Phi^m(\alpha), \Phi^n(\alpha) \in V$, *we have* $n - m > c^m$.

We prove Theorem 11.5.0.2 in Section 11.8 using the technical results derived in Sections 11.6 and 11.7.

We show next how to deduce Theorem 11.1.0.9 from Theorem 11.5.0.2.

PROOF OF THEOREM 11.1.0.9. Since α is not preperiodic, then $\mathcal{O}_\Phi(\alpha)$ is infinite, and so alternative (i) from Theorem 11.5.0.2 does not hold. Indeed, otherwise we would have that V contains the Zariski closure W of $\mathcal{O}_{\Phi^M}(\Phi^\ell(\alpha))$ (see Theorem 11.5.0.2). But then V contains W which is a positive dimensional periodic subvariety (see Proposition 3.1.2.14); this is a contradiction. So, indeed alternative (ii) from Theorem 11.5.0.2 must hold.

With the same notation as in Theorem 11.5.0.2, in order to derive the conclusion of Theorem 11.1.0.9 it suffices to prove that for each $i \in \{0, \ldots, M - 1\}$ we have

$$(11.5.0.3) \quad \#\{0 \leq k \leq n \colon n \equiv i \pmod{M} \text{ and } \Phi^n(\alpha) \in V(\mathbb{C})\} = o\left(\log^{(m)}(n)\right),$$

for any $m \in \mathbb{N}$. We let $n_1 < n_2 < \cdots$ be the set of all nonnegative integers n satisfying both conditions:

$$n \equiv i \pmod{M} \text{ and } \Phi^n(\alpha) \in V(\mathbb{C}).$$

Then alternative (ii) in Theorem 11.5.0.2 yields that for all $j \geq m$ we have

$$\log^{(m)}(n_j) > c^{c^{c^{\cdots}}},$$

where in the above inequality there are $j - m$ exponentials. Using the fact that $c > 1$, then for each $\epsilon > 0$, we obtain that there exists a positive number

$$L := L(m, c, \epsilon)$$

such that for all $j > L$, we have

$$\log^{(m)}(n_j) > c^{c^{c^{\cdots}}} > \frac{j}{\epsilon}.$$

Hence (11.5.0.3) holds, as desired.   □

The proof of Theorem 11.1.0.9 yields that assuming alternative (i) in Theorem 11.5.0.2 does not occur (for any given $M \in \mathbb{N}$), if n_i is the i-th integer n such that $\Phi^n(\alpha) \in V(\mathbb{C})$, then n_i grows much faster than $\exp^{(m)}(i)$, where $\exp^{(m)}$ is the m-th iterate of the exponential function. Using a similar argument as in the proof of Theorem 11.4.2.1 based on Proposition 11.4.3.1, we can deduce the following result regarding very sparse sets which may appear in an unlikely counterexample to the Dynamical Mordell-Lang Conjecture.

THEOREM 11.5.0.4. *Let* $f_1, \ldots, f_N \in \mathbb{C}(z)$ *be rational functions, and let* $\Phi = (f_1, \ldots, f_N)$ *denote the map on* $(\mathbb{P}^1)^N$ *given by the coordinatewise action. Let*

$$\alpha = (\alpha_1, \ldots, \alpha_N) \in (\mathbb{P}^1)^N(\mathbb{C}),$$

and let $V \subseteq (\mathbb{P}^1)^N$ *be a subvariety. Then the set*

$$S_V := \{n \in \mathbb{N}_0 \colon \Phi^n(\alpha) \in V(\mathbb{C})\}$$

is a union of finitely many arithmetic progressions along with a set T_V *with the property that*

$$(11.5.0.5) \qquad \#\{0 \le i \le n \colon i \in T_v\} = o\left(\log^{(m)}(n)\right),$$

for all $m \in \mathbb{N}$.

PROOF. We let $\mathcal{V}$ be the set of all subvarieties V of $X := (\mathbb{P}^1)^N$ with the property that there exists an endomorphism Φ of X (as above) and there exists a point $\alpha \in X(\mathbb{C})$ such that the set

$$S_V := \{n \in \mathbb{N}_0 \colon \Phi^n(\alpha) \in V(\mathbb{C})\}$$

is not a union of at most finitely many arithmetic progressions along with a set T_V satisfying condition (11.5.0.5) above. We assume $\mathcal{V}$ is non-empty and then we let $V \in \mathcal{V}$ be of minimal dimension. We may assume that V is irreducible (since an irreducible component of it has to be in $\mathcal{V}$ anyway). Also we assume the point α is not preperiodic, since otherwise the Dynamical Mordell-Lang Conjecture holds (see Proposition 3.1.2.9). By definition, the set

$$S_V := \{n \in \mathbb{N}_0 \colon \Phi^n(\alpha) \in V(\mathbb{C})\}$$

does not satisfy condition (11.5.0.5) above. Then Theorem 11.1.0.9 yields that S_V contains an infinite arithmetic progression $\{\ell + nM\}_{n \ge 0}$. We split our analysis in two cases:

Case 1. $V \subseteq \Phi^{-M}(V)$.

In this case, whenever $i \in S_V$ then $i + Mn \in S_V$ for all $n \ge 0$; hence S_V is a finite union of arithmetic progressions, which is a contradiction.

Case 2. $W := V \cap \Phi^{-M}(V)$ is a proper subvariety of V.

Then $W \notin \mathcal{V}$ (by the minimality of $\dim(V)$) and moreover, for each $j = 0, \ldots, M - 1$, we have that each

$$(11.5.0.6) \qquad W_j := \overline{\Phi^j(W)} \text{ has dimension less than } \dim(V).$$

Furthermore $\Phi^{j+\ell+nM}(\alpha) \in W_j$ for each $j = 0, \ldots, M-1$. So, apart from a finite set, S_V is the union of the sets $S_j \cdot M + j + \ell$, for $j = 0, \ldots, M-1$, where

$$S_j := \{n \in \mathbb{N}_0 : \Phi^{j+\ell+nM}(\alpha) \in (W_j \cap V)(\mathbb{C})\}.$$

However, since $W_j \cap V \notin \mathcal{V}$ (using (11.5.0.6) and the minimality of $\dim(V)$ among all subvarieties contained in $\mathcal{V}$), we get that each S_j is a union of at most finitely many arithmetic progressions along with a set T_j satisfying condition (11.5.0.5). But then $V \in \mathcal{V}$, a contradiction. □

Now, for any subvariety $V \subseteq (\mathbb{P}^1)^N$, if one removes from V all positive dimensional subvarieties which are fixed by Φ^M, then the intersection of that set with $\mathcal{O}_\Phi(\alpha)$ should be a set very sparse as described by alternative (ii) in Theorem 11.5.0.2. Indeed, once one removes the positive dimensional subvarieties of V which are fixed by Φ^M, then alternative (i) cannot occur on the intersection of this set with $\mathcal{O}_\Phi(\alpha)$. Furthermore, all the positive dimensional subvarieties of V fixed by Φ^M are contained in the closed subset of V given by

$$\bigcap_{k \in \mathbb{N}_0} \left(\Phi^{kM}\right)^{-1}(V).$$

Hence Theorem 11.5.0.2 says that if the Dynamical Mordell-Lang Conjecture were to fail, then any such counterexample would produce a set

$$\{n \in \mathbb{N}_0 : \Phi^n(\alpha) \in V(\mathbb{C})\}$$

which is very sparse. This suggests that finding such a counterexample is very unlikely, and hence the Dynamical Mordell-Lang Conjecture should hold at least for endomorphisms of $(\mathbb{P}^1)^N$.

When our points and maps are defined over a number field K, we may phrase this discussion in terms of Weil heights; see Section 2.6 for background on heights. If α is not preperiodic, then the Weil height $h(\Phi^n(\alpha))$ grows *at least* as fast as $\deg_{\min}(\Phi)^n$, where

$$\deg_{\min}(\Phi) := \min_j \deg(f_j).$$

Indeed, for any point

$$P := (x_1, \ldots, x_N) \in (\mathbb{P}^1)^N(\overline{\mathbb{Q}}),$$

we let

$$h(P) := h(x_1) + \cdots + h(x_N),$$

where $h(\cdot)$ is the usual Weil height of points in $\mathbb{P}^1(\overline{\mathbb{Q}})$. Also, as proven in Proposition 2.6.4.2, for a map ψ of degree $d > 1$, we have that

$$|h(\psi(x)) - h(x)| \text{ is uniformly bounded on } \mathbb{P}^1(\overline{\mathbb{Q}}).$$

Hence $h(\Phi^n(\alpha)) \gg \deg_{\min}(\Phi)^n$, as claimed. Thus, we obtain the following dynamical analogue of Theorem 11.5.0.1.

COROLLARY 11.5.0.7 ([**BGKT10**]). *Let α, Φ, and V be as in Theorem 11.5.0.2, and let n_k denote the k-th integer n such that $\Phi^n(\alpha) \in V(\mathbb{C})$. Assume that α and Φ are defined over some number field K, that $\deg_{\min}(\Phi) \geq 2$, that α_i is not preperiodic for f_i for each i, and that the set*

$$\mathcal{S} := \{n \in \mathbb{N}_0 : \Phi^n(\alpha) \in V\}$$

does not contain any infinite arithmetic progressions. Then $h(\Phi^{n_i}(P))$ grows faster than $\exp^{(k)}(i)$ for any $k \geq 1$.

This growth is much more rapid than that of the "gap principles" of Mumford [**Mum65**] and Davenport-Roth [**DR55**]. If C is a curve of genus greater than 1, Mumford showed (see Theorem 11.5.0.1) that there are constants $a, b > 0$ such that if we order the K-rational points of C (for some number field K) according to Weil height, then the ℓ-th point has Weil height at least $e^{a+b\ell}$. Of course, later Faltings [**Fal83**] proved Mordell's conjecture therefore showing that there are at most finitely many K-rational points on C. In his proof, Mumford embedded points of C into $\mathbb{R}^d$ (where d is the rank of the Mordell-Weil group of the Jacobian); *Mumford's gap principle* roughly states that there is a constant $c > 1$ such that if $v_1, v_2 \in \mathbb{R}^d$ are the images of two points on the curve lying in a small sector, then either $|v_1| > c \cdot |v_2|$ or $|v_2| > c \cdot |v_1|$. More precisely (see [**HS00**, Theorem B.6.6 (b)]), Mumford's gap principle yields that there exist positive constants c_1 and c_2 such that for any two points $P, Q \in C(K)$, we have

$$\widehat{h}(P - Q) \geq c_1 \cdot \left(\widehat{h}(P) + \widehat{h}(Q) \right) - c_2.$$

Similarly, in our Theorem 11.5.0.2, two indices n_1, n_2 lying in the same congruence class modulo N can be considered analogous to two vectors v_1, v_2 lying in a small sector. In fact, in [**BGKT10**, Theorem 4.1], one shows that the pair of constants (N, C) in the conclusion of Theorem 11.5.0.2 may be replaced by the pair $(eN, C^{e-\epsilon})$, for any positive integer e and any positive real number $\epsilon > 0$. Hence, by the same analogy to Mumford's gap principle, one proves that "the smaller the angles" between two indices, "the larger the gap" between them.

Our proof of Theorem 11.5.0.2 uses p-adic dynamics by combining all possible p-adic parametrizations of orbits under rational maps (see Section 6.1). First we find a suitable prime number p such that V, Φ, and α are defined over $\mathbb{Q}_p$, and Φ has good reduction modulo p. Then, using Lemmas 6.2.1.1 and 6.2.2.1, we find a positive integer M, and for each $\ell = 0, \ldots, M-1$ and for each H in the vanishing ideal of V, we construct a p-adic power series $G_{H,\ell}(z_0, z_1, \ldots, z_m)$ such that for n sufficiently large, we have

(11.5.0.8) $\Phi^{\ell+nM}(\alpha) \in V$ if and only if $G_{H,\ell}(n, p^n, p^{2^n}, \ldots, p^{m^n}) = 0$ for all H.

We then show that either $G_{H,\ell}$ is identically zero for all H (which implies conclusion (i) of Theorem 11.5.0.2), or the integers n with $\Phi^{\ell+nM}(\alpha) \in V(\mathbb{C})$ grow as in conclusion (ii).

For each prime number p, we also construct an example (see Proposition 11.10.0.1 and also, [**BGKT10**, Proposition 7.1]) of a power series $f \in \mathbb{Z}_p[[z]]$ such that for an infinite increasing sequence $\{n_k\}_{k\geq 1} \subseteq \mathbb{N}$ we have

$$f(p^{n_k}) = n_k, \text{ and moreover } n_{k+1} < n_k + p^{2n_k}$$

for each $k \geq 1$. This example shows that Theorem 11.5.0.2 cannot be improved to finding a proof of the Dynamical Mordell-Lang Conjecture merely by sharpening our p-adic methods; some new technique would be required for a full proof of Conjecture 1.5.0.1 even in the case of endomorphisms of $(\mathbb{P}^1)^N$. Our example shows that the set S_V in Theorem 11.4.2.1 may be very sparse (but infinite) in the context of rigid analytic spaces.

If V is a curve defined over a number field, then we can prove the following more precise result.

THEOREM 11.5.0.9 ([**BGKT10**]). *Let* P, Φ, *and* V *be as in Theorem* 11.5.0.2. *Assume further that* V *is an irreducible curve that is not periodic, and that both* V *and* P *are defined over a number field* K. *Then for any* $\epsilon > 0$, *there are infinitely many primes* p *and associated constants*

$$C = C(p) > p - \epsilon \ and \ N = N(p) = O(p^{2[K:\mathbb{Q}]})$$

with the following property: For any integers $n > m \geq 0$ *and* $\ell \in \{1, 2, \ldots, N\}$, *if* m *is sufficiently large and if*

$$\Phi^{\ell+mN}(P) \ and \ \Phi^{\ell+nN}(P) \in V,$$

then $n - m > C^m$.

Theorem 11.5.0.9 was obtained by the last two authors together with Benedetto and Kurlberg while writing [**BGKT10**], but it was ultimately not included in the final version of [**BGKT10**].

We remark that the Dynamical Mordell-Lang Conjecture is still open in the case considered by Theorem 11.5.0.9, i.e.:

- the ambient variety X is $(\mathbb{P}^1)^N$,
- the subvariety $V \subseteq X$ is a curve; and
- the endomorphism Φ is given by coordinatewise action of one-variable rational maps f_i.

Xie's theorem [**Xieb**] (see Theorem 5.10.0.6) covers only the case each f_i is a polynomial map.

We proceed as follows: in Sections 11.6 and 11.7 we construct the p-adic analytic functions $G_{H,\ell}$ (see (11.5.0.8)); we conclude the proof of Theorem 11.5.0.2 in Section 11.8, and then in Section 11.9 we prove Theorem 11.5.0.9. Finally, in Section 11.10 we prove Proposition 11.10.0.1 which yields the limitation in our current p-adic approach for extending Theorem 11.5.0.2 to a proof of Conjecture 1.5.0.1 for endomorphisms of $(\mathbb{P}^1)^N$.

11.6. Reductions in the proof of Theorem 11.5.0.2

We continue with the notation from Theorem 11.5.0.2. First note that we lose no generality if we replace α by an iterate of it under Φ since the orbit of $\Phi^\ell(\alpha)$ under Φ differs by only finitely many points from the orbit of α under Φ (see also Proposition 3.1.2.4).

Secondly, it suffices to prove Theorem 11.5.0.2 if we replace Φ by Φ^k for some suitable $k \in \mathbb{N}$ and replace α by $\Phi^j(\alpha)$ for $j = 0, \ldots, k - 1$. Indeed, for each $j = 0, \ldots, k - 1$, we let

$$(M_j, C_j) \in \mathbb{N} \times \mathbb{R}_+$$

be the corresponding constants appearing in the conclusion of Theorem 11.5.0.2 for each pair (endomorphism and starting point)

$$(\Phi^k, \Phi^j(\alpha))$$

and then we let

$$M := k \cdot \mathrm{lcm}\,[M_0, \ldots, M_{k-1}] \ \text{and} \ C := \min_{0 \leq j \leq k-1} C_j.$$

Then Theorem 11.5.0.2 holds for the pair (Φ, α) with the constants (M, C).

Thirdly, we note that we may assume that each α_i is not preperiodic for f_i. Indeed, otherwise if α_N is preperiodic for f_N, say, then we may reduce N to $N-1$ by replacing V with the subvariety $\bigcup_i W_i$ of $(\mathbb{P}^1)^{N-1}$, where for each $i \in \mathbb{N}_0$,

$$W_i = (\pi \mid_V)^{-1} (f_N^i(\alpha_N)),$$

where $\pi \mid_V$ is the restriction to V of the projection map

$$\pi : (\mathbb{P}^1)^N \longrightarrow (\mathbb{P}^1)^{N-1}$$

on the first $N-1$ coordinates. Note that $\mathcal{O}_{f_N}(\alpha_N)$ is finite since α_N is f_N-preperiodic and thus $\bigcup_i W_i$ is a union of finitely many subvarieties.

We may assume V is a hypersurface in $(\mathbb{P}^1)^N$ since any subvariety is an intersection of hypersurfaces, and once we know that Theorem 11.5.0.2 holds for the subvarieties V_1 and V_2 with corresponding constants (M_1, C_1) and (M_2, C_2), then Theorem 11.5.0.2 holds for the subvariety $V_1 \cap V_2$ with the constants (M, C), where

$$M = \operatorname{lcm}[M_1, M_2] \text{ and } C = \min\{C_1, C_2\}.$$

So, we assume from now on that V is the Zariski closure in $(\mathbb{P}^1)^N$ of a hypersurface in $\mathbb{A}^N$ given by the zero set of a polynomial $H \in \mathbb{C}[x_1, \ldots, x_N]$.

There exists a finitely generated $\mathbb{Z}$-algebra R such that each f_i is defined over R, and also $\alpha \in (\mathbb{P}^1)^N (R)$. We choose a prime p such that there exists an embedding into $\mathbb{Z}_p$ of R such that each f_i has good reduction modulo p (the argument is similar to the one presented in Subsection 4.4.1; see Propositions 4.4.1.3 and 4.4.1.4). At the expense of enlarging the ring R we may assume R contains the $(m-1)$-st root of the coefficient c_m of each leading term of degree m appearing in an expansion at each one of the finitely many super-attracting points for each f_j (see the super-attracting case in Lemma 6.2.1.1 (2) with its corresponding notation). For more details, we refer the interested reader to Step (i) from [**BGKT10**, Section 4].

Each residue class of α_j modulo p is preperiodic under the induced action of f_j, i.e., there exist nonnegative integers ℓ_j and positive integers k_j such that the residue class of $f_j^{\ell_j}(\alpha_j)$ is fixed by $f_j^{k_j}$ modulo p. So, at the expense of replacing α by $\Phi^\ell(\alpha)$, where

$$\ell := \max_j \ell_j,$$

and replacing Φ by Φ^k, where

$$k = \operatorname{lcm}[k_1, \ldots, k_N],$$

we may assume that the residue class of α is fixed by the action of Φ modulo p. In Section 11.7 we use the three possible p-adic parametrizations for the orbit of α_j under f_j (see Lemmas 6.2.1.1 and 6.2.2.1) to construct the aforementioned p-adic analytic function $G_{H,\ell}(z_0, z_1, z_2, \ldots, z_m)$ for which

(11.6.0.10) $\Phi^{\ell+nM}(\alpha) \in V$ if and only if $G_{H,\ell}(n, p^n, p^{2^n}, \ldots, p^{m^n}) = 0.$

11.7. Construction of a suitable p-adic analytic function

The construction of the desired p-adic analytic function $G_{H,\ell}(z_0, z_1, z_2, \ldots, z_m)$ satisfying (11.6.0.10) follows along the steps (ii)—(vii) outlined in the proof of [**BGKT10**, Theorem 1.4, Section 4].

We continue with the reductions from Section 11.6; in particular, the residue class of each α_j is fixed by f_j modulo p. If

$$|f_j'(\alpha_j)|_p = 1,$$

then Corollary 6.2.2.2 yields the existence of a positive integer κ_j and of p-adic analytic functions

$$u_{j,\ell} \in \mathbb{Q}_p[[z_0]] \text{ for } \ell = 0, \dots, \kappa_j - 1$$

such that for each $n \in \mathbb{N}_0$ we have

$$f_j^{n\kappa_j + \ell}(\alpha_j) = u_{j,\ell}(n).$$

Note that $u_{j,\ell}$ is slightly changed by a linear factor from the corresponding u_j from Corollary 6.2.2.2.

If

$$0 < |f_j'(\alpha_j)|_p < 1,$$

then Corollary 6.2.1.2 yields the existence of a p-adic analytic function u_j such that for all $n \in \mathbb{N}_0$ we have

$$f_j^n(\alpha_j) = u_j\left(\lambda_j^n\right), \text{ where } \lambda_j := f_j'(\alpha_j).$$

On the other hand, $\lambda_j = p^{e_j} \cdot \epsilon_j$, where ϵ_j is a unit in $\mathbb{Z}_p$ and $e_j \in \mathbb{N}$. We let $\kappa_j \in \{1, \dots, p-1\}$ such that

$$|\epsilon_j^{\kappa_j} - 1|_p \leq \frac{1}{p}.$$

Hence the function

$$z \mapsto \left(\epsilon_j^{\kappa_j}\right)^z$$

is p-adic analytic (see Lemma 2.3.4.2) and therefore there exist p-adic analytic functions

$$u_{j,\ell} \in \mathbb{Q}_p[[z_0, z_1]] \text{ for each } \ell = 0, \dots, \kappa_j - 1$$

such that

$$f_j^{n\kappa_j + \ell} = u_{j,\ell}(n, p^n).$$

If

$$f_j'(\alpha_j) = 0,$$

then Corollary 6.2.1.2 yields the existence of an integer $m_j \geq 2$, and of a number $\gamma_j \in p\mathbb{Z}_p$ (defined as the corresponding number c from the conclusion of Corollary 6.2.1.2 (ii)) such that for all $n \in \mathbb{N}_0$ we have

$$f_j^n(\alpha_j) = u_j\left(\gamma_j^{m_j^n}\right).$$

Again we write $\gamma_j = p^{e_j} \cdot \epsilon_j$, where ϵ_j is a p-adic unit. There exist then $\kappa_{j,0} \in \mathbb{N}_0$ and $\kappa_j \in \mathbb{N}$ such that

$$\left|\epsilon_j^{m_j^{\kappa_{j,0}}} - \epsilon_j^{m_j^{\kappa_{j,0}+\kappa_j}}\right|_p \leq \frac{1}{p}.$$

If $p \nmid m_j$, then for each $\ell = 0, \dots, \kappa_j - 1$ there exists a p-adic analytic function

$$u_{j,\ell} \in \mathbb{Z}_p[[z_0, z_{m_j}]]$$

such that for each integer $n \geq \kappa_{j,0}/\kappa_j$ we have that

$$f_j^{n\kappa_j + \ell}(\alpha_j) = u_{j,\ell}\left(n, p^{m_j^n}\right).$$

Now, if $p \mid m_j$, then one still finds similar power series $u_{j,\ell}$ only that they will involve three variables z_0, z_1, z_{m_j} and thus

$$f_j^{n\kappa_j + \ell}(\alpha_j) = u_{j,\ell}\left(n, p^n, p^{m_j^n}\right).$$

For more details on this construction, see [**BGKT10**, Step (iv), Theorem 1.4].

We let

$$\kappa := \prod_j \kappa_j,$$

with κ_j defined for each coordinate $j = 1, \ldots, N$. Also we let

$$m := \max_j m_j$$

with each m_j defined as previously in this section for each α_j which lies in a super-attracting cycle for the action of f_j. Hence for each $\ell = 0, \ldots, \kappa - 1$ there exists a p-adic analytic function

$$G_{H,\ell} \in \mathbb{Q}_p[[z_0, z_1, z_2, \ldots, z_m]]$$

such that

(11.7.0.1) $\Phi^{\ell + n\kappa}(\alpha) \in V$ if and only if $G_{H,\ell}(n, p^n, p^{2^n}, \ldots, p^{m^n}) = 0.$

If $G_{H,\ell} = 0$ identically, then $\Phi^{n\kappa + \ell} \in V$ for all $n \geq 0$ thus showing that alternative (i) holds in Theorem 11.5.0.2. So, from now on, assume that

$$G_{H,\ell} \text{ is not identically zero.}$$

In Section 11.8 we show that the set of $n \in \mathbb{N}$ such that the p-adic analytic function $G_{H,\ell}$ vanishes at $\left(n, p^n, p^{2^n}, \ldots, p^{m^n}\right)$ is very sparse (as prescribed in Theorem 11.5.0.2 (ii)). First we set up additional notation for $G_{H,\ell}$. We may write

$$G_{H,\ell}\left(z_0, p^n, p^{2^n}, \ldots, p^{m^n}\right) = \sum_{w \in \mathbb{N}^m} g_w(z_0) p^{f_w(n)}$$

where for any m-tuple $w = (a, b_2, \ldots, b_m) \in \mathbb{N}^m$ and $n \geq 0$, define $f_w : \mathbb{N} \to \mathbb{N}$ by

(11.7.0.2) $$f_w(n) = an + \sum_{j=2}^m b_j j^n,$$

while $g_w \in \mathbb{Q}_p[[z]]$ is p-adic analytic. For each $w := (w_1, \ldots, w_m) \in \mathbb{N}^m$ we let $|w| := w_1 + \cdots w_m$. As shown in [**BGKT10**, Step (vii), Section 4] there exists a positive real number B such that all coefficients of g_w have absolute value at most $p^{B|w|}$, for every $w \in \mathbb{N}^m$.

We order $\mathbb{N}_0^m$ using a lexicographic ordering reading right-to-left. That is,

$$(b_1, \ldots, b_m) \prec (b_1', \ldots, b_m')$$

if either $b_m < b_m'$, or $b_m = b_m'$ but $b_{m-1} < b_{m-1}'$, or $b_m = b_m'$ and $b_{m-1} = b_{m-1}'$ but $b_{m-2} < b_{m-2}'$, etc. Note that this order $\prec$ gives a well-ordering of $\mathbb{N}_0^m$. Then we may write

$$G_{H,\ell} \in \mathbb{Q}_p[[z_0, z_1, \ldots, z_m]]$$

uniquely as

(11.7.0.3) $$G_{H,\ell}(z_0, z_1, \ldots, z_m) = \sum_{w \in \mathbb{N}^m} g_w(z_0) z^w,$$

where $g_w \in \mathbb{Q}_p[[z_0]]$, and for $w = (a, b_2, \ldots, b_m) \in \mathbb{N}_0^m$, z^w denotes

$$z^w = z_1^a z_2^{b_2} z_3^{b_3} \cdots z_m^{b_m}.$$

Assume $G_{H,\ell}$ is not identically equal to 0. Then there exists $v \in \mathbb{N}^m$ minimal with respect to $\prec$ such that g_v is a nonzero p-adic analytic function. By Lemma 2.3.6.1 there exist (at most) finitely many zeros of g_v in $\overline{D}(0,1)$ (where we see this closed disk inside $\mathbb{Z}_p$; so, $\overline{D}(0,1) = \mathbb{Z}_p$). So, there exists $s \in (0,1]$ such that for all $x \in \overline{D}(0,1)$, there exists at most one zero of g_v in $D(x,s)$. We let $k \in \mathbb{N}$ such that $p^{-k} < s$; then we can cover $\overline{D}(0,1) = \mathbb{Z}_p$ by the finitely many disks $D(i,s)$ with $i = 0, \ldots, p^k - 1$.

11.8. Conclusion of the proof of Theorem 11.5.0.2

The following technical result (Lemma 11.8.0.1; see also [**BGKT10**, Lemma 3.1]) finishes the proof of Theorem 11.5.0.2; note that the previously constructed function $G_{H,\ell}$ (see Section 11.7) satisfies the technical hypotheses of the following result.

LEMMA 11.8.0.1 ([**BGKT10**]). *Let*

$$G(z_0, z_1, z_2, \ldots, z_m) \in \mathbb{Q}_p[[z_0, z_1, z_2, \ldots, z_m]]$$

be a non-trivial power series in $m + 1 \geq 1$ variables. Write

$$G = \sum_w g_w(z_0) z^w$$

as in Equation (11.7.0.3), and let $v \in \mathbb{N}^m$ be the minimal index with respect to $\prec$ such that $g_v \neq 0$. Assume that g_v converges on $\overline{D}(0,1)$, and let s be a positive real number such that for all $x \in \mathbb{Z}_p$, g_v does not vanish at more than one point of the disk $\overline{D}(x,s)$. Assume also that there exists $B > 0$ such that for each $w \succ v$, all coefficients of g_w have absolute value at most $p^{B|w|}$.

Then there exists $C > 1$ with the following property: If $x \in \overline{D}(0,1)$, and if $\{n_i\}_{i \geq 1}$ is a strictly increasing sequence of positive integers such that for each $i \geq 1$,

(a) $|n_i - x|_p \leq s$, *and*

(b) $G\left(n_i, p^{n_i}, p^{2^{n_i}}, p^{3^{n_i}}, \ldots, p^{m^{n_i}}\right) = 0$,

then $n_{i+1} - n_i > C^{n_i}$ for all sufficiently large i.

We observe that the rate of growth from Lemma 11.8.0.1 of the p-adic absolute value of the coefficients of each g_w appearing in $G_{H,\ell}$ follows from the construction of $G_{H,\ell}$; see Section 11.7.

Before proceeding to the proof of Lemma 11.8.0.1, we show how to deduce the conclusion of Theorem 11.5.0.2 from Lemma 11.8.0.1 (our argument follows [**BGKT10**, Step (viii), Section 4]).

PROOF OF THEOREM 11.5.0.2. We apply Lemma 11.8.0.1 (with the radius s established in the previous Section) to $G_{H,\ell}$, and let $C_0 > 1$ be the constant from the conclusion of Lemma 11.8.0.1 and let

$$C := C_0^{p^k - 1} > 1,$$

where k is related to s as in Section 11.7, i.e., $p^{-k} < s$. Also, we let

$$M := p^k \cdot \kappa;$$

we recall from (11.7.0.1) that $\Phi^{n\kappa+\ell}(\alpha) \in V$ if and only if $G_{H,\ell}\left(n, p^n, p^{2^n}, \ldots, p^{m^n}\right) = 0$.

Unless conclusion (ii) of Theorem 11.5.0.2 holds with the values C and M defined in the above paragraph, there is $\ell \in \{0, \ldots, M-1\}$ and there are infinitely many pairs (n, n') of positive integers such that:

(a) $\Phi^{\ell+nM}(\alpha), \Phi^{\ell+n'M}(\alpha) \in V$, and
(b) $0 < n' - n \le C^n$.

For any fixed $n \ge 1$, there are only finitely many choices of n' for which condition (b) holds; thus, there are pairs (n, n') with n arbitrarily large satisfying these two conditions. Using the pigeonhole principle, we obtain that there are

$$\ell_1 \in \{0, \ldots, \kappa-1\} \text{ and } j \in \{0, \ldots, p^k - 1\},$$

and there are infinitely many pairs (n_1, n_1') such that

(1) $\Phi^{\ell_1+n_1\kappa}(\alpha), \Phi^{\ell_1+n_1'\kappa}(\alpha) \subset V$,
(2) $n_1 \equiv n_1' \equiv j \pmod{p^k}$, and
(3) $0 < \frac{n_1'-n_1}{p^k} \le C^{\frac{n_1-j}{p^k}}$.

We navigate between conditions (a)—(b) above and conditions (1)—(3) as follows:

- $\ell = \ell_1 + j\kappa$;
- $n_1 = np^k + j$ and $n_1' = n'p^k + j$.

Recalling that

$$C = C_0^{p^k-1} > 1 \text{ and } j \ge 0,$$

condition (3) becomes

(3') $0 < n_1' - n_1 \le p^k C_0^{(n_1-j)(1-p^{-k})} \le C_0^{n_1}$,

for n_1 sufficiently large (more precisely, for $n_1 \ge \frac{kp^k \log p}{\log C_0}$). However, conditions (1), (2) and (3'), along with (11.7.0.1) contradict the conclusion of Lemma 11.8.0.1. Hence Theorem 11.5.0.2 must hold. $\square$

We conclude this section by proving the technical Lemma used in the proof of Theorem 11.5.0.2; we reproduce the proof of [**BGKT10**, Lemma 3.1].

PROOF OF LEMMA 11.8.0.1. If $g_w = 0$ for all $w \ne v$, then $G = g_v(z_0)z^v$. By hypothesis (b), then, the one-variable nonzero power series $g_v(z_0)$ vanishes at all points of the sequence $\{n_i\}_{i\ge1}$, a contradiction; hence, no such sequence exists. In particular, if $m = 0$, then G is a non-trivial power series in the one variable z_0, and therefore G vanishes at only finitely many points n_i. Thus, we may assume that g_w is nonzero for some $w \succ v$.

For any $w, w' \in \mathbb{N}^m$, note that $w \prec w'$ if and only if $f_w(n)$ grows more slowly than $f_{w'}(n)$ as $n \to \infty$.

CLAIM 11.8.0.2. *For any $A > 0$, there is an integer $M = M(v, A) \ge 0$ such that for each $w \succ v$ and $n \ge M$,*

$$f_w(n) - f_v(n) \ge n + A(|w| - |v| - 1).$$

PROOF OF CLAIM 11.8.0.2. Write $v = (a, b_2, \ldots, b_m)$, and choose $M \ge A$ large enough so that

$$j^x \ge (a+1)x + \sum_{k=2}^{j-1} b_k k^x$$

for all $x \geq M$, and for all $j = 2, \ldots, m$. Write

$$w = (a', b'_2, \ldots, b'_m).$$

Then

$$f_w(n) - f_v(n) = (a' - a)n + (b'_2 - b_2)2^n + \cdots + (b'_m - b_m)m^n.$$

We consider two cases:

Case 1. If $b'_k = b_k$ for each $k = 2, \ldots, m$, then $a' > a$, and therefore

$$f_w(n) - f_v(n) - n = (a' - a - 1)n \geq (a' - a - 1)A = A(|w| - |v| - 1)$$

for $n \geq M$, because $M \geq A$.

Case 2. Otherwise, there exists $k = 2, \ldots, m$ such that $b'_k > b_k$. Let j be the largest such k, so that $b'_k = b_k$ for $k > j$. Then

$$f_w(n) - f_v(n) - A|w| + A|v| = (a' - a)(n - A)$$

$$+ \sum_{k=2}^{j-1} (b'_k - b_k)(k^n - A) + (b'_j - b_j)(j^n - A)$$

$$\geq -an - \sum_{k=2}^{j-1} b_k k^n + j^n - A$$

$$\geq n - A,$$

where the first inequality is because $n \geq A$ and $b'_j - b_j \geq 1$, and the second is because $n \geq M$. The proof of Claim 11.8.0.2 is now complete. □

By hypothesis (b), for any i such that $n_i \geq M(v, B)$, we have

$$(11.8.0.3) \qquad |g_v(n_i)|_p = \left| \sum_{w \succ v} g_w(n_i) p^{f_w(n_i) - f_v(n_i)} \right|_p \leq p^{-n_i + B|v| + B},$$

where the inequality is by Claim 11.8.0.2, the fact that $|n_i|_p \leq 1$, and the fact that the absolute values of all coefficients of g_w are at most $p^{B|w|}$. Let $y \in \overline{D}(x, s) \cap \mathbb{Z}_p$ be a limit point of the sequence $\{n_i\}_{i \geq 1}$. Then by inequality (11.8.0.3), we have $g_v(y) = 0$. Thus, g_v can be written as

$$g_v(z) = \sum_{i \geq \delta} c_i (z - y)^i,$$

where $\delta \geq 1$ and $c_\delta \neq 0$. In fact, we must have

$$|c_\delta|_p s^\delta > |c_i|_p s^i \text{ for all } i > \delta;$$

otherwise, inspection of the Newton polygon shows that g_v would have a zero besides y in $\overline{D}(x, s)$. Thus, for i sufficiently large (i.e., such that $n_i \geq M(v, B)$ with the notation as in Claim 11.8.0.2), we have

$$|c_\delta (n_i - y)^\delta|_p = |g_v(n_i)|_p \leq p^{-n_i + O(1)},$$

by hypothesis (a) and inequality (11.8.0.3), and hence

$$(11.8.0.4) \qquad |n_i - y|_p \leq |c_\delta|_p^{-\frac{1}{\delta}} p^{-\frac{n_i}{\delta} + O(1)}.$$

It follows that

$$(11.8.0.5) \qquad n_{i+1} \equiv n_i \mod p^{\lfloor \frac{n_i}{\delta} - O(1) \rfloor}.$$

Hence, if we choose C such that $1 < C < p^{\frac{1}{\delta}}$, we have

$$n_{i+1} - n_i > C^{n_i}$$

for i sufficiently large, as desired. Finally, note that C depends only on G since there exist at most finitely many zeros of g_v in $\mathbb{Z}_p$, and thus finitely many possible limit points for the sequence $\{n_i\}_i$ in $\mathbb{Z}_p$ (and thus finitely many possibilities both for δ and for the $O(1)$-constant in (11.8.0.4) and in (11.8.0.5)). $\square$

We note that Lemma 11.8.0.1 holds also if G is defined over a finite extension K of $\mathbb{Q}_p$; the only significant change is that the constant C will depend also on the ramification index e of $K/\mathbb{Q}_p$.

11.9. Curves

In this section we prove Theorem 11.5.0.9; we thank Rob Benedetto and Par Kurlberg for giving us the permission of using the proof that was discovered while writing [**BGKT10**].

The proof of Theorem 11.5.0.9 is simpler than the proof of Theorem 11.5.0.2, but it requires an additional ingredient that is only available over number fields, namely, the existence of a suitable indifferent cycle in at least one of the variables (which one obtains over number fields by Lemma 7.1.1.1, which in turn, relies on [**Sil93**, Theorem 2.2]). Because of the p-adic example which we will present in Proposition 11.10.0.1 (which shows that for rigid analytic spaces, the set S_V in Theorem 11.4.2.1 may be infinite, but very sparse), it seems likely that a proof of Conjecture 1.5.0.1 would also have to involve extra information beyond what is used in the proof of Theorem 11.5.0.2. Thus, although Theorem 11.5.0.9 only applies to curves, it may well be that the techniques used to prove it are better adapted to a general proof of Conjecture 1.5.0.1.

To prove Theorem 11.5.0.9 we need a sharper version of Lemma 6.2.2.1, giving an upper bound on k. We first recall the following special case of [**Bel06**, Theorem 3.3] (see also Lemma 6.2.2.1).

THEOREM 11.9.0.1. *Let $p > 3$ be prime, let $K_p/\mathbb{Q}_p$ be a finite unramified extension, and let $\mathcal{O}_p$ denote the ring of integers in K_p. Let*

$$g(z) = a_0 + a_1 z + a_2 z^2 + \cdots \in \mathcal{O}_p[[z]]$$

be a power series with $|a_0|_p, |a_1 - 1|_p < 1$ and for each $i \geq 2$, $|a_i|_p \leq p^{1-i}$. Then for any $z_0 \in \mathcal{O}_p$, there is a power series $u \in \mathcal{O}_p[[z]]$ mapping $\overline{D}(0,1) \subset K_p$ into itself such that

$$u(0) = z_0, \text{ and } u(z+1) = g(u(z)).$$

REMARK 11.9.0.2. In [**Bel06**], Theorem 11.9.0.1 is only stated for $K_p = \mathbb{Q}_p$, but the proof goes through essentially unchanged for any finite unramified extension of $\mathbb{Q}_p$. This is another instance of the p-adic arc lemma; see Chapter 4, especially Theorem 4.4.2.1.

We can now give an explicit bound on k. However, we give up any claims on the size of the image of u. In fact, if z_0 is a periodic point, the map u is constant. On the other hand, if z_0 is not periodic, then the derivative of u is non-vanishing at zero, and hence u is a local bijection.

PROPOSITION 11.9.0.3. *Let $p > 3$ be prime, let K_p and $\mathcal{O}_p$ be as in Theorem 11.9.0.1, let $h(z) \in \mathcal{O}_p[[z]]$ be a power series, and let $z_0 \in \mathcal{O}_p$. Suppose that*

$$|h(z_0) - z_0|_p < 1 \text{ and } |h'(z_0)|_p = 1.$$

Then there is an integer $1 \le k \le p^{[K_p:\mathbb{Q}_p]}$ and a power series $u \in \mathcal{O}_p[[z]]$ mapping $\overline{D}(0, 1) \subset K_p$ into itself such that

$$u(0) = z_0 \text{ and } h^k(u(z)) = u(z + 1).$$

In particular, $h^{nk}(z_0) = u(n)$ for all $n \ge 0$.

PROOF. Let $q = p^{[K_p:\mathbb{Q}_p]}$ denote the cardinality of the residue field of $\mathcal{O}_p$. Conjugating by a translation we may assume that $z_0 = 0$. Let

$$g(z) := h(pz)/p = b_0 + b_1 z + b_2 z^2 + \cdots \in \mathcal{O}_p[[z]]$$

We find that $|b_0|_p \le 1, |b_1|_p = 1$, and $|b_i|_p \le p^{1-i}$ for each $i \ge 2$. By considering the iterates of the map

$$z \mapsto b_0 + b_1 z,$$

we have

$$g^k(z) \equiv z \pmod{p} \text{ for some } 1 \le k \le q.$$

Hence,

$$g^k(z) = a_0 + a_1 z + a_2 z^2 + \cdots$$

satisfies the hypotheses of Theorem 11.9.0.1, giving a power series $\tilde{u} \in \mathcal{O}_p[[z]]$ mapping $\overline{D}(0, 1)$ into itself, with $\tilde{u}(0) = 0$ and $\tilde{u}(z + 1) = g^k(\tilde{u}(z))$. It follows that $u(z) = p\tilde{u}(z)$ has the desired properties. □

Now we are ready to prove Theorem 11.5.0.9.

PROOF OF THEOREM 11.5.0.9. For simplicity we assume that $X = \mathbb{P}^1 \times \mathbb{P}^1$, and that $V \subset X$ is an irreducible curve; the argument can be easily modified to include the general case (by considering projections on two coordinates at a time; see also Section 5.10). If x_i is preperiodic under f_i for either $i = 1$ or $i = 2$, the result is trivial. If both f_1 and f_2 are of degree one, V can be shown to be be periodic, either by the Skolem-Mahler-Lech theorem, or by the main result of [**Bel06**]. Thus, possibly after permuting indices, we may assume that the degree of f_1 is greater than 1. Define

$$\pi_1 : V \to \mathbb{P}^1 \text{ by } (z_1, z_2) \to z_1.$$

By taking a periodic cycle $D = \{d_1, \ldots, d_a\}$ of f_1 of sufficiently large cardinality a, defined over some number field L, we assume that

- D is not super-attracting (i.e., no d_i is a critical point of f_1);
- all points $(\alpha_1, \alpha_2) \in \pi_1^{-1}(D) \cap V$ are smooth points on V; and
- for (z_1, z_2) near (α_1, α_2), we have

(11.9.0.4) $$z_1 - \alpha_1 = \gamma_\alpha \cdot (z_2 - \alpha_2) + O\left((z_2 - \alpha_2)^2\right),$$

for some $\gamma_\alpha \ne 0$.

Note that only finitely many points violate these conditions.

Since f_1 is not prepreiodic, by [**Sil93**, Theorem 2.2] (or Lemma 7.1.1.1), we can find infinitely many primes p such that

$$|f_1^n(x_1) - d_1|_p < 1 \text{ for some } n,$$

where $|\cdot|_p$ denotes the extension of the p-adic absolute value on $\mathbb{Q}$ to L. We may of course assume that $L/\mathbb{Q}$ is unramified at p and that

$$|\gamma_\alpha|_p = |(f_1^a)'(d_1)|_p = 1$$

for all sufficiently large p, as there are only finitely many p not fitting these conditions. In particular, the orbit of x_1 under f_1 ends up in a domain of quasiperiodicity.

If the orbit of x_2 under f_2 also has quasiperiodic behavior, then V is periodic by [**BGKT12**, Theorem 3.4] (see also Theorem 6.2.3.1). Otherwise, the orbit of x_2 ends up in an attracting or super-attracting domain. The arguments in these two cases are very similar, and we shall only give details for the attracting case. Hence, assume that $f_2^n(x_2)$ tends to an attracting cycle

$$E = \{e_1, e_2, \ldots, e_b\},$$

with multiplier λ_2 satisfying $0 < |\lambda_2|_p < 1$. Since λ_2 and E are defined over K_p, and $K_p/\mathbb{Q}_p$ is unramified, we have

$$|\lambda|_p \le 1/p.$$

Note that $b \le p^{[K_p:\mathbb{Q}_p]} + 1 \le p^{[K:\mathbb{Q}]} + 1$. Let $N = \operatorname{lcm}(a, b)$, so that

$$N \le a \cdot (p^{[K:\mathbb{Q}]} + 1) = O(p^{[K:\mathbb{Q}]}).$$

Choose representatives $\{\alpha_{ij} : 1 \le i \le a,\ 1 \le j \le b\}$ for $\mathbb{Z}/N\mathbb{Z}$ such that

$$|f_1^{\alpha_{ij}+Nn}(x_1) - d_i|_p < 1, \quad |f_2^{\alpha_{ij}+Nn}(x_2) - e_j|_p < 1$$

for n sufficiently large. At the cost of increasing a (and hence N) by a factor bounded by $p^{[K:\mathbb{Q}]}$, by Proposition 11.9.0.3 and Lemma 6.2.1.1 there exist p-adic power series A_i, B_j, such that

(11.9.0.5) $$f_1^{\alpha_{ij}+Nn}(x_1) - d_i = A_i(k), \quad f_2^{\alpha_{ij}+Nn}(x_2) - e_j = B_j(\lambda_2^k)$$

for n sufficiently large. If $n > m$ and $\phi^{mN+\alpha_{ij}}(P), \phi^{nN+\alpha_{ij}}(P) \in V$, then (11.9.0.4) and (11.9.0.5) yield that

$$|A_i(n) - A_i(m)|_p = O(|\lambda_2|_p^m),$$

since we had $|\gamma_\alpha|_p = 1$ in (11.9.0.4). Hence $n \equiv m \pmod{p^{m-O_p(1)}}$, where the $O_p(1)$ depends on the derivative of A_i. Thus, if we take $C < p$, we find that $n \ge m + C^m$ for m sufficiently large. $\qquad\square$

11.10. An analytic counterexample to a p-adic formulation of the Dynamical Mordell-Lang Conjecture

In this section we present the example from [**BGKT10**, Section 7] which shows that there are indeed nonzero p-adic analytic functions

$$G \in \mathbb{Z}_p[[z_0, z_1, z_2, \ldots, z_m]]$$

such that the set of $n \in \mathbb{N}_0$ for which

$$G\left(n, p^n, p^{2^n}, \ldots, p^{m^n}\right) = 0$$

is infinite and very sparse. More precisely, we will show that there exists a p-adic analytic function $f \in \mathbb{Z}_p[[z]]$ such that the set of $n \in \mathbb{N}_0$ such that $p^n = f(n)$ is infinite, but very sparse.

PROPOSITION 11.10.0.1 ([**BGKT10**]). *For any prime $p \geq 2$ and for any positive integer n_1, there is an increasing sequence $\{n_j\}_{j \geq 2}$ of positive integers and a power series $f(z) \in \mathbb{Z}_p[[z]]$ such that*

$$f(p^{n_j}) = n_j \quad and \quad n_j + p^{n_j} \leq n_{j+1} \leq n_j + p^{n_1 + \cdots + n_j}$$

for all $j \geq 1$. Moreover, $n_1 + \cdots + n_{j-1} \leq n_j$, and hence

$$n_j + p^{n_j} \leq n_{j+1} \leq n_j + p^{2n_j}.$$

PROOF. We will inductively construct the sequence $\{n_j : j \geq 2\}$ of positive integers and a sequence $\{f_j(z) : j \geq 1\}$ of polynomials $f_j \in \mathbb{Z}_p[z]$, with $\deg(f_j) = j - 1$. The power series f will be then

$$f = \lim_{j \to \infty} f_j.$$

Let f_1 be the constant polynomial equal to n_1. Then, for each $j \geq 1$, suppose we are already given

$$f_1, \ldots, f_j \text{ and } n_1, \ldots, n_j$$

such that

$$f_k(p^{n_i}) = n_i$$

for each i, k with $1 \leq i \leq k \leq j$. Choose n_{j+1} to be the unique integer such that

$$n_j + 1 \leq n_{j+1} \leq n_j + p^{n_1 + \cdots + n_j}$$

and

$$(11.10.0.2) \qquad |n_{j+1} - f_j(0)|_p \leq |p|_p^{n_1 + \cdots + n_j}.$$

Note that because $f_j \in \mathbb{Z}_p[z]$ and $f_j(p^{n_j}) = n_j$, we have

$$|f_j(0) - n_j|_p = |f_j(0) - f_j(p^{n_j})|_p \leq |p|_p^{n_j},$$

and therefore $|n_{j+1} - n_j|_p \leq |p|_p^{n_j}$, implying that

$$n_{j+1} \geq n_j + p^{n_j}$$

and that

$$n_{j+1} \geq n_1 + n_2 + \cdots + n_j,$$

as claimed in the proposition. Define

$$g_j(z) := (z - p^{n_1})(z - p^{n_2}) \cdots (z - p^{n_j}),$$

and set

$$c_j := \frac{n_{j+1} - f_j(p^{n_{j+1}})}{g_j(p^{n_{j+1}})} \in \mathbb{Q}_p,$$

and

$$f_{j+1}(z) := f_j(z) + c_j g_j(z) \in \mathbb{Q}_p[z].$$

We claim that $|c_j|_p \leq 1$. Indeed, we have

$$(11.10.0.3) \qquad |f_j(0) - f_j(p^{n_{j+1}})|_p \leq |p|_p^{n_{j+1}},$$

because $f_j \in \mathbb{Z}_p[[z]]$. Therefore,

$$|n_{j+1} - f_j(p^{n_{j+1}})|_p \leq \max\{|n_{j+1} - f_j(0)|_p, |f_j(0) - f_j(p^{n_{j+1}})|_p\}$$
$$\leq \max\{|p|_p^{n_1 + \cdots + n_j}, |p|_p^{n_{j+1}}\}$$
$$= |p|_p^{n_1 + \cdots + n_j} = |g_j(p^{n_{j+1}})|_p,$$

where the second inequality follows from equations (11.10.0.2) and (11.10.0.3). It follows immediately that $|c_j|_p \leq 1$, as claimed.

Clearly, $f_{j+1}(p^{n_i}) = n_i$ for all $i = 1, \ldots, j+1$. Since $c_j \in \mathbb{Z}_p$, we obtain that $f_{j+1} \in \mathbb{Z}_p[z]$, completing the induction. In fact, because (for any fixed $m \geq 0$) the size of the z^m-coefficient of g_j goes to zero as $j \to \infty$, it follows that

$$\lim_{j \to \infty} f_j$$

converges coefficient-wise to some power series $f \in \mathbb{Z}_p[[z]]$. Since every f_j also lies in $\mathbb{Z}_p[[z]]$, it follows that the convergence $f_j \to f$ is uniform on $p\mathbb{Z}_p$. Hence,

$$f(p^{n_i}) = n_i \text{ for all } i \geq 1,$$

as desired. $\qquad\qquad\qquad\qquad\qquad\qquad\qquad\qquad\qquad\qquad\qquad\qquad\qquad$ $\square$

In particular the example constructed in Proposition 11.10.0.1 shows that for the endomorphism

$$\Phi : \mathbb{A}^2 \longrightarrow \mathbb{A}^2 \text{ given by } \Phi(x, y) = (px, y + 1),$$

there exists an infinite (very sparse) set of $n \in \mathbb{N}$ such that $\Phi^n(1, 0)$ lies on the p-adic analytic curve $X = f(Y)$. Hence there exists *no* p-adic analytic version of the Dynamical Mordell-Lang Conjecture.

11.11. Approximating an orbit by a p-adic analytic function

We have seen in Chapter 4 that for an étale endomorphism of a quasiprojective variety X, one can find a suitable p-adic parametrization of each orbit; this leads to the proof of the Dynamical Mordell-Lang Conjecture for all étale endomorphisms (see Theorem 4.3.0.1). The p-adic arc lemma was used also in Chapter 7 to prove other instances of Conjecture 1.5.0.1. On the other hand, we saw in Chapter 8 that is not always reasonable to expect that the p-adic arc lemma can be employed to construct a p-adic analytic parametrization of any orbit of any endomorphism.

On the other hand, in Section 11.7 we saw that for endomorphisms of $(\mathbb{P}^1)^N$ one can always find *almost* a p-adic arc lemma; the construction of the function $G_{\ell,H}$ from Sections 11.7 and 11.8 is the appropriate substitute for a p-adic analytic parametrization of an orbit for arbitrary endomorphisms of $(\mathbb{P}^1)^N$. Our analysis from Sections 11.6 to 11.8 leads to proving that, if it were for Conjecture 1.5.0.1 to fail for a subvariety $V \subset (\mathbb{P}^1)^N$, then the set

$$S(V, \Phi, \alpha) := \{n \in \mathbb{N}_0 : \Phi^n(\alpha) \in V\}$$

is *very sparse* (see Theorem 11.5.0.2). In this section, building on our previous work from Chapter 4, we show that often one can find a p-adic analytic function which *approximates* the orbit of a point. Our result leads to showing that the gap principle from Theorem 11.5.0.2 extends beyond endomorphisms of $(\mathbb{P}^1)^N$; see Theorem 11.11.3.1 for an application for our method.

11.11.1. A p-adic approximating function. Let p be a prime number, and let K be a complete valued field with respect to a valuation satisfying

$$|p| = \frac{1}{p}.$$

Let R be the valuation ring of K. For an integer $d \geq 1$, we let

$$R[\mathbf{x}] := R[x_1, \ldots, x_d],$$

and for $f \in R[\mathbf{x}]$ we let $\|f\|$ be the supremum of the absolute values of the coefficients of f. We let $R\langle \mathbf{x} \rangle$ be the completion of $R[\mathbf{x}]$ with respect to $\| \cdot \|$; this is called the *Tate algebra* and it consists of all power series in $\mathbf{x}$ with the property that the absolute values of its coefficients tends to 0.

We let c be a positive real number. For two power series $F, G \in R\langle \mathbf{x} \rangle$, we write

$$F \equiv G \pmod{p^c}$$

if each coefficient a_α of $F - G$ satisfies $|a_\alpha| \le p^{-c}$. More generally, for two d-tuples of power series

$$\mathcal{F} := (F_1, \ldots, F_d) \text{ and } \mathcal{G} := (G_1, \ldots, G_d)$$

contained in $R\langle \mathbf{x} \rangle^d$, we write

$$\mathcal{F} \equiv \mathcal{G} \pmod{p^c} \text{ if } F_i \equiv G_i \pmod{p^c}$$

for each i. Finally, for each $n \in \mathbb{N}$ we denote by $\mathcal{F}^n$ the composition of $\mathcal{F}$ with itself n times. Since the coefficients of each F_i converge to 0, the composition $\mathcal{F}^n$ is well-defined.

THEOREM 11.11.1.1. *Let E be a d-by-d idempotent matrix. If $f \in R\langle x_1, \ldots, x_d \rangle^d$ satisfies*

$$f(\mathbf{x}) \equiv E\mathbf{x} \pmod{p^c}$$

for some $c > 1/(p-1)$, then there exists $g \in R\langle x_1, \ldots, x_d, z \rangle$ such that

$$\|g(\mathbf{x}, n) - f^n(\mathbf{x})\| \le p^{-nc}$$

for each $n \in \mathbb{Z}_{\ge 0}$.

PROOF. This proof is similar to Poonen's proof [**Poo14**]. We let Δ denote the linear operator defined by

$$(\Delta h)\mathbf{x} = h(f(\mathbf{x})) - h \circ E(\mathbf{x}).$$

Then Δ maps $R\langle \mathbf{x} \rangle^d$ into $p^c R\langle \mathbf{x} \rangle^d$. We define

$$g(\mathbf{x}, z) := \sum_{s \ge 1} \frac{(z-1)(z-2) \cdots (z-s+1)}{(s-1)!}$$

$$\times \sum_{j_0, j_s \ge 0, j_1, \ldots, j_{s-1} \ge 1} \Delta^{j_0} E \Delta^{j_1} E \cdots \Delta^{j_{s-1}} E \Delta^{j_s}(\mathbf{x}).$$

Notice that the coefficient of $\frac{(z-1)(z-2) \cdots (z-s+1)}{(s-1)!}$ is a sum of terms of the form

$$\Delta^{j_0} E \Delta^{j_1} E \cdots \Delta^{j_{s-1}} E \Delta^{j_s}(\mathbf{x})$$

and by assumption we must have at least $s - 1$ copies of Δ in each term and thus the coefficient is in $p^{(s-1)c} R\langle \mathbf{x} \rangle^d$ and so $g(\mathbf{x}, z)$ converges in $R\langle \mathbf{x}, z \rangle^d$ with respect to $\| \cdot \|$ (note also that $z \mapsto \frac{z(z-1) \cdots (z-s+1)}{s!}$ is a p-adic analytic function for $z \in \mathbb{Z}_p$). Observe that for $n \in \mathbb{Z}_{\ge 0}$ we have

$$f^n(\mathbf{x}) = (E + \Delta)^n(\mathbf{x}) = \sum_{s=0}^{n} \sum_{i_0 + \cdots + i_s = n-s} E^{i_0} \Delta E^{i_1} \Delta \cdots E^{i_{s-1}} \Delta E^{i_s}(\mathbf{x}).$$

For $s < n$, the condition $i_0 + \cdots + i_s = n - s$ implies that $i_j \geq 1$ for some j. Moreover, $E^2 = E$ and hence

$$\sum_{i_0 + \cdots + i_s = n-s} E^{i_0} \Delta E^{i_1} \Delta \cdots E^{i_{s-1}} \Delta E^{i_s}$$

$$= \sum_{\substack{t \geq 1 \\ }} \sum_{\substack{j_0, j_t \geq 0, j_1, \ldots, j_{t-1} \geq 1 \\ j_0 + \cdots + j_t = s}} \binom{n-1}{t-1} \Delta^{j_0} E \Delta^{j_1} E \cdots \Delta^{j_{t-1}} E \Delta^{j_t}(\mathbf{x}).$$

It follows that all terms appearing in $f^n(\mathbf{x}) - g(\mathbf{x}, n)$ have at least n copies of Δ appearing and hence $f^n(\mathbf{x}) - g(\mathbf{x}, n) \in p^{nc} R\langle \mathbf{x} \rangle^d$ for $n \in \mathbb{Z}_{\geq 0}$. $\square$

11.11.2. The geometric setting. Power series in a Tate algebra $R\langle \mathbf{x} \rangle$ appear naturally in a geometric setting as described in Subsection 4.4.1 (see also [**BGT10**]). So, consider a quasiprojective variety X defined over a field K of characteristic 0 endowed with an endomorphism Φ, and let $x \in X(K)$. As always, we are interested in describing the intersection $Y(K) \cap \mathcal{O}_\Phi(x)$ for some subvariety Y of X, i.e., according to the Dynamical Mordell-Lang Conjecture we would like to conclude that

$S := S_Y := \{n \in \mathbb{N}_0 \colon \Phi^n(x) \in Y(K)\}$ is a finite union of arithmetic progressions.

Assume X is a smooth variety of dimension d. Arguing as in Propositions 4.4.1.3 and 4.4.1.4 we find a suitable prime number p and a smooth $\mathbb{Z}_p$-model $\mathcal{X}$ such that:

- Φ extends to an endomorphism of $\mathcal{X}$ sharing the same properties that the original endomorphism of X may have (unramified, flat, etc.); and
- x extends to a section $\alpha \in \mathcal{X}(\mathbb{Z}_p)$.

Using Propositions 3.1.2.4 and 3.1.2.5, we may replace x by $\Phi^m(x)$ (for some $m \in \mathbb{N}_0$) and replace Φ by Φ^ℓ, and therefore assume that the reduction $\bar{x}$ of x modulo p (in other words, $\bar{x}$ is the intersection of the section α with the special fiber of $\mathcal{X}$) is fixed by the induced action of the reduction of Φ modulo p. We let $U_{\bar{x}}$ be the p-adic neighborhood of $\mathcal{X}(\mathbb{Z}_p)$ containing all points β with the same reduction $\bar{x}$ modulo p. Arguing exactly as in the proof of [**BGT10**, Proposition 2.2] (see also Subsection 4.4.1) we then obtain that there is a p-adic analytic isomorphism

$$\iota : U_{\bar{x}} \longrightarrow \mathbb{Z}_p^d,$$

such that there are power series $F_1, \ldots, F_d \in \mathbb{Z}_p[[z_1, \ldots, z_d]]$ with the properties that

(i) each F_i converges on $\mathbb{Z}_p^d$;

(ii) for all $(\beta_1, \ldots, \beta_d) \in \mathbb{Z}_p^d$, we have

(11.11.2.1) $\iota(\Phi(\iota^{-1}(\beta_1, \ldots, \beta_d))) = (F_1(\beta_1, \ldots, \beta_d), \ldots, F_d(\beta_1, \ldots, \beta_d))$; and

(iii) each F_i is congruent to a linear polynomial modulo p (in other words, all the coefficients of terms of degree greater than one are in the maximal ideal of $\mathbb{Z}_p$). Moreover, for each i, we have

$$F_i(z_1, \ldots, z_d) = \frac{1}{p} \cdot H_i(pz_1, \ldots, pz_d),$$

for some $H_i \in \mathbb{Z}_p[[z_1, \ldots, z_d]]$.

Denoting $\vec{\beta} = (\beta_1, \ldots, \beta_d)$ and $\iota\Phi\iota^{-1}$ as $\mathcal{F}_\Phi$, we thus have

$$(11.11.2.2) \qquad \mathcal{F}_\Phi(\vec{\beta}) \equiv C_\Phi + L_\Phi(\vec{\beta}) \pmod{p}$$

for some $C_\Phi \in \mathbb{Z}_p^d$ and some $d \times d$ matrix L_Φ with coefficients in $\mathbb{Z}_p$. Since the reduction of L_Φ modulo p is a matrix with entries in $\mathbb{F}_p$, then a power of it is idempotent. So, at the expense of replacing again Φ by an iterate (see Proposition 3.1.2.5) we assume

$$L_\Phi^2 \equiv L_\Phi \pmod{p}.$$

From now on assume

$$p \geq 3.$$

An idempotent matrix $M \in \mathrm{M}_d(\mathbb{Z}/p\mathbb{Z})$ can be diagonalized using a matrix in $\mathrm{SL}_d(\mathbb{Z}/p\mathbb{Z})$ since all of its eigenvalues are in $\{0, 1\}$. Since diagonal idempotent matrices in $\mathrm{M}_d(\mathbb{Z}/p\mathbb{Z})$ clearly lift to diagonal idempotent matrices in $\mathrm{M}_d(\mathbb{Z})$ and since the natural map

$$\mathrm{SL}_d(\mathbb{Z}) \to \mathrm{SL}_d(\mathbb{Z}/p\mathbb{Z})$$

is surjective, we see that there is an idempotent matrix $E \in \mathrm{M}_d(\mathbb{Z})$ such that

$$L \equiv E \pmod{p}.$$

The key thing here is that $E^2 = E$—not just when we look mod p.

Now, if $C_\Phi \equiv 0 \pmod{p}$, then we are in position to apply Theorem 11.11.1.1. We sketch below possible applications of Theorem 11.11.1.1 to the Dynamical Mordell-Lang Conjecture.

11.11.3. Applications. We present an application of Theorem 11.11.1.1 in a very explicit case which cannot be proven using Theorem 4.3.0.1. Also, we note that the case we consider is not covered by other known instances of the Dynamical Mordell-Lang Conjecture, such as the ones contained in [**Xieb**].

THEOREM 11.11.3.1. *Let $p \geq 3$ be a prime number, let $N \geq 2$ be an integer, let*

$$\Phi : \mathbb{A}^2 \longrightarrow \mathbb{A}^2$$

be an endomorphism defined over $\mathbb{Q}_p$, let $V \subseteq \mathbb{A}^N$ be a subvariety defined over $\mathbb{Q}_p$ and let $\alpha \in \mathbb{A}^N(\mathbb{Q}_p)$ be a point. Assume the following conditions are met:

(1) *$\alpha := (\alpha_1, \ldots, \alpha_N)$ and each $\alpha_i \in p\mathbb{Z}_p$;*
(2) *V contains no positive dimensional periodic subvariety;*
(3) *the endomorphism Φ is given by*

$$(x_1, \ldots, x_N) \mapsto (f_1(x_1, \ldots, x_N), \ldots, f_N(x_1, \ldots, x_N))$$

for some polynomials $f_i \in \mathbb{Z}_p[x_1, \ldots, x_N]$ and furthermore for each $i = 1, \ldots, N$ we have

$$f_i(x_1, \ldots, x_N) \equiv \sum_{j=1}^{N} a_{i,j} x_j \pmod{p}$$

for some $a_{i,j} \in \mathbb{Z}_p$;
(4) *for any $n \in \mathbb{N}$, the orbit $\mathcal{O}_{\Phi^n}(\alpha)$ does not converge p-adically to a periodic point of Φ lying on V.*

Then the set $S_V := \{n \in \mathbb{N}_0 \colon \Phi^n(\alpha) \in V(\mathbb{Q}_p)\}$ has the property that

$$\#\{i \le n \colon i \in S_V\} = o\left(\log^{(m)}(n)\right),$$

for any $m \in \mathbb{N}$ (where $\log^{(m)}$ is the m-th iterated logarithmic function).

We already know (see Theorem 11.1.0.7) that S_V has Banach density 0. However, Theorem 11.11.3.1 yields a much stronger statement than the one from Theorem 11.1.0.7 or Theorem 11.1.0.8; note that even in the case V is a curve, Theorem 11.1.0.8 yields only that

$$\#\{i \le n \colon i \in S_V\} = O\left(\frac{n}{\log(n)}\right).$$

Before proceeding to the proof of Theorem 11.11.3.1 we make several remarks regarding the hypotheses of our result. First of all, as previously discussed (see Section 4.4.2), we can always assume there exists a suitable prime number p such that V, Φ and α are defined over $\mathbb{Z}_p$ (as long as V and Φ are defined over a field of characteristic 0). We added the extra condition that

$$\alpha_i \equiv 0 \pmod{p} \text{ for each } i = 1, \ldots, N$$

for the coordinates of α just so we simplify the technical conditions needed later, i.e. the fact that Φ fixes α modulo p. This condition is not necessary, and can easily be achieved at the expense of replacing both α and Φ by iterates.

In condition (2), we assumed that V contains no positive dimensional periodic subvariety. As proven in Subsection 3.1.3, Conjecture 3.1.3.2 is equivalent to Conjecture 1.5.0.1, and thus under condition (2), one expects (assuming the Dynamical Mordell-Lang Conjecture) that the set S_V is finite. We are proving that, if it is infinite, then it is very sparse.

In condition (3), we needed to assume there are no constant terms for the polynomial maps f_i modulo p in order to apply Theorem 11.11.1.1. Even though it is a technical restrictive condition, often it holds, and more importantly, it is immediate to check whether it is satisfied in each explicit example of an endomorphism Φ of $\mathbb{A}^N$.

Finally, in condition (4), we ask that (even passing to an iterate of Φ) the orbit $\mathcal{O}_\Phi(\alpha)$ does not converge p-adically to a periodic point contained on V. So, condition (4) is automatically satisfied if V contains no periodic point (which may happen quite often if V is a curve). Essentially, condition (4) is equivalent with asking that the point α does not lie in an attracting periodic cycle modulo p. In Chapter 8 we showed heuristics supporting the idea that for general endomorphisms of $\mathbb{A}^m$ (or of $\mathbb{P}^m$) it may not be possible to avoid the ramification locus of Φ modulo any prime p, and thus the orbit of α might always land in an attracting periodic cycle modulo each prime number. Hence, condition (4) is in reality the most restrictive condition out of the four conditions imposed in the hypotheses of Theorem 11.11.3.1.

PROOF OF THEOREM 11.11.3.1. Clearly, it suffices to assume V is irreducible. Also, we may assume the set S_V is infinite; otherwise the conclusion of Theorem 11.11.3.1 is trivial. Now, if S_V is infinite, then condition (4) from the hypotheses can be strengthen by asking that for any $k \in \mathbb{N}$, the orbit $\mathcal{O}_{\Phi^k}(\alpha)$ does not converge p-adically to a point in $\mathbb{A}^N$. Indeed, if

$$\mathcal{O}_{\Phi^k}(\alpha) \text{ converges to } \beta \in \mathbb{A}^N(\mathbb{Q}_p),$$

then for each $\ell = 0, \ldots, k - 1$ we have that

(11.11.3.2) $\mathcal{O}_{\Phi^k}(\Phi^\ell(\alpha))$ converges to $\Phi^\ell(\beta) \in \mathbb{A}^N(\mathbb{Q}_p).$

Since S_V is infinite, then by the pigeonhole principle, there exists some $\ell \in \{0, \ldots, k - 1\}$ such that

(11.11.3.3) $\Phi^{nk+\ell}(\alpha) \in V(\mathbb{Q}_p)$ for infinitely many $n \in \mathbb{N}_0.$

Combining (11.11.3.2) and (11.11.3.3), we get that

(11.11.3.4) $\Phi^\ell(\beta) \in V(\mathbb{Q}_p).$

But then (11.11.3.4) and (11.11.3.2) contradict condition (4) in Theorem 11.11.3.1.

We let $A := (a_{i,j})$ be the corresponding N-by-N matrix. At the expense of replacing Φ by an iterate Φ^k we may assume the reduction of A modulo p is idempotent. (Note that we are allowed to replace Φ by a power of it and conclusion of Theorem 11.11.3.1 is unchanged.)

If A modulo p is invertible, then we can apply Theorem 4.3.0.1 and conclude that S_V is finite.

On the other hand, because of condition (4) in Theorem 11.11.3.1, we know that A modulo p cannot be nilpotent; otherwise $\Phi^n(\alpha)$ converges to the origin in $\mathbb{A}^N$ (note that, as explained in the begining of our proof, condition (4) can be relaxed to asking that there is no accumulation point for $\mathcal{O}_{\Phi^k}(\alpha)$ for any $k \in \mathbb{N}$). Then arguing as in Section 11.11.2, at the expense of replacing Φ by a suitable conjugate (through a matrix with entries in $\mathbb{Z}_p$—hence this conjugation preserves condition (1) in Theorem 11.11.3.1) we may assume there exists an idempotent matrix E with entries in $\mathbb{Z}_p$ such that

$$A \equiv E \pmod{p}.$$

Thus the hypotheses of Theorem 11.11.1.1 are met for Φ (with $c = 1$) and then there exists a p-adic analytic function

$$G : \mathbb{Z}_p \longrightarrow \mathbb{Z}_p^N$$

such that for each $n \in \mathbb{N}_0$ we have

(11.11.3.5) $\|\Phi^n(\alpha) - G(n)\| \le p^{-n},$

where $\|(x_1, \ldots, x_N)\| := \max\{|x_1|_p, \ldots, |x_N|_p\}$, while in general, we let

$$(a_1, \ldots, a_N) - (b_1, \ldots, b_N) := (a_1 - b_1, \ldots, a_N - b_N)$$

for any two points $(a_1, \ldots, a_N)$ and $(b_1, \ldots, b_N)$ in $\mathbb{A}^N(\mathbb{Z}_p)$. Now, since each f_i has coefficients in $\mathbb{Z}_p$ (and also that $\alpha, G(n) \in \mathbb{Z}_p^N$) we obtain that

(11.11.3.6) $\|\Phi^{n+1}(\alpha) - \Phi(G(n))\| \le p^{-n}.$

Combining (11.11.3.5) and (11.11.3.6) we obtain that

(11.11.3.7) $\|\Phi(G(n)) - G(n + 1)\| \le p^{-n}.$

On the other hand, the function $H(z) := \Phi(G(z)) - G(z + 1)$ given by

$$H(z) := (H_1(z), \ldots, H_N(z))$$

satisfies the following properties:

(1) $H_i : \mathbb{Z}_p \longrightarrow \mathbb{Z}_p$ is p-adic analytic for each $i = 1, \ldots, N$; and
(2) $|H_i(n)|_p \le p^{-n}$ for each $n \ge 1$ and each $i = 1, \ldots, N$.

Since $\mathbb{N}$ is dense in $\mathbb{Z}_p$ we obtain that H_i is identically equal to 0; so

(11.11.3.8) $\Phi(G(n)) = G(n+1)$ for all n.

There are now two cases.

Case 1. G is not constant.

There are then two additionally subcases.

Subcase (i). There exists $N \in \mathbb{N}$ such that for all $n \ge N$, we have $G(n) \in V(\mathbb{Q}_p)$.

Since the orbit of $G(N)$ under Φ is exactly the set $\{G(n)\}_{n \ge N}$ (see (11.11.3.8)), Proposition 3.1.2.14 yields that V contains a positive dimensional periodic subvariety (we also use here that G is not constant). This contradicts condition (2) from Theorem 11.11.3.1.

Subcase (ii). There exist infinitely many positive integers n such that $G(n) \notin V(\mathbb{Q}_p)$.

We let $\{F_1, \ldots, F_m\}$ be a finite set of polynomials in $\mathbb{Z}_p[x_1, \ldots, x_N]$ which generate the vanishing ideal of V. Then for each $i = 1, \ldots, m$, we let

$$L_i := F_i \circ G : \mathbb{Z}_p \longrightarrow \mathbb{Z}_p,$$

which is a p-adic analytic function. By our assumption from Subcase (ii), there exists some $i_0 \in \{1, \ldots, m\}$ such that

$$L := L_{i_0} \text{ is nonzero.}$$

CLAIM 11.11.3.9. *For each $n \in S_V$, we have $|L(n)|_p \le p^{-n}$.*

PROOF. Using (11.11.3.5) and Taylor expansion (noting that each coefficient of F_{i_0} is in $\mathbb{Z}_p$ and that $F_{i_0}(\Phi^n(\alpha)) = 0$ if $n \in S_V$), we obtain the desired conclusion. $\square$

Since L is not identically equal to 0, then L has at most finitely many zeros in $\mathbb{Z}_p$. We let $s \in (0, 1]$ (as in Lemma 11.8.0.1) such that for each $x \in \mathbb{Z}_p$ there is at most one zero of L in $\overline{D}(x, s)$. We let $k \in \mathbb{N}$ such that $p^{-k} < s$ and we cover $\mathbb{Z}_p$ by the disks $\overline{D}(i, p^{-k})$ with $i = 0, \ldots, p^k - 1$. For each $i \in \{0, \ldots, p^k - 1\}$, we let

$$S_{V,i} := \{n \in S_V : n \equiv i \pmod{p^k}\};$$

then

(11.11.3.10) $$S_V = \bigcup_{i=0}^{p^k - 1} S_{V,i}.$$

CLAIM 11.11.3.11. *Let $i \in \{0, \ldots, p^k - 1\}$. If there are no zeros of L in $D(i, p^{-k})$, then $S_{V,i}$ is finite.*

PROOF. Indeed, otherwise there is an accumulation point $\gamma \in D(i, p^{-k})$ for the elements of $S_{V,i}$. Using Claim 11.11.3.9 we obtain that $L(\gamma) = 0$, contradiction. $\square$

CLAIM 11.11.3.12. *Let $i \in \{0, \ldots, p^k - 1\}$ and assume $S_{V,i}$ is an infinite set. If we list the elements in $S_{V,i}$ in increasing order as $\{n_j\}_{j \ge 1}$, then there exists $C > 1$ such that for all sufficiently large j, we have*

$$n_{j+1} - n_j \ge C^{n_j}.$$

PROOF. The proof is similar to the proof of Lemma 11.8.0.1. We let $\gamma \in D\left(i, p^{-k}\right)$ (be the unique point) such that $L(\gamma) = 0$ (see also Claim 11.11.3.11). Letting δ be the order of the zero γ for L, and arguing similarly as in the proof of Lemma 11.8.0.1 (see (11.8.0.4) and (11.8.0.5)) we obtain (for sufficiently large j) that

$$|n_j - \gamma|_p \leq p^{-\frac{n_j}{\delta} + O(1)} \text{ and thus, } |n_{j+1} - n_j|_p \leq p^{-\frac{n}{\delta} + O(1)}.$$

Hence Claim 11.11.3.12 holds with $C := p^{\frac{1}{\delta} - \epsilon}$ for any $\epsilon > 0$.   □

Using Claims 11.11.3.11 and 11.11.3.12 (and (11.11.3.10)) coupled with a simple counting argument (as in the proof of Theorem 11.1.0.9) we obtain that

$$(11.11.3.13) \qquad \#\{i \leq n \colon i \in S_V\} = o\left(\log^{(m)}(n)\right),$$

for each $m \in \mathbb{N}$, as desired.

Case 2. G is constant.

So, letting $\beta \in \mathbb{Z}_p^N$ such that $G(n) = \beta$ (identically), (11.11.3.5) yields that the orbit $\mathcal{O}_\Phi(\alpha)$ converges p-adically to β (which must be a point on V since there exist infinitely many points of the orbit landing on V). This contradicts assumption (4) above.   □

CHAPTER 12

Denis-Mordell-Lang Conjecture

12.1. Denis-Mordell-Lang Conjecture

12.1.1. The beginning of the Dynamical Mordell-Lang Conjecture.
In this chapter we discuss a variant of the Dynamical Mordell-Lang Conjecture in the context of Drinfeld modules, which is a conjecture of Denis [**Den92a**]. Actually, Denis' Conjecture (see Conjecture 12.1.3.1) was the starting point for the Dynamical Mordell-Lang Conjecture. Informally speaking (for the exact connection with Denis' question stated in Conjecture 12.1.3.1, see Corollary 12.1.7.1), Denis [**Den92a**] asked for a description of the set

$$S_{V,\Phi,\alpha} := \{n \in \mathbb{N}_0 \colon \Phi^n(\alpha) \in V(K)\},$$

where Φ is an endomorphism of $\mathbb{A}^N$ given by the coordinatewise action of some additive one-variable polynomials $\varphi_1, \ldots, \varphi_N$ defined over a field K of characteristic p, while $\alpha \in \mathbb{A}^N(K)$ and $V \subseteq \mathbb{A}^N$ is a subvariety defined over K. We recall that a polynomial φ is *additive* if

$$\varphi(x+y) = \varphi(x) + \varphi(y)$$

for any x and y. Over a field of characteristic 0, all additive polynomials are of the form

$$x \mapsto cx \text{ for some constant } c.$$

Over a field K of characteristic p, there are many more additive polynomials. Indeed, any polynomial

$$\varphi(z) := \sum_{i=0}^{m} c_i z^{p^i},$$

for some coefficients $c_i \in K$ is additive. On the other hand, not any additive polynomials φ_i appearing in the definition of the endomorphism Φ of $\mathbb{A}^N$ lead to the set $S(V,\Phi,\alpha)$ being a finite union of arithmetic progressions; Example 3.4.5.1 shows that if each φ_i is linear, then $S(V,\Phi,\alpha)$ might be infinite, but very sparse.

If Φ is an arbitrary endomorphism of $\mathbb{A}^N$, one cannot expect that the Dynamical Mordell-Lang principle holds; the set $S(V,\Phi,\alpha)$ can have a more complicated structure. In Chapter 13 we discuss a version of the Dynamical Mordell-Lang Conjecture for any endomorphism of a variety defined over a field of positive characteristic. In the present Chapter we confine ourselves to discussing Denis' original conjecture which is formulated solely in the context of additive polynomials.

Motivated by the deep analogy between abelian varieties defined over fields of characteristic 0 and Drinfeld modules, Denis [**Den92a**] conjectured (essentially) that if Φ is an endomorphism of $\mathbb{A}^N$ given by the diagonal action of a Drinfeld module (which is an additive polynomial of special type; for more details, see Definition 12.1.2.1), then $S(V,\Phi,\alpha)$ should be a finite union of arithmetic progressions.

It was precisely Denis' question (see Conjecture 12.1.3.1) along with the classical Mordell-Lang Conjecture which motivated the authors of [**GT09**] to conjecture that the same Dynamical Mordell-Lang principle should hold quite generally for endomorphisms of quasiprojective varieties defined over a field of characteristic 0.

12.1.2. Drinfeld modules. The main results that we present in this chapter are from [**GT08b**]; for some results we include the proofs from [**GT08b**] in entirety, and in other cases we give only a sketch. The results of [**GT08b**] offer a partial answer to Denis' conjecture from [**Den92a**]. First, we define a Drinfeld module (of generic characteristic); for more details on Drinfeld modules see [**Gos96**].

DEFINITION 12.1.2.1. Let p be a prime number and let K be a field containing $\mathbb{F}_p(t)$. We call the ring homomorphism

$$\varphi : \mathbb{F}_p[t] \longrightarrow \operatorname{End}_K(\mathbb{G}_a)$$

a *Drinfeld module (of generic characteristic)* if for each $f(t) \in \mathbb{F}_p[t]$, $\varphi_f := \varphi(f)$ is a (separable) endomorphism of $\mathbb{G}_a$ defined over K such that $\varphi'_f(z) = f$, and moreover φ_t is not a linear endomorphism.

If $\varphi_t(z) = tz + \sum_{i=1}^r a_i z^{p^i}$, then we call r the rank of φ. Note that by our hypothesis, φ_t is non-linear, and therefore $r \geq 1$.

If there exists some nonzero $f \in \mathbb{F}_p[t]$ such that φ_f is not separable, then φ is called a Drinfeld module of *special characteristic*. There is a rich theory for Drinfeld modules of special characteristic (again we refer the reader to [**Gos96**]), but in this book we restrict our attention to Drinfeld modules of generic characteristic—hence we will simply call them *Drinfeld modules*.

We also define naturally the notion of *submodule* under a Drinfeld module action.

DEFINITION 12.1.2.2. Let $g \in \mathbb{N}$, and let $\varphi_1, \ldots, \varphi_g$ be Drinfeld modules defined over a field K. We say that $\Gamma \subset \mathbb{A}^g(K)$ is a $(\varphi_1, \ldots, \varphi_g)(\mathbb{F}_p[t])$-submodule if it is mapped into itself by the coordinatewise action on $\mathbb{A}^g$ given by

$$(\varphi_1, \ldots, \varphi_g).$$

More generally, a Drinfeld module is a ring homomorphism from a ring A of functions defined on a $\mathbb{F}_p$-curve C which are regular away from a given point η of C. In our Definition 12.1.2.1,

$$A = \mathbb{F}_p[t], \text{ and thus } (C, \eta) = (\mathbb{P}^1, \infty).$$

All the results we will be stating in this chapter are equivalent to the more general definition of a Drinfeld module, since a ring A as above is always a finite integral extension of $\mathbb{F}_p[t]$ and thus a subset S of K has a structure of finitely generated $\varphi(\mathbb{F}_p[t])$-submodule if and only if it is a finitely generated $\varphi(A)$-submodule. In particular, the notion of *torsion* for a Drinfeld module is the same regardless of which definition we use.

DEFINITION 12.1.2.3. Let

$$\varphi : \mathbb{F}_p[t] \longrightarrow \operatorname{End}_K(\mathbb{G}_a)$$

be a Drinfeld module. For each $a \in \mathbb{F}_p[t] \setminus \{0\}$, we denote by $\varphi[a]$ the set of all $x \in \overline{K}$ such that $\varphi_a(x) = 0$. Each such point x is called *torsion* for φ, and we denote by φ_{tor} (which is called the *torsion submodule of φ*) the set of all torsion points for φ.

12.1.3. Denis' Conjecture. The following conjecture was raised by Denis in [**Den92a**].

CONJECTURE 12.1.3.1 (Denis-Mordell-Lang [**Den92a**]). *Let K be a field of characteristic p, let $g \in \mathbb{N}$, let*

$$\Phi_1, \dots, \Phi_g : \mathbb{F}_p[t] \longrightarrow \operatorname{End}_K(\mathbb{G}_a)$$

be Drinfeld modules, let $\Gamma \subseteq \mathbb{G}_a^g(K)$ be a finitely generated $(\Phi_1, \dots, \Phi_g)(\mathbb{F}_p[t])$-submodule (where Φ_i acts on the i-th coordinate of $\mathbb{G}_a^g$), and let $V \subseteq \mathbb{A}^g$ be an affine K-subvariety. Then

$$V(K) \cap \Gamma$$

is a union of at most finitely many $(\Phi_1, \dots, \Phi_g)(\mathbb{F}_p[t])$-submodules of Γ.

REMARK 12.1.3.2. If one were to extend the above conjecture to Drinfeld modules of special characteristic, then the conclusion would not hold in its full generality, in the sense that one can *only* expect that the intersection has the structure of finitely generated subgroups invariant under some infinite subring of $\mathbb{F}_p[t]$, as proven in [**Ghi05**].

12.1.4. Questions generalizing Denis-Mordell-Lang Conjecture. Note that Denis [**Den92a**] formulated Conjecture 12.1.3.1 more generally for T-modules, which includes the case of product of Drinfeld modules as stated in Conjecture 12.1.3.1.

Furthermore, Denis also asked Conjecture 12.1.3.1 for the division hull of Γ, i.e., when Γ is replaced by $\Gamma \otimes_{\mathbb{F}_p[t]} \mathbb{F}_p(t)$. In particular, when $\Gamma = \{0\}$, and thus its division hull is the set of torsion points under the action

$$(x_1, \dots, x_g) \mapsto (\Phi_1(x_1), \dots, \Phi_g(x_g)),$$

this more general problem reduces to the description of the intersection between any affine subvariety of $\mathbb{A}^g$ with the torsion submodule under the Drinfeld module action. This latter problem is called the Denis-Manin-Mumford conjecture since it is modelled after the classical Manin-Mumford Conjecture for abelian varieties. The Denis-Manin-Mumford conjecture was proven by Scanlon in [**Sca02**] when

$$(12.1.4.1) \qquad \Phi_1 = \dots = \Phi_g.$$

Quite surprisingly, Demangos [**Dem**] found counterexamples to the Denis-Manin-Mumford conjecture if (12.1.4.1) does not hold, i.e., if the Drinfeld modules Φ_i are distinct. In [**Dem**], Demangos formulated a new version of Denis-Manin-Mumford conjecture which takes into account the families of counterexmaples found by him. Also, it is worth pointing out that Demangos' new conjecture is consistent with the behaviour exhibited by Drinfeld modules of special characteristic for the Denis-Mordell-Lang conjecture; for more details, see [**Ghi05**]. Loosely speaking, Demangos' counterexamples arise from having Drinfeld modules of different rank which satisfy a skew-commutation relation; so, Demangos' examples are reminiscent of the necessity of imposing the polarizability condition in the Dynamical Manin-Mumford Conjecture that we will be discussing in Section 14.1.

12.1.5. Known results towards Conjecture 12.1.3.1. There are only a few cases when the Denis-Mordell-Lang conjecture is known to hold (see [**Ghi05, Ghi10, GT08b**]). The first two cited results hold in the case when

$$\varphi := \Phi_1 = \dots = \Phi_g,$$

and moreover, φ is not isomorphic to a Drinfeld module defined over a field of transcendence degree 1 over $\mathbb{F}_p$; in particular,

$$\operatorname{trdeg}_{\mathbb{F}_p} K \geq 2.$$

The results of [**Ghi05, Ghi10**] can be viewed as the "function field" version of Conjecture 12.1.3.1; in addition, they hold only for the case when V does not contain a translate of a positive dimensional algebraic subgroup of $\mathbb{G}_a^g$. The method of proof from [**Ghi05**] follows the general strategy employed by Hrushovski to deduce the classical Manin-Mumford Conjecture by reduction to positive characteristic through a specialization argument (for more details, see [**Hru99**]).

On the other hand, [**GT08b**, Theorem 2.5] is valid for Drinfeld modules Φ_i not necessarily equal to each other, which are defined over finite extensions of $\mathbb{F}_p(t)$.

THEOREM 12.1.5.1 ([**GT08b**]). *Let K be a finite extension of $\mathbb{F}_p(t)$, let*

$$\Phi_1 : \mathbb{F}_p[t] \to \operatorname{End}_K(\mathbb{G}_a), \ldots, \Phi_g : \mathbb{F}_p[t] \to \operatorname{End}_K(\mathbb{G}_a)$$

be Drinfeld modules, let $\alpha := (\alpha_1, \ldots, \alpha_g) \in \mathbb{G}_a^g(K)$, and let $\Gamma \subseteq \mathbb{G}_a^g(K)$ be the cyclic $(\Phi_1, \ldots, \Phi_g)$-submodule generated by $(\alpha_1, \ldots, \alpha_g)$. If $V \subseteq \mathbb{G}_a^g$ is an affine subvariety defined over K, then $V(K) \cap \Gamma$ is a finite union of cosets of $(\Phi_1, \ldots, \Phi_g)$-submodules of Γ.

REMARK 12.1.5.2. Moreover, each submodule of Γ whose coset appears in the above intersection is of the form $B_i(K) \cap \Gamma$, where each B_i is an *algebraic* $(\Phi_1, \ldots, \Phi_g)$-*submodule* of $\mathbb{G}_a^g$, i.e., B_i is an algebraic subgroup of $\mathbb{G}_a^g$ invariant under the action

$$(x_1, \ldots, x_g) \mapsto (\Phi_1(x_1), \ldots, \Phi_g(x_g)).$$

Indeed, if $(b + H) \subseteq V(K)$ is a coset of a submodule H of Γ, then

$$(b + D) \subseteq V,$$

where D is the Zariski closure of H. Since H is a $(\Phi_1, \ldots, \Phi_g)$-submodule, each D is mapped into itself by the $(\Phi_1, \ldots, \Phi_g)$-action. Hence, it is a finite union of translates $(b_i + B_i)$ of algebraic $(\Phi_1, \ldots, \Phi_g)$-submodules B_i of $\mathbb{G}_a^g$ (see [**Den92a**, Lemme 4]). Therefore, we may write

$$(b + H) \subseteq \bigcup_i (c_i + (B_i(K) \cap \Gamma)) \subseteq V(K),$$

where $c_i \in (b_i + B_i(K)) \cap \Gamma$ for each i.

From now on, in order to simplify the notation, we write Φ-*submodule* instead of $(\Phi_1, \ldots, \Phi_g)$-submodule.

12.1.6. The method of proof for Theorem 12.1.5.1. The idea behind the proof of our Theorem 12.1.5.1 is very similar to the use of the p-adic arc lemma in Chapter 4. Assuming that an affine variety $V \subseteq \mathbb{G}_a^g$ has infinitely many points in common with a cyclic Φ-submodule Γ, we can then find a suitable submodule $\Gamma_0 \subseteq \Gamma$ whose coset lies in V. Indeed, applying the logarithmic map to Γ_0 (associated to a suitable place v; for the technical details, see Section 12.2) yields a line in the vector space $\mathbb{C}_v^g$.

Each polynomial f that vanishes on V, then gives rise to an analytic function F on this line (by composing with the exponential function); this idea was used already several times for the Dynamical Mordell-Lang Conjecture and it originated in [**GT08b**]. Since we assumed there are infinitely many points in $V \cap \Gamma$, the zeros of F must have an accumulation point on this line, which means that F vanishes identically on the line. This means that there is an entire translate of Γ_0 contained in the zero locus of f.

As mentioned before, the inspiration for this idea comes from the Skolem-Mahler-Lech method discussed in Section 2.5 (see also Chapter 4). Also, the authors of [**GT08b**] took inspiration from the method employed by Chabauty in [**Cha41**] (and later refined by Coleman in [**Col85**]) to study the intersection of a curve C of genus g, embedded in its Jacobian J, with a finitely generated subgroup of J of rank less than g. Finally, our technique also bears a resemblance to Skolem's method for treating Diophantine equations (see [**BS66**, Chapter 4.6]).

Note that for a Drinfeld module Φ (of generic characteristic), Φ_t is an étale map; however the geometric approach from Chapter 4 fails in this case since we work in positive characteristic. For example, note that the self-map

$$\Psi : \mathbb{A}^2 \longrightarrow \mathbb{A}^2 \text{ given by } \Psi(x,y) = (tx, (1-t)y)$$

is also étale, and so one might think that the Dynamical Mordell-Lang principle would also apply for the intersection of the plane line V given by the equation

$$x + y = 1$$

defined over $\mathbb{F}_p(t)$ with the orbit of $(1,1)$ under Ψ. However, as seen in Example 3.4.5.1, the set of $n \in \mathbb{N}_0$ such that

$$\Psi^n(1,1) \in V(\mathbb{F}_p(t))$$

is the set of all powers of p, thus contradicting the Dynamical Mordell-Lang principle. The reason for which a similar v-adic parametrization works for Drinfeld modules is *not* because they are a family of étale maps, but because there exists a *global* analytic parametrization with respect to the place at infinity (very similar to the classical exponential map associated to abelian varieties in characteristic 0) which induces a *local* v-adic analytic parametrization for all *finite* places v (i.e., places which do not lie over the place at infinity from $\mathbb{F}_p(t)$; for more details regarding places of $\mathbb{F}_p(t)$, see Subsection 12.2.1).

Finally, we note that one needs to use a v-adic analytic parametrization with respect to finite places v because with respect to such a place (as opposed to the place at infinity from $\mathbb{F}_p(t)$) the elements of $\mathbb{F}_p[t]$ accumulate near 0, and therefore we can apply Lemma 2.3.6.1 which states that an analytic function vanishing on a non-discrete set of points must be identically equal to 0.

So, similar to the p-adic arc lemma and the proof of Theorem 4.4.1.1, the proof of Theorem 12.1.5.1 relies on constructing a v-adic analytic parametrization of the cyclic module Γ with respect to a suitable place v of K; this will be done in Section 12.2. Then we finish the proof of Theorem 12.1.5.1 in Section 12.3.

12.1.7. A dynamical variant of Theorem 12.1.5.1. Since a Drinfeld module is a family of polynomial actions on the affine line, we can extract from Theorem 12.1.5.1 a Dynamical Mordell-Lang statement for endomorphisms of the additive group scheme in characteristic p.

COROLLARY 12.1.7.1. *Let p be a prime number, let K be a finite extension of $\mathbb{F}_p(t)$, let*

$$\Psi_1, \ldots, \Psi_g \in \mathrm{End}_K(\mathbb{G}_a)$$

such that Ψ_i *is non-linear and*

$$\Psi_i'(z) = t \text{ for each } i = 1, \ldots, g,$$

let $\alpha \in K^g$, *and let* $V \subseteq \mathbb{A}^g$ *be an affine K-subvariety. We let*

$$\Psi := (\Psi_1, \ldots, \Psi_g)$$

be the endomorphism of $\mathbb{G}_a^g$ given by the coordinatewise action of the Ψ_i's. Then the set of $n \in \mathbb{N}_0$ such that

$$\Psi^n(\alpha) \in V(K)$$

is a union of finitely many arithmetic progressions.

PROOF. For each $i = 1, \ldots, g$ we let $\Phi_i : \mathbb{F}_p[t] \longrightarrow \text{End}_K(\mathbb{G}_a)$ be the Drinfeld module defined by

$$(\Phi_i)_t = \Psi_i.$$

Then Theorem 12.1.5.1 yields that V intersects the cyclic

$$(\Phi_1, \ldots, \Phi_g)(\mathbb{F}_p[t])\text{-submodule } \Gamma \text{ generated by } \alpha$$

in a union of at most finitely many cosets of submodules of Γ. Clearly, each submodule of Γ is also cyclic (since $\mathbb{F}_p[t]$ is a PID), and so the above cosets of Γ are of the form

$$\Gamma_0 := \{(\Phi_1(a + bc), \ldots, \Phi_g(a + bc))(\alpha) : c \in \mathbb{F}_p[t]\}$$

for some given $a, b \in \mathbb{F}_p[t]$. Hence

$$\Psi^n(\alpha) \in \Gamma_0$$

if and only if there exists $c_n \in \mathbb{F}_p[t]$ such that for each $i = 1, \ldots, g$ we have that

$$(12.1.7.2) \qquad \Phi_i(t^n - a - bc_n)(\alpha_i) = 0,$$

where $\alpha := (\alpha_1, \ldots, \alpha_g)$. Let J_i be the ideal of $\mathbb{F}_p[t]$ which kills α_i under the action of Φ_i (if α_i is non-torsion, then $J_i = (0)$). Then (12.1.7.2) is equivalent to

$$(12.1.7.3) \qquad t^n \equiv a \pmod{\tilde{J}_i},$$

where $\tilde{J}_i := J_i + (b)$. Clearly the solutions n to (12.1.7.3) form an arithmetic progression (possibly constant, or empty). Since the intersection of arithmetic progressions is also an arithmetic progression, we conclude our proof. $\square$

12.2. Preliminaries on function field arithmetic

In this section, we set up the basic notation and also introduce some technical results used in the proof of Theorem 12.1.5.1. Some of the contents of this section overlap with Chapter 2, particularly the construction of absolute values for a set of inequivalent places for a given function field.

12.2.1. Valuations. The contents of this subsection are from [**GT08b**].

Let $M_{\mathbb{F}_p(t)}$ be the set of inequivalent places on $\mathbb{F}_p(t)$; for more details, see Subsection 2.3.6. We denote by v_∞ the place in $M_{\mathbb{F}_p(t)}$ such that

$$v_\infty\left(\frac{f}{g}\right) = \deg(g) - \deg(f)$$

for every nonzero $f, g \in A = \mathbb{F}_p[t]$. We let M_K be the set of valuations on K. Then M_K is a set of valuations which satisfies a product formula (see [**Lan83**] and also Section 2.6). Thus

- for each nonzero $z \in K$, there are finitely many $v \in M_K$ such that $|z|_v \neq 1$; and

- for each nonzero $z \in K$, we have $\prod_{v \in M_K} |z|_v = 1$.

DEFINITION 12.2.1.1. Each place in M_K which lies over v_∞ is called an *infinite place*. Each place in M_K which does not lie over v_∞ is called a *finite place*.

When we fix some infinite place of K, we simply denote it by $\infty \in M_K$.

For $v \in M_K$ we let K_v be the completion of K with respect to v. Let $\mathbb{C}_v$ be the completion of an algebraic closure of K_v. Then $|\cdot|_v$ extends to a unique absolute value on all of $\mathbb{C}_v$. We fix an embedding

$$\iota : \overline{K} \longrightarrow \mathbb{C}_v.$$

For $z \in \overline{K}$, we denote $|\iota(z)|_v$ simply as $|z|_v$, by abuse of notation.

12.2.2. Logarithms and exponentials associated to a Drinfeld module. The contents of this subsection are from [**GT08b**] and provide the technical details for the v-adic analytic parametrization of a cyclic module under the diagonal action of g Drinfeld modules acting on $\mathbb{A}^g$.

Let $v \in M_K$. According to [**Gos96**, Proposition 4.6.7], there exists a unique formal power series $\exp_{\Phi,v} \in \mathbb{C}_v\{\tau\}$ such that for every $a \in \mathbb{F}_p[t]$, we have

$$(12.2.2.1) \qquad\qquad \Phi_a = \exp_{\Phi,v} a \exp_{\Phi,v}^{-1}.$$

In addition, the coefficient of the linear term in $\exp_{\Phi,v}(X)$ is equal to 1. We let $\log_{\Phi,v}$ be the formal power series $\exp_{\Phi,v}^{-1}$, which is the inverse of $\exp_{\Phi,v}$.

If $v = \infty$ is an infinite place, then $\exp_{\Phi,\infty}(z)$ is convergent for all $z \in \mathbb{C}_\infty$ (see [**Gos96**, Theorem 4.6.9]). There exists a sufficiently small ball B_∞ centered at the origin such that $\exp_{\Phi,\infty}$ is an isometry on B_∞ (see [**GT08a**, Lemma 3.6]). Hence, $\log_{\Phi,\infty}$ is convergent on B_∞. Moreover, the restriction of $\log_{\Phi,\infty}$ on B_∞ is an analytic isometry (see also [**Gos96**, Proposition 4.14.2]).

If v is a finite place, then $\exp_{\Phi,v}$ is convergent on a sufficiently small ball $B_v \subseteq \mathbb{C}_v$ (this follows identically as the proof of the analyticity of $\exp_{\Phi,\infty}$ from [**Gos96**, Theorem 4.6.9]). Similarly as in the above paragraph, at the expense of replacing B_v by a smaller ball, we may assume $\exp_{\Phi,v}$ is an isometry on B_v. Hence, also $\log_{\Phi,v}$ is an analytic isometry on B_v.

For every place $v \in M_K$, for every $z \in B_v$ and for every polynomial $a \in \mathbb{F}_p[t]$, we have (see (12.2.2.1))

$$(12.2.2.2) \qquad a \log_{\Phi,v}(z) = \log_{\Phi,v}(\Phi_a(z)) \text{ and } \exp_{\Phi,v}(az) = \Phi_a(\exp_{\Phi,v}(z)).$$

By abuse of language, $\exp_{\Phi,\infty}$ and $\exp_{\Phi,v}$ will be called exponentials, while $\log_{\Phi,\infty}$ and $\log_{\Phi,v}$ will be called logarithms.

12.2.3. Integrality and reduction. The following definition is essentially Definition 6.1.1.1 in the context of Drinfeld modules.

DEFINITION 12.2.3.1. A Drinfeld module Φ has *good reduction* at a place v if for each nonzero $a \in \mathbb{F}_p[t]$, all coefficients of Φ_a are v-adic integers and the leading coefficient of Φ_a is a v-adic unit. If Φ does not have good reduction at v, then we say that Φ has *bad reduction* at v.

It is immediate to see that Φ has good reduction at v if and only if all coefficients of Φ_t are v-adic integers, while the leading coefficient of Φ_t is a v-adic unit. All infinite places of K are places of bad reduction for Φ. We also note that our definition for places of good reduction is not invariant under isomorphisms of Drinfeld modules. Finally, the notion of good reduction for Drinfeld modules is equivalent to the notion of good reduction for the polynomial Φ_t as defined in Subsection 6.1.1.

12.3. Proof of our main result

In this section we prove Theorem 12.1.5.1 using the background introduced in Section 12.2. We follow the proof from [**GT08b**] by presenting the main ingredients of the argument from that paper.

Let K be a finite extension of $\mathbb{F}_p(t)$, let $\Phi_1, \ldots, \Phi_g$ be Drinfeld modules defined over K, and let Φ denote the action of

$$(\Phi_1, \ldots, \Phi_g)$$

on $\mathbb{G}_a^g$. Also, let $(\alpha_1, \ldots, \alpha_g) \in \mathbb{G}_a^g(K)$ and let Γ be the cyclic Φ-submodule of $\mathbb{G}_a^g(K)$ generated by

$$\alpha := (\alpha_1, \ldots, \alpha_g).$$

Unless otherwise stated, $V \subseteq \mathbb{G}_a^g$ is an affine subvariety defined over K.

The following easy combinatorial result is proved in [**GT08b**, Lemma 3.1].

LEMMA 12.3.0.1 ([**GT08b**]). *Let Γ be a cyclic Φ-submodule of $\mathbb{G}_a^g(K)$, let Γ_0 be a non-trivial Φ-submodule of Γ, and let $S \subseteq \Gamma$ be an infinite set. Suppose that for every infinite subset $S_0 \subseteq S$, there exists a coset C_0 of Γ_0 such that*

$$C_0 \cap S_0 \neq \emptyset \text{ and } C_0 \subseteq S.$$

Then S is a finite union of cosets of Φ-submodules of Γ.

PROOF. The main observation in our proof is that a cyclic, infinite Φ-submodule is isomorphic to $\mathbb{F}_p[t]$. Thus Γ_0 is isomorphic to a non-trivial ideal I of $\mathbb{F}_p[t]$. Since $\mathbb{F}_p[t]/I$ is finite, there are finitely many cosets of Γ_0 in Γ. Thus S contains at most finitely many cosets of Γ_0. So let

$$\{y_i + \Gamma_0\}_{i=1}^{\ell}$$

be all of the cosets of Γ_0 that are contained in S. Suppose that

$$(12.3.0.2) \qquad S_0 := S \setminus \bigcup_{i=1}^{\ell} (y_i + \Gamma_0) \text{ is infinite.}$$

If S_0 is infinite, then there is a coset of Γ_0 that is contained in S but is not one of the cosets $(y_i + \Gamma_0)$ (because it has a non-empty intersection with S_0). This contradicts the fact that

$$\{y_i + \Gamma_0\}_{i=1}^{\ell}$$

are *all* the cosets of Γ_0 that are contained in S. Therefore S_0 must be finite. Since any finite subset of Γ is a finite union of cosets of the trivial submodule of Γ, this completes the proof of Lemma 12.3.0.1. $\square$

We will also use the following result in the proof of Theorem 12.1.5.1.

LEMMA 12.3.0.3 ([**GT08b**]). *Let*

$$\theta : A \to K\{\tau\} \text{ and } \psi : A \to K\{\tau\}$$

be Drinfeld modules, let v be a place of good reduction for θ and ψ, let $x, y \in \mathbb{C}_v$, and let $r_v \in (0, 1)$ and

$$B_v := \{z \in \mathbb{C}_v \colon |z|_v < r_v\}$$

be a sufficiently small ball centered at the origin with the property that both $\log_{\theta,v}$ and $\log_{\psi,v}$ are analytic isometries on B_v. Then for every polynomials $P, Q \in \mathbb{F}_p[t]$ such that

$$(\theta_P(x), \psi_P(y)) \in B_v \times B_v \text{ and } (\theta_Q(x), \psi_Q(y)) \in B_v \times B_v,$$

we have

$$\log_{\theta,v}(\theta_P(x)) \cdot \log_{\psi,v}(\psi_Q(y)) = \log_{\theta,v}(\theta_Q(x)) \cdot \log_{\psi,v}(\psi_P(y)).$$

PROOF. This is [**GT08b**, Lemma 3.2]. The proof is a simple application of (12.2.2.2).   □

The following result is an immediate corollary of Lemma 12.3.0.3.

COROLLARY 12.3.0.4. *With the notation as in Theorem 12.1.5.1, assume in addition that*

$$\alpha_1 \notin (\Phi_1)_{\text{tor}}.$$

Let v be a place of good reduction for each ϕ_i. Suppose $B_v \subset \mathbb{C}_v$ is a small ball (of radius less than 1) centered at the origin such that each $\log_{\Phi_i,v}$ is an analytic isometry on B_v. Then for each $i \in \{2, \dots, g\}$, the fractions

$$\lambda_i := \frac{\log_{\Phi_i,v}\left((\Phi_i)_P(\alpha_i)\right)}{\log_{\Phi_1,v}\left((\Phi_1)_P(\alpha_1)\right)}$$

are independent of the choice of the nonzero polynomial $P \in \mathbb{F}_p[t]$ for which we have

$$\Phi_P(\alpha_1, \dots, \alpha_g) \in B_v^g.$$

We are ready to prove the main result of this chapter.

PROOF OF THEOREM 12.1.5.1. We may assume $V(K) \cap \Gamma$ is infinite (otherwise the conclusion of Theorem 12.1.5.1 is obviously satisfied). Assuming $V(K) \cap \Gamma$ is infinite, we will show that there exists a non-trivial Φ-submodule $\Gamma_0 \subseteq \Gamma$ such that each infinite subset of points S_0 in $V(K) \cap \Gamma$ has a non-empty intersection with a coset C_0 of Γ_0, and moreover, $C_0 \subseteq V(K) \cap \Gamma$. Theorem 12.1.5.1 will then follow immediately from Lemma 12.3.0.1.

First we observe that Γ is not a torsion Φ-submodule. Otherwise Γ is finite, contradicting our assumption that $V(K) \cap \Gamma$ is infinite. Hence, from now on, we assume without loss of generality that α_1 is not a torsion point for Φ_1.

We fix a finite set of polynomials $\{f_1, \dots, f_\ell\} \subseteq K[z_1, \dots, z_g]$ which generate the vanishing ideal of V.

Let $v \in M_K$ be a place of K which is of good reduction for all Φ_i (for $1 \leq i \leq g$). In addition, we assume each x_i is integral at v (for $1 \leq i \leq g$). Then for each $P \in \mathbb{F}_p[t]$, we have

$$\Phi_P(\alpha_1, \dots, \alpha_g) \in \mathbb{G}_a^g(\mathfrak{o}_v),$$

where $\mathfrak{o}_v$ is the ring of v-adic integers in K_v (the completion of K at v). Since $\mathfrak{o}_v$ is a compact space (we use the fact that K is a function field of transcendence degree 1 and thus it has a finite residue field at v), we conclude that every infinite sequence of points

$$\Phi_P(\alpha_1, \ldots, \alpha_g) \in V(K) \cap \Gamma$$

contains a convergent subsequence in $\mathfrak{o}_v^g$. Using Lemma 12.3.0.1, it suffices to show that there exists a non-trivial Φ-submodule $\Gamma_0 \subseteq \Gamma$ such that every convergent sequence of points in $V(K) \cap \Gamma$ has a non-empty intersection with a coset C_0 of Γ_0, and moreover,

$$C_0 \subseteq V(K) \cap \Gamma.$$

Now, let S_0 be an infinite subsequence of distinct points in $V(K) \cap \Gamma$ which converges v-adically to

$$(\alpha_{0,1}, \ldots, \alpha_{0,g}) \in \mathfrak{o}_v^g,$$

let $0 < r_v < 1$, and let

$$B_v := \{z \in \mathbb{C}_v \colon |z|_v < r_v\}$$

be a small ball centered at the origin on which each of the logarithmic functions $\log_{\Phi_i,v}$ is an analytic isometry (for $1 \le i \le g$). Since

$$(\alpha_{0,1}, \ldots, \alpha_{0,g}) \text{ is the limit point for } S_0,$$

there exist $d \in \mathbb{F}_p[t]$ and an infinite subsequence $\{\Phi_{d+P_n}(\alpha)\}_{n \ge 0} \subseteq S_0$ (with $P_n = 0$ if and only if $n = 0$), such that for each $n \ge 0$, we have

$$(12.3.0.5) \qquad \left| (\Phi_i)_{d+P_n} (\alpha_i) - \alpha_{0,i} \right|_v < \frac{r_v}{2} \quad \text{for each } 1 \le i \le g.$$

We will show that there exists an algebraic group Y_0, independent of S_0 and invariant under Φ, such that

$$\Phi_d(\alpha_1, \ldots, \alpha_g) + Y_0 \subseteq V, \text{ and moreover, } \Phi_{d+P_n}(\alpha_1, \ldots, \alpha_g) \in \Phi_d(\alpha_1, \ldots, \alpha_g) + Y_0,$$

for all P_n. Thus the submodule

$$\Gamma_0 := Y_0(K) \cap \Gamma$$

will satisfy the hypothesis of Lemma 12.3.0.1 for the infinite subset $V(K) \cap \Gamma \subseteq \Gamma$; this yields the conclusion of Theorem 12.1.5.1.

Using (12.3.0.5) for $n = 0$ (we recall that $P_0 = 0$), and then for arbitrary n, we see that

$$(12.3.0.6) \qquad \left| (\Phi_i)_{P_n} (\alpha_i) \right|_v < \frac{r_v}{2} \text{ for each } 1 \le i \le g.$$

Hence $\log_{\Phi_i,v}$ is well-defined at $(\Phi_i)_{P_n} (\alpha_i)$ for each $i \in \{1, \ldots, g\}$ and for each $n \ge 1$. Moreover, the fact that

$$\left((\Phi_i)_{P_n+d} (\alpha_i) \right)_{n \ge 1} \text{ converges to a point in } \mathfrak{o}_v$$

yields that $\left((\Phi_i)_{P_n} (\alpha_i) \right)_{n \ge 1}$ converges to a point which is contained in B_v (see (12.3.0.6)).

Without loss of generality, we may assume

$$(12.3.0.7) \qquad |\log_{\Phi_1,v} ((\Phi_1)_{P_1} (\alpha_1))|_v = \max_{i=1}^{g} |\log_{\Phi_i,v} ((\Phi_i)_{P_1} (\alpha_i))|_v.$$

In (12.3.0.7), we used the fact that the maximum cannot be attained at a torsion point α_i, because the logarithm vanishes precisely on the torsion points (actually, the only torsion point contained in B_v is 0 because $\log_{\Phi_i,v}$ is an analytic isometry on B_v for each i).

Using the result of Corollary 12.3.0.4, we conclude that for each $i \in \{2, \ldots, g\}$, the following fraction is independent of n and of the sequence $\{P_n\}_n$:

$$(12.3.0.8) \qquad \lambda_i := \frac{\log_{\Phi_i, v}\left((\Phi_i)_{P_n}(\alpha_i)\right)}{\log_{\Phi_1, v}\left((\Phi_1)_{P_n}(\alpha_1)\right)}.$$

Since α_1 is not a torsion point for Φ_1, the denominator of λ_i (12.3.0.8) is nonzero. From Equation (12.3.0.7), we may then conclude that $|\lambda_i|_v \leq 1$ for each i.

The fact that λ_i is independent of the sequence $\{P_n\}_{n \geq 1}$ will be used later to show that the Φ-submodule Γ_0 that we construct is independent of the sequence $\{P_n\}_{n \geq 1}$.

For each $n \geq 1$ and each $i \in \{2, \ldots, g\}$, we have

$$(12.3.0.9) \qquad \log_{\Phi_i, v}\left((\Phi_i)_{P_n}(\alpha_i)\right) = \lambda_i \cdot \log_{\Phi_1, v}\left((\Phi_1)_{P_n}(\alpha_1)\right).$$

For each i, applying the exponential function $\exp_{\Phi_i, v}$ to both sides of (12.3.0.9) yields

$$(12.3.0.10) \qquad (\Phi_i)_{P_n}(\alpha_i) = \exp_{\Phi_i, v}\left(\lambda_i \cdot \log_{\Phi_1, v}\left((\Phi_1)_{P_n}(\alpha_1)\right)\right).$$

Since $\Phi_{d+P_n}(\alpha_1, \ldots, \alpha_g) \in V(K)$, for each $j \in \{1, \ldots, \ell\}$ we have

$$(12.3.0.11) \qquad f_j\left(\Phi_{d+P_n}(\alpha_1, \ldots, \alpha_g)\right) = 0 \text{ for each } n.$$

For each $j \in \{1, \ldots, \ell\}$ we let $f_{d,j} \in K[z_1, \ldots, z_g]$ be defined by

$$(12.3.0.12) \qquad f_{d,j}(z_1, \ldots, z_g) := f_j\left(\Phi_d(\alpha_1, \ldots, \alpha_g) + (z_1, \ldots, z_g)\right).$$

We let $V_d \subseteq \mathbb{G}_a^g$ be the affine subvariety defined by the equations

$$f_{d,j}(z_1, \ldots, z_g) = 0 \text{ for each } j \in \{1, \ldots, \ell\}.$$

Using (12.3.0.11) and (12.3.0.12), we see that for each $j \in \{1, \ldots, \ell\}$ we have

$$(12.3.0.13) \qquad f_{d,j}\left(\Phi_{P_n}(\alpha_1, \ldots, \alpha_g)\right) = 0$$

for each n, and so,

$$(12.3.0.14) \qquad \Phi_{P_n}(\alpha_1, \ldots, \alpha_g) \in V_d(K).$$

For each $j \in \{1, \ldots, \ell\}$, we let $F_{d,j}(u)$ be the analytic function defined on B_v by

$$F_{d,j}(z) := f_{d,j}\left(z, \exp_{\Phi_2, v}\left(\lambda_2 \log_{\Phi_1, v}(z)\right), \ldots, \exp_{\Phi_g, v}\left(\lambda_g \log_{\Phi_1, v}(z)\right)\right).$$

Using (12.3.0.7) and the fact that $\log_{\Phi_1, v}$ is an analytic isometry on B_v, we see that for each $z \in B_v$, we have

$$(12.3.0.15) \qquad |\lambda_i \cdot \log_{\Phi_1, v}(z)|_v = |\lambda_i|_v \cdot |\log_{\Phi_1, v}(z)|_v \leq |z|_v < r_v.$$

Equation (12.3.0.15) shows that $\lambda_i \cdot \log_{\Phi_1, v}(z) \in B_v$, and so, $\exp_{\Phi_i, v}\left(\lambda_i \cdot \log_{\Phi_1, v}(z)\right)$ is well-defined.

Using (12.3.0.10) and (12.3.0.13) we obtain that for every $n \geq 1$, we have

$$(12.3.0.16) \qquad F_{d,j}\left((\Phi_1)_{P_n}(\alpha_1)\right) = 0.$$

Thus $\left((\Phi_1)_{P_n}(\alpha_1)\right)_{n \geq 1}$ is a sequence of zeros for the analytic function $F_{d,j}$ which has an accumulation point in B_v. Lemma 2.3.6.1 then implies that

$$F_{d,j} = 0,$$

and so, for each $j \in \{1, \ldots, \ell\}$, we have

$$(12.3.0.17) \qquad f_{d,j}\left(z, \exp_{\Phi_2,v}\left(\lambda_2 \log_{\Phi_1,v}(z)\right), \ldots, \exp_{\Phi_g,v}\left(\lambda_g \log_{\Phi_1,v}(z)\right)\right) = 0.$$

For each $z \in B_v$, we let

$$Q_z := \left(z, \exp_{\Phi_2,v}\left(\lambda_2 \log_{\Phi_1,v}(z)\right), \ldots, \exp_{\Phi_g,v}\left(\lambda_g \log_{\Phi_1,v}(z)\right)\right) \in \mathbb{G}_a^g(\mathbb{C}_v).$$

Then (12.3.0.17) implies that

$$(12.3.0.18) \qquad Q_z \in V_d \text{ for each } z \in B_v.$$

Let Y_0 be the Zariski closure of $\{Q_z\}_{z \in B_v}$. Then $Y_0 \subseteq V_d$. Note that Y_0 is independent of the sequence $\{P_n\}_n$ (because the λ_i are independent of the sequence $\{P_n\}_n$, according to Corollary 12.3.0.4).

We claim that for each $z \in B_v$ and for each $P \in \mathbb{F}_p[t]$, we have

$$(12.3.0.19) \qquad \Phi_P(Q_z) = Q_{(\Phi_1)_P(z)}.$$

Note that for each $z \in B_v$, we also have that $(\Phi_1)_P(z) \in B_v$ for each $P \in \mathbb{F}_p[t]$, because each coefficient of Φ_1 is a v-adic integer. To see that (12.3.0.19) holds, we use (12.2.2.2), which implies that for each $i \in \{2, \ldots, g\}$ we have

$$\begin{aligned}
\exp_{\Phi_i,v}\left(\lambda_i \log_{\Phi_1,v}\left((\Phi_1)_P(z)\right)\right) &= \exp_{\Phi_i,v}\left(\lambda_i \cdot P \cdot \log_{\Phi_1,v}(z)\right) \\
&= \exp_{\Phi_i,v}\left(P \cdot \lambda_i \log_{\Phi_1,v}(z)\right) \\
&= (\Phi_i)_P\left(\exp_{\Phi_i,v}\left(\lambda_i \log_{\Phi_1,v}(z)\right)\right).
\end{aligned}$$

Hence, (12.3.0.19) holds, and so Y_0 is invariant under Φ. Furthermore, since all of the $\exp_{\Phi_i,v}$ and $\log_{\Phi_i,v}$ are additive functions, we have

$$Q_{z_1+z_2} = Q_{z_1} + Q_{z_2} \text{ for every } z_1, z_2 \in B_v.$$

Hence Y_0 is an algebraic group, which is also a Φ-submodule of $\mathbb{G}_a^g$. Moreover, Y_0 is defined independently of Γ.

Let $\Gamma_0 := Y_0(K) \cap \Gamma$. Since Y_0 is invariant under Φ, we have that Γ_0 is a submodule of Γ. Since $Y_0 \subseteq V_d$, it follows that

$$\Phi_d(\alpha_1, \ldots, \alpha_g) + Y_0 \subseteq V,$$

and moreover,

$$\Phi_{d+P_n}(\alpha_1, \ldots, \alpha_g)\}_n \subset \Phi_d(\alpha_1, \ldots, \alpha_g) + Y_0.$$

In particular, the (infinite) translate C_0 of Γ_0 by $\Phi_d(\alpha_1, \ldots, \alpha_g)$ is contained in $V(K) \cap \Gamma$. Hence, every infinite sequence of points in $V(K) \cap \Gamma$ has a non-trivial intersection with a coset C_0 of (the non-trivial Φ-submodule) Γ_0, and moreover, $C_0 \subseteq V(K) \cap \Gamma$. Applying Lemma 12.3.0.1 thus finishes the proof of Theorem 12.1.5.1. $\qquad\square$

In the course of our proof of Theorem 12.1.5.1 we also proved the following statement.

THEOREM 12.3.0.20 ([**GT08b**]). *Let Γ be an infinite cyclic Φ-submodule of $\mathbb{G}_a^g$. Then there exists an infinite Φ-submodule $\Gamma_0 \subseteq \Gamma$ such that for every affine subvariety $V \subseteq \mathbb{G}_a^g$, if*

$$V(\overline{K}) \cap \Gamma \text{ is infinite,}$$

then $V(\overline{K}) \cap \Gamma$ contains a coset of Γ_0.

PROOF. This is [**GT08b**, Theorem 3.5]. With the notation as in the proof of Theorem 12.1.5.1, there exists a positive dimensional algebraic group Y_0, invariant under Φ, and depending only on Γ and v (but not on V), such that a translate of Y_0 by a point in Γ lies in V. Moreover,

$$\Gamma_0 := Y_0(\overline{K}) \cap \Gamma \text{ is infinite.}$$

Hence Γ_0 satisfies the conclusion of Theorem 12.3.0.20. $\qquad\square$

In particular, Theorem 12.3.0.20 shows that if the intersection $V(K) \cap \Gamma$ is infinite, then there exists a uniform bound for the number of cosets of (maximal) submodules of Γ which are contained in V.

CHAPTER 13

Dynamical Mordell-Lang Conjecture in positive characteristic

In this chapter we discuss a version of the Dynamical Mordell-Lang Conjecture for endomorphisms of varieties defined over fields of characteristic p—see Conjecture 13.2.0.1. It is somewhat surprising to see that there are *almost* no partial results towards Conjecture 13.2.0.1. On the other hand, this phenomenon mirrors the history of the classical Mordell-Lang Conjecture which was proven in the function field case of positive characteristic by Hrushovski [**Hru96**] a few years *after* the result was established over $\mathbb{C}$ by Faltings [**Fal91**]. Furthermore, Hrushovski's proof was very much different than Faltings' proof thus showing the intrinsic difficulties posed by the function field arithmetic in positive characteristic. We expect that similarly, the characteristic p Dynamical Mordell-Lang question posed in Conjecture 13.2.0.1 is *very* difficult.

We begin the chapter by presenting in Section 13.1 Hrushovski's result [**Hru96**] for the characteristic p classical Mordell-Lang problem, and then state the refinement proven by Moosa and Scanlon [**MS03, MS04**] in the case of semiabelian varieties defined over a finite field. This leads us naturally in Section 13.2 to the formulation of the characteristic p version of the Dynamical Mordell-Lang Conjecture. We continue in Section 13.3 by presenting a special case of Conjecture 13.2.0.1— essentially, this is the only significant case known of the conjecture (besides the case of translations on a semiabelian variety). We conclude this chapter with what might seem to be a detour from the Dynamical Mordell-Lang problem, but we consider the problems discussed in Section 13.4 be relevant to the main theme of our book. So, in Section 13.4 we go back to the classical Skolem-Mahler-Lech problem of describing the set

$$(13.0.0.1) \qquad S_{\mathbf{a}} := \{n \in \mathbb{N}_0 : a_n = 0\},$$

for some linear recurrence sequence $\{a_n\}$. When the sequence is defined over a field of characteristic 0, the problem was completely solved in Section 2.5 and its proof was the building block in developing the p-adic arc lemma which was employed in Chapter 4 for proving the Dynamical Mordell-Lang Conjecture for étale maps. However, when the field of definition for the sequence $\{a_n\}$ has positive characteristic, then the set $S_{\mathbf{a}}$ from (13.0.0.1) is no longer a finite union of arithmetic progressions. The structure of the set $S_{\mathbf{a}}$ is similar to the one encountered in the characteristic p case of the Mordell-Lang problem, as proven by Moosa and Scanlon [**MS03**]. However, the methods we employ in Section 13.4 are quite different than the algebraic geometric and model theoretic techniques used in [**MS03**]. Instead, we use *automata theory*. For the results of Section 13.4 we follow both [**Der07**] (where Derksen reproved using the automata theory a special case of the Mordell-Lang theorem established by Moosa and Scanlon [**MS03**]), and also [**AB12**] where a similar problem is solved in the context of algebraic power series. More precisely, the main result of [**AB12**] is to show that for a power series

$$F(z) = \sum_{n=0}^{\infty} c_n z^n$$

defined over a field K of positive characteristic with the property that $F(z)$ is algebraic over $K(z)$, the set

$$\{n \in \mathbb{N}_0 : c_n = 0\}$$

is *p-automatic*. We believe that the results of Section 13.4 besides their intrinsic interest also could be of use for a further study of the characteristic p Dynamical Mordell-Lang Conjecture.

13.1. The Mordell-Lang Conjecture over fields of positive characteristic

As previously discussed there are counterexamples (see Examples 3.4.5.1 and 11.1.0.6) to an immediate translation of the Dynamical Mordell-Lang conjecture in characteristic p. Even the classical Mordell-Lang conjecture, which was the main motivation for Conjecture 1.5.0.1 does not hold identically in positive characteristic. Hrushovski [**Hru96**] gave a complete description of the *special* subvarieties for the Mordell-Lang conjecture in characteristic p; for an interpretation of the Mordell-Lang problem in terms of *special points* and *special subvarieties*, see Subsection 3.4.3.

THEOREM 13.1.0.1 (Hrushovski [**Hru96**]). *Let X be a semiabelian variety defined over an algebraically closed field K of characteristic p, let $\Gamma \subset X(K)$ be a finitely generated subgroup, and let $V \subset X$ be an irreducible subvariety. If*

$$V(K) \cap \Gamma \text{ is Zariski dense in } V,$$

then there exists

- *a semiabelian subvariety $X_1 \subseteq X$ defined over K,*
- *a semiabelian variety X_0 defined over $\overline{\mathbb{F}}_p$,*
- *a point $\alpha \in X_1(K)$, a subvariety $V_0 \subseteq X_0$ defined over $\overline{\mathbb{F}}_p$, and*
- *an algebraic group endomorphism $h : X_1 \longrightarrow X_0$ defined over K*

such that $V = \alpha + h^{-1}(V_0)$.

However, Theorem 13.1.0.1 leaves open the description of the intersection of the subvariety V of with the finitely generated subgroup Γ. If X is defined over a finite field (i.e., $K = \overline{\mathbb{F}}_p$ in Theorem 13.1.0.1), Moosa and Scanlon [**MS04, MS03**] gave a concrete description of the intersection (see Theorem 13.1.0.3). To describe their result, we first need a definition from [**MS04**].

DEFINITION 13.1.0.2. Let X be a semiabelian variety defined over a finite field $\mathbb{F}_q$ (where q is a power of the prime number p), let F be the corresponding Frobenius map for the finite field $\mathbb{F}_q$ which extends thus to an endomorphism of X, let K be an algebraically closed field of characteristic p, and let $\Gamma \subseteq X(K)$ be a finitely generated subgroup.

(a) By a *sum of F-orbits* in Γ we mean a set of the form

$$S(a_1, \ldots, a_m; \delta_1, \ldots, \delta_m) := \sum_{j=1}^{m} \left\{ F^{n\delta_j} a_j : n \in \mathbb{N}_0 \right\} \subseteq \Gamma$$

where $a_1, \ldots, a_m \in X(K)$ and $\delta_1, \ldots, \delta_m$ are positive integers.
(b) An F-set in Γ is a set of the form $b + C + \Gamma'$ where $b \in \Gamma$, C is a sum of F-orbits in Γ, and $\Gamma' \subseteq \Gamma$ is a subgroup.

As shown in Example 11.1.0.6, the intersection of a subvariety of $\mathbb{G}_m^n$ (defined over a field K of characteristic p) with a finitely generated subgroup of $\mathbb{G}_m^n(K)$ may contain sums of orbits under the Frobenius endomorphism. Moosa and Scanlon [**MS04, MS03**] show that for all subvarieties V of semiabelian varieties X defined over $\mathbb{F}_q$, their intersection with a finitely generated subgroup of $X(K)$ (where K is any field containing $\mathbb{F}_q$) is a finite union of F-sets.

THEOREM 13.1.0.3 (Moosa-Scanlon [**MS04**]). *Let $\mathbb{F}_q$, K, X, Γ be as given in Definition 13.1.0.2, and let $V \subseteq X$ be a subvariety defined over K. If Γ is invariant under F, then $V(K) \cap \Gamma$ is a union of at most finitely many F-sets in Γ.*

We note, very importantly, that in Theorem 13.1.0.3 the semiabelian variety X is *assumed* to be defined over $\mathbb{F}_q$, *but V is not necessarily defined over $\overline{\mathbb{F}}_q$*. If X is not defined over a finite field, there are only some conjectural statements made regarding the intersection $V(K) \cap \Gamma$ (see [**MS03**]).

The proof of Theorem 13.1.0.3 uses the model theoretic interpretation of the Mordell-Lang Conjecture as in the proof of Hrushovski [**Hru96**] for the positive characteristic Mordell-Lang Conjecture. For a thorough model theoretic treatment of the Mordell-Lang conjecture we refer the reader to [**Pil98**].

Ghioca [**Ghi08b**] extended the result of Theorem 13.1.0.3 to all finitely generated subgroups Γ. The proof from [**Ghi08b**] uses a combinatorial argument coupled with the use of the classical Mordell-Lang conjecture for tori. Independently, using arguments from automata theory, Derksen [**Der07**] proved Theorem 13.1.0.3 for linear subvarieties V of a tori $X = \mathbb{G}_m^g$. Later, Derksen and Masser [**DM12**] gave an effective proof of Derksen's result using Diophantine techniques.

13.2. Dynamical Mordell-Lang Conjecture over fields of positive characteristic

Theorem 13.1.0.3 suggests the following conjecture for the Dynamical Mordell-Lang problem in characteristic p.

CONJECTURE 13.2.0.1 (Ghioca-Scanlon). *Let X be a quasiprojective variety defined over a field K of characteristic p. Let $\alpha \in X(K)$, let $V \subseteq X$ be a closed subvariety defined over K, and let*

$$\Phi : X \longrightarrow X$$

be an endomorphism defined over K. Then the set of integers $n \in \mathbb{N}_0$ such that $\Phi^n(\alpha) \in V(K)$ is a finite union of finitely many arithmetic progressions, and finitely many sets of the form

$$(13.2.0.2) \qquad \left\{ \sum_{j=1}^m c_j p^{k_j n_j} : n_j \in \mathbb{N}_0 \text{ for each } j = 1, \ldots m \right\},$$

for some $c_j \in \mathbb{Q}$, and some $k_j \in \mathbb{N}_0$.

As showed in Example 11.1.0.6, it is possible to construct examples so that sets of the form

$$\{ p^m + p^n : m, n \in \mathbb{N} \}$$

appear in (13.2.0.2) corersponding to an intersection of a subvariety with a given orbit. We also note that we already know by Theorem 11.1.0.7 that apart from finitely many arithmetic progressions, the set

$$S_V := \{n \in \mathbb{N}_0 \colon \Phi^n(\alpha) \in V(K)\}$$

has Banach density 0. However, Conjecture 13.2.0.1 predicts a much more precise structure for the set S_V. Already, in the special case of semiabelian varieties defined over $\overline{\mathbb{F}}_p$, Conjecture 13.2.0.1 leads to deep Diophantine questions as we will explain in Section 13.3. Also in Section 13.3 we prove an instance of Conjecture 13.2.0.1 when $X = \mathbb{G}_m^N$ and Φ is an algebraic group endomorphism.

13.3. Dynamical Mordell-Lang Conjecture for tori in positive characteristic

The setup for this section is as follows:
- K is an algebraically closed field of characteristic p,
- $V \subseteq \mathbb{G}_m^N$ is a subvariety defined over K,
- $\alpha := (\alpha_1, \ldots, \alpha_N) \in \mathbb{G}_m^N(K)$, and
- $\Phi \colon \mathbb{G}_m^N \longrightarrow \mathbb{G}_m^N$ is an algebraic group endomorphism defined over K.

Our goal is to describe $V(K) \cap \mathcal{O}_\Phi(\alpha)$.

Since $\Phi \in \mathrm{End}(\mathbb{G}_m^N)$, we know that Φ acts on any point $(x_1, \ldots, x_N) \in \mathbb{G}_m^N$ by

$$\Phi(x_1, \ldots, x_N) = \left(\prod_{j=1}^{N} x_j^{a_{1,j}}, \cdots, \prod_{j=1}^{N} x_j^{a_{N,j}} \right),$$

where the matrix $(a_{i,j})$ is the Jacobian of Φ at the identity of $\mathbb{G}_m^N$. So, we let

$$\Gamma := \Gamma_0^N \subseteq \mathbb{G}_m^N(K),$$

where $\Gamma_0 \subseteq \mathbb{G}_m(K)$ is the subgroup generated by $\alpha_1, \ldots, \alpha_N$. Clearly then $\mathcal{O}_\Phi(\alpha) \subseteq \Gamma$. Since Γ is finitely generated (and even invariant under the Frobenius endomorphism of $\mathbb{G}_m^N$ corresponding to $\mathbb{F}_p$, which is simply the p-th powering map), Theorem 13.1.0.3 yields that

$$V(K) \cap \Gamma \text{ is a finite union of } F\text{-sets.}$$

Thus the problem reduces to understanding the intersection (inside Γ) between an F-set U and $\mathcal{O}_\Phi(\alpha)$. We discuss next a special case in which we can completely describe the intersection $U \cap \mathcal{O}_\Phi(\alpha)$.

Since Φ is an algebraic group endomorphism of $\mathbb{G}_m^N$, it is integral over $\mathbb{Z}$ (which is seen as a subring of $\mathrm{End}(\mathbb{G}_m^N)$); more precisely there exists $m \in \mathbb{N}$ and $c_0, \ldots, c_{m-1} \in \mathbb{Z}$ such that

(13.3.0.1)
$$\Phi^m = \sum_{i=0}^{m-1} c_i \Phi^i,$$

where the sum in (13.3.0.1) is made with respect to the natural group operation of $\mathbb{G}_m^N$. Furthermore, the polynomial

$$f(z) := z^m - \sum_{i=0}^{m-1} c_i z^i$$

is the minimal polynomial for the matrix $(a_{i,j})$; so in particular $m \leq N$. If all the nonzero roots of the polynomial $f(z)$ are distinct, then we can describe completely the intersection $U \cap \mathcal{O}_{\Phi}(\alpha)$.

PROPOSITION 13.3.0.2. *With the notation for Φ, α, Γ, U and f as in the beginning of this section, assume in addition that the nonzero roots of f are distinct. Then the set of all $n \in \mathbb{N}_0$ such that $\Phi^n(\alpha) \in U$ is a finite union of arithmetic progressions.*

PROOF. Without loss of generality we may assume all roots of f are nonzero since otherwise (if $z = 0$ is a root of order k for f), then we may simply replace m by $m - k$ and disregard $\Phi^i(\alpha)$ for $i = 0, \ldots, k-1$, and then obtain a linear recurrence relation of order $m - k$ which is valid starting with the k-th iterate of Φ in place of (13.3.0.1). In other words, this amounts to replacing α by $\Phi^k(\alpha)$ which does not change the desired conclusion, as shown by Proposition 3.1.2.4.

Now, since for each $n \in \mathbb{N}_0$ we have

$$\Phi^{n+m}(\alpha) = \sum_{i=0}^{m-1} c_i \Phi^{n+i}(\alpha),$$

(where each time the sums are taken inside $\mathbb{G}_m^N$) we conclude that there exist linear recurrence sequences $\{b_{i,n}\}_{n \in \mathbb{N}_0}$ (for $i = 0, \ldots, m-1$) such that for each $n \in \mathbb{N}_0$ we have

$$\Phi^n(\alpha) = \sum_{i=0}^{m-1} b_{i,n} \Phi^i(\alpha),$$

and moreover, each sequence $\{b_{i,n}\}_n$ satisfies the same linear recurrence relation:

$$b_{i,n+m} = \sum_{j=0}^{m-1} c_j b_{i,n+j}.$$

So, letting $r_1, \ldots, r_m$ be the distinct (nonzero) roots of the polynomial f and using Proposition 2.5.1.4, we obtain that there exist constants $d_{i,j} \in \overline{\mathbb{Q}}$ such that for each $i = 0, \ldots, m-1$ we have

$$b_{i,n} = \sum_{j=0}^{m-1} d_{i,j} r_j^n, \text{ for each } n \in \mathbb{N}_0.$$

On the other hand, each element in U is of the form

$$\beta_0 + p^{\delta_1 k_1} \beta_1 + \cdots p^{\delta_s k_s} \beta_s + \gamma_0,$$

for some $k_i \in \mathbb{N}_0$, where $\beta_i \in \Gamma$ are given (we can enlarge Γ so that it contains all the points β_i above), and $\delta_1, \ldots, \delta_s \in \mathbb{N}$ are given, and also $\gamma_0 \in U_0$, where U_0 is a given subgroup of Γ. It follows that

$$\Phi^n(\alpha) \in U$$

if and only if there exist $k_1, \ldots, k_s \in \mathbb{N}_0$ such that

$$(13.3.0.3) \qquad \left(\sum_{i=0}^{m-1} b_{i,n} \Phi^i(\alpha) \right) - \left(\beta_0 + \sum_{i=1}^{s} p^{\delta_i k_i} \beta_i \right) \in U_0.$$

Since Γ is a finitely generated group, its torsion subgroup is finite, and the same is true for U_0. Hence, at the expense of replacing (13.3.0.3) by finitely many similar conditions (the only difference being a different β_0 each time, altered by a torsion point in Γ) we may assume Γ (and thus also U_0) is torsion-free.

We fix a $\mathbb{Z}$-basis $\gamma_1, \ldots, \gamma_\ell$ for Γ and express each $\Phi^i(\alpha)$ for $i = 0, \ldots, m-1$, and each β_i in terms of this basis. Then (13.3.0.3) is equivalent to a simultaneous solution to finitely many equations which are of the form

$$(13.3.0.4) \qquad \sum_{i=0}^{m-1} f_i r_i^n + \sum_{j=1}^{s} g_j p^{\delta_j k_j} = g_0$$

or

$$(13.3.0.5) \qquad \sum_{i=0}^{m-1} f_i r_i^n + \sum_{j=1}^{s} g_j p^{\delta_j k_j} \equiv g_0 \pmod{M},$$

for some constants $f_i, g_j \in \overline{\mathbb{Q}}$ and some positive integer M. The derivation of the two equations (13.3.0.4) and (13.3.0.5) follows verbatim using the argument from the proof of [**Ghi08b**, Claim 3.4].

Now, it is elementary to see that the set of $n \in \mathbb{N}_0$ for which there exist some $k_i \in \mathbb{N}_0$ such that (13.3.0.5) holds is a finite union of arithmetic progressions because the set of residue classes of $r_i^n \pmod{M}$ is preperiodic.

On the other hand, (13.3.0.4) also yields a set of solutions n which is a finite union of arithmetic progressions by an easy application of Laurent's theorem [**Lau84**] (the classical Mordell-Lang conjecture for finitely generated subgroups of algebraic tori; see Theorem 3.4.1.1). This concludes the proof of Proposition 13.3.0.2. $\qquad \square$

So, as proven in Proposition 13.3.0.2, when the roots of the minimal polynomial for the (algebraic group) endomorphism Φ of $\mathbb{G}_m^N$ are all distinct, then the set of integers $n \in \mathbb{N}_0$ such that $\Phi^n(\alpha) \in V$ is simply a finite union of arithmetic progression, and thus one does not need the more complicated sets from the conclusion of Conjecture 13.2.0.1.

If we drop the hypothesis that the (nonzero) roots of the minimal polynomial $f(z)$ for the endomorphism Φ are distinct, then the problem is *much* harder. Indeed, the difference is that the linear recurrence relations satisfied by the sequences $\{b_{i,n}\}_{n \in \mathbb{N}_0}$ from the proof of Proposition 13.3.0.2 do not have distinct characteristic roots, and so, one obtains (see Proposition 2.5.1.4) that the formula for the general element $b_{i,n}$ is of the form

$$\sum_{j=1}^{k} P_j(n) r_j^n,$$

where the r_j's are the distinct roots of $f(z)$, while $P_j \in \overline{\mathbb{Q}}[z]$. Hence, going through the same argument as in the proof of Proposition 13.3.0.2 one obtains that the set of solutions n correspond to equations of the form

$$(13.3.0.6) \qquad \sum_{i=0}^{m-1} f_i(n) r_i^n + \sum_{j=1}^{s} g_j p^{\delta_j k_j} = g_0$$

or

$$(13.3.0.7) \qquad \sum_{i=0}^{m-1} f_i(n) r_i^n + \sum_{j=1}^{s} g_j p^{\delta_j k_j} \equiv g_0 \pmod{M},$$

for some constants $g_j \in \overline{\mathbb{Q}}$ and $M \in \mathbb{N}$, and polynomials $f_i \in \overline{\mathbb{Q}}[z]$. Now, Equation (13.3.0.7) yields the solutions n are in a set which is a finite union of arithmetic progressions because the values $f_i(n) \pmod{M}$ are also preperiodic as n ranges over all nonnegative integers. On the other hand, the polynomial-exponential Equation (13.3.0.6) is *very difficult*. For example, even the simpler case

$$(13.3.0.8) \qquad f(n) = \sum_{j=1}^{s} g_j p^{k_j}$$

is unknown in general (see [**BBM13, CZ00, CZ13**] for more details). We would expect (cf. Conjecture 13.2.0.1) that the set of $n \in \mathbb{N}_0$ for which there exist $k_1, \ldots, k_s \in \mathbb{N}_0$ such that (13.3.0.8) holds would always be of the form given in (13.2.0.2). For example, even the special case

$$(13.3.0.9) \qquad n^2 = \sum_{j=1}^{s} g_j p^{k_j}$$

is not known unless $g_1 = 1$ (where $k_1 < \cdots < k_s$) and $s \leq 4$. However, we would always expect that solutions n to (13.3.0.9) are of the form

$$\sum_{i=1}^{\ell} f_i p^{m_i}$$

for some $\ell \in \mathbb{N}$ and constants f_i. Informally, Equation (13.3.0.9) says that if n^2 has at most s nonzero p-adic digits, then n has at most ℓ nonzero p-adic digits (where ℓ is bounded solely on s).

13.4. The Skolem-Mahler-Lech Theorem in positive characteristic

We recall that the Skolem-Mahler-Lech theorem gives a concrete description of the zero sets of linear recurrence sequences over fields of characteristic zero. Furthermore, in Chapter 4, we showed the connection between this theorem and the Dynamical Mordell Lang theorem, which can be seen as proposing a sweeping generalization of the Skolem-Mahler-Lech result.

As we have noted many times, in positive characteristic these results do not hold. In particular, Example 3.4.5.1 can be suitably modified to give an example of a linear recurrence sequence over $\mathbb{F}_p(t)$ whose zero set is precisely the powers of p. In this section, we give a presentation of Derksen's analogue of the Skolem-Mahler-Lech theorem in positive characteristic, which shows that the zero sets of linear recurrences over a field of characteristic $p > 0$ are given by finite unions of (possibly finite) arithmetic progressions along with what he calls *p-normal* sets, which roughly speaking are sets that are built from powers of p, exactly as in (13.2.0.2) (see also the definition of F-sets from Definition 13.1.0.2 which generalizes the p-normal sets).

Unlike earlier results from this chapter, Derksen's methods have the advantage of being *effective*. In particular, he provides algorithms which allow one to completely determine whether a linear recurrence sequence over a field of positive characteristic takes the value zero and moreover, it allows one to describe the zero set completely. This is in stark contrast with the zero characteristic case, where it is currently unknown whether or not it is decidable to determine if an integer-valued linear recurrence sequence has a zero. The bulk of Derksen's argument rests with using *finite-state automata* and we give an overview of these ideas. To present Derksen's ideas, we must give the notion of an automatic sequence and an automatic set.

DEFINITION 13.4.0.1. Let p be a prime number, let Δ be a finite set, and let

$$f : \mathbb{N}_0 \to \Delta.$$

For each $j \in \{0, 1, \ldots, p-1\}$, we define a map

$$e_j : \mathbb{N}_0 \to \mathbb{N}_0 \text{ by } e_j(n) = pn + j$$

and we let Σ denote the semigroup generated by the collection of all e_j under composition. We say that f is a *p-automatic sequence* if the set of distinct sequences in

$$\{f \circ e \ : \ e \in \Sigma\}$$

is a finite set. We say that a subset $S \subseteq \mathbb{N}_0$ is a *p-automatic set* if the characteristic sequence of S is a p-automatic sequence.

We note that this is not the conventional definition of an automatic sequence, which is generally defined in terms of sequences whose n-th term is produced via a finite-state machine that accepts as input the base-p expansion of n. Nevertheless, the definition we give can be seen to be equivalent (cf. Allouche and Shallit [**AS03**, Theorem 6.6.2]).

THEOREM 13.4.0.2 (Derksen [**Der07**]). *Let $f(n)$ be a sequence satisfying a recurrence over a field K of characteristic p. Then the set of natural numbers n such that $f(n) = 0$ is a p-automatic set.*

In fact, Derksen gives a further refinement of Theorem 13.4.0.2, which we shall describe in Subsection 13.4.2 (see Theorem 13.4.2.3). The main ingredient of his proof, however, is the fact that the zero set is p-automatic. In light of the equivalences given in Proposition 2.5.1.4, we note that this can be recast in terms of describing the set of zero coefficients of the power series expansion of a rational function over a field of positive characteristic. We prove an extension of this result for algebraic power series in Subsection 13.4.1, following the exposition from [**AB12**].

13.4.1. Algebraic power series.

DEFINITION 13.4.1.1. Let K be a field. We say that a power series

$$F(t) = \sum_{n=0}^{\infty} f(n)t^n \in K[[t]]$$

is *algebraic* if there exists a natural number d and rational functions

$$\phi_0(t), \ldots, \phi_{d-1} \in K(t)$$

such that

$$F(t)^d + \sum_{j=0}^{d-1} \phi_j(t)F(t)^j \ = \ 0.$$

REMARK 13.4.1.2. It is clear that rational power series form a subset of *algebraic power series*. In fact, Theorem 13.4.0.2, when recast in terms of coefficients of rational functions, holds at the level of algebraic power series.

It is interesting to note that the Skolem-Mahler-Lech theorem in characteristic 0 has no analogue for multivariate rational functions. For instance,

$$G(x, y) = \sum_{m,n} (2^m - n^2) x^m y^n$$

is a bivariate rational power series and its zero set is

$$\{(m, n) \ : \ m \equiv 0 \ (\text{mod} \ 2), n = 2^{m/2}\}.$$

Thus we cannot expect the zero set to be given in terms of arithmetic progressions. Remarkably, in positive characteristic an analogue of Derksen's result holds for multivariate rational power series—in fact it even holds for multivariate algebraic power series!

In this subsection, we give a proof of the following result (see [**AB12**]).

THEOREM 13.4.1.3. *Let K be a field of characteristic $p > 0$ and let $F(x) \in K[[x]]$ be the power series expansion of an algebraic function over $K(x)$. Then the set of n in $\mathbb{N}_0$ for which the coefficient of x^n in $F(x)$ is zero is p-automatic.*

As noted in Remark 13.4.1.2, Theorem 13.4.1.3 can in fact be done at the level of multivariate algebraic power series [**AB12**], but we do not prove this in the book. In order to prove Theorem 13.4.1.3 we need to introduce some notation. We also recall that a field K of characteristic $p > 0$ is *perfect* if the map $x \mapsto x^p$ is surjective on K.

DEFINITION 13.4.1.4. Let p be a prime number and let K be a perfect field of characteristic p. For a power series

$$F(x) = \sum_{n=0}^{\infty} f(n) x^n \in K[[x]]$$

we define the j-th *Cartier operator*

$$E_j(F(x)) := \sum_{n=0}^{\infty} (f \circ e_j(n))^{1/p} x^n$$

for $j \in \{0, 1, \ldots, p-1\}$. We let Ω denote the semigroup generated by the collection of E_j under composition and we let $\Omega(F)$ denote the K-vector space spanned by all power series of the form $E \circ F$ with $E \in \Omega$.

We point out that if $G \in \Omega(F)$ then $E \circ G \in \Omega(F)$ for all $E \in \Omega$. Cartier operators are particularly useful in that they can be used to decompose power series over fields of positive characteristic.

REMARK 13.4.1.5. Let p be a prime number an let K be a perfect field of characteristic p. For a power series

$$F(x) = \sum_{n=0}^{\infty} f(n) x^n \in K[[x]]$$

we have a decomposition

$$F(x) = \sum_{j \in \{0,\ldots,p-1\}} x^j E_j(F(x))^p.$$

Using Proposition 2.5.1.4, it is straightforward to show that if $F(x) \in K[[x]]$ is the power series expansion of a rational function, then $\Omega(F)$ is finite-dimensional. More generally, a result due to Sharif and Woodcock [**SW88**, Corollary 5.4] gives a characterization of the algebraic power series over a perfect field of positive characteristic in terms of this property.

THEOREM 13.4.1.6 (Sharif and Woodcock [**SW88**]). *Let p be a prime number and let K be a perfect field of characteristic p. A power series*

$$F(x) \in K[[x]]$$

is algebraic over $K(x)$ if and only if $\Omega(F)$ is a finite-dimensional K-vector space.

One can rephrase the theorem of Sharif and Woodcock in terms of the coefficients of an algebraic power series.

LEMMA 13.4.1.7. *Let p be a prime number, let K be a perfect field of characteristic p, and let $f : \mathbb{N}_0 \to K$ be a sequence with the property that*

$$F(x) := \sum_{n=0}^{\infty} f(n)x^n \in K[[x]]$$

is the power series expansion of an algebraic function over $K(x)$. Then there exists a positive integer m and there exist maps

$$f_1, \ldots, f_m : \mathbb{N}_0 \to K$$

such that:

(1) *$F_i(x) := \sum_{n=0}^{\infty} f_i(n)x^n \in \Omega(F)$ with $1 \le i \le m$ form a basis for $\Omega(F)$ as a K-vector space;*
(2) *$F_1 = F$;*
(3) *if $g : \mathbb{N}_0 \to K$ has the property that*

$$G(x) := \sum_{n \in \mathbb{N}_0} g(n)x^n \in \Omega(F),$$

then $g \circ e_j \in K f_1^p + \cdots + K f_m^p$ for $j \in \{0, \ldots, p-1\}$.

PROOF. Since $F(x)$ is algebraic, $\dim_K(\Omega(F))$ is finite by Theorem 13.4.1.6. It follows that there exist maps

$$f_1, \ldots, f_m : \mathbb{N} \to K$$

such that the m power series

$$F_i(x) := \sum_{n=0}^{\infty} f_i(n)x^n \in \Omega(F)$$

with $i \in \{1, \ldots, m\}$ form a basis for $\Omega(F)$ as a K-vector space. Let

$$g : \mathbb{N}_0 \to K$$

be such that

$$G(x) := \sum_{n=0}^{\infty} g(n)x^n \in \Omega(F).$$

Then

$$(13.4.1.8) \qquad G(x) = \sum_{j \in \{0,\ldots,p-1\}} x^j E_j(G(x))^p.$$

Let $j \in \{0, 1, \ldots, p-1\}$. By assumption,

$$E_j(G(x)) \in K F_1(x) + \cdots + K F_m(x)$$

and hence

$$E_j(G(x))^p \in K F_1(x)^p + \cdots + K F_m(x)^p.$$

Considering the coefficient of x^{pn+j} in Equation (13.4.1.8), we see $g \circ e_j(n)$ is equal to the coefficient of x^{pn} in $E_j(G(x))^p$, which is in

$$K f_1(n)^p + \cdots + K f_m(n)^p,$$

as desired. $\qquad\qquad\qquad\qquad\qquad\qquad\qquad\qquad\qquad\qquad\qquad\qquad\qquad\square$

We are almost ready to prove Theorem 13.4.1.3. Before doing so, we fix some notation. Given a finitely generated field extension K_0 of $\mathbb{F}_p$ we let $K_0^{\langle p \rangle}$ denote the subfield consisting of all elements of the form c^p with $c \in K_0$. Given $\mathbb{F}_p$-vector subspaces V and W of K_0 we let VW denote the $\mathbb{F}_p$-subspace of K_0 spanned by all products of the form vw with $v \in V, w \in W$ and we let $V^{\langle p \rangle}$ denote the $\mathbb{F}_p$-vector subspace consisting of all elements of the form v^p with $v \in V$. Since K_0 is a finitely generated field extension of $\mathbb{F}_p$, K_0 is a finite-dimensional $K_0^{\langle p \rangle}$-vector space. If we fix a basis

$$K_0 = \bigoplus_{i=1}^{r} K_0^{\langle p \rangle} h_i$$

then we have *projections* $\pi_1, \ldots, \pi_r : K_0 \to K_0$ defined by

$$(13.4.1.9) \qquad c = \sum_{i=1}^{r} \pi_i(c)^p h_i.$$

REMARK 13.4.1.10. For $1 \le i \le r$ and $a, b, c \in K_0$ we have

$$\pi_i(c^p a + b) = c\pi_i(a) + \pi_i(b).$$

The last ingredient of the proof is a technical result due to Derksen.

PROPOSITION 13.4.1.11 (Derksen [**Der07**]). *Let K_0 be a finitely generated field extension of $\mathbb{F}_p$ and let*

$$\pi_1, \ldots, \pi_r : K_0 \to K_0$$

be as in Equation (13.4.1.9). If V is a finite-dimensional $\mathbb{F}_p$-vector subspace of K_0, then there exists a finite-dimensional $\mathbb{F}_p$-vector subspace W of K_0 containing V such that

$$\pi_i(WV) \subseteq W \text{ for } 1 \le i \le r.$$

PROOF. This is [**Der07**, Proposition 5.2]. $\qquad\qquad\qquad\qquad\qquad\qquad\square$

PROOF OF THEOREM 13.4.1.3. By enlarging K if necessary, we may assume that K is perfect. By Lemma 13.4.1.7 we can find maps

$$f_1, \ldots, f_m : \mathbb{N}_0 \to K$$

such that:

(1) $F_i(x) := \sum_{n=0}^{\infty} f_i(n) x^n \in \Omega(F)$, $i = 1, \ldots, m$, form a basis for $\Omega(F)$ as a K-vector space;

(2) $F_1 = F$;

(3) if $g : \mathbb{N}_0 \to K$ has the property that

$$G(x) := \sum_{n=0}^{\infty} g(n)x^n \in \Omega(F),$$

then $g \circ e_j \in K f_1^p + \cdots + K f_m^p$ for $j \in \{0, \ldots, p-1\}$.

It follows that there exist elements $\lambda_{i,j}^{(\ell)}$ for $i, j \in \{1, \ldots, m\}$ and $\ell \in \{0, 1, \ldots, p-1\}$ such that

$$(13.4.1.12) \qquad\qquad f_i \circ e_\ell = \sum_{j=1}^{m} \lambda_{i,j}^{(\ell)} f_j^p.$$

Since $F_1, \ldots, F_m$ are algebraic power series, we have that there is a finitely generated field extension of $\mathbb{F}_p$ such that all coefficients of $F_1, \ldots, F_m$ are contained in this field extension. It follows that the subfield K_0 of K, generated by the coefficients of $F_1(x), \ldots, F_m(x)$ and the elements $\lambda_{i,j}^{(\ell)}$ for $i, j \in \{1, \ldots, m\}$ and $\ell \in \{0, 1, \ldots, p-1\}$, is a finitely generated field extension of $\mathbb{F}_p$.

Since K_0 is a finite-dimensional $K_0^{\langle p \rangle}$ space, we can fix a basis $\{h_1, \ldots, h_r\}$; that is,

$$K_0 = \bigoplus_{i=1}^{r} K_0^{\langle p \rangle} h_i.$$

This then gives *projections* $\pi_1, \ldots, \pi_r : K_0 \to K_0$ defined by

$$(13.4.1.13) \qquad\qquad c = \sum_{i=1}^{r} \pi_i(c)^p h_i.$$

We let V denote the finite-dimensional $\mathbb{F}_p$-vector subspace of K_0 spanned by the elements $\lambda_{i,j}^{(\ell)}$ for $1 \leq i, j \leq d$ and $\ell \in \{0, 1, \ldots, p-1\}$ and by 1. In particular, by Equation (13.4.1.12) for $1 \leq i, j \leq d$ and $\ell \in \{0, 1, \ldots, p-1\}$ we have

$$(13.4.1.14) \qquad\qquad f_i \circ e_\ell \in \sum_{j=1}^{m} V f_j^p.$$

By Proposition 13.4.1.11 there is a finite-dimensional $\mathbb{F}_p$-vector subspace W of K_0 that contains V and has the property that $\pi_i(WV) \subseteq W$ for $1 \leq i \leq r$. Let

$$U = W f_1 + W f_2 + \cdots + W f_m \subseteq \{g \ : g : \mathbb{N}_0 \to K_0\}.$$

Then $|U| \leq |W|^m < \infty$, since we are working with vector spaces over finite fields. If $\ell \in \{1, \ldots, r\}$, $i \in \{1, \ldots, m\}$, and $j \in \{0, 1, \ldots, p-1\}$ then by equation (13.4.1.14) and Remark 13.4.1.10 we have

$$\begin{aligned} \pi_\ell(W f_i \circ e_j) \quad &\subseteq \quad \pi_\ell(WV f_1^p + \cdots + WV f_m^p) \\ &\subseteq \quad \sum_{k=1}^{m} \pi_\ell(WV) f_k \\ &\subseteq \quad \sum_{k=1}^{m} W f_k \\ &= \quad U. \end{aligned}$$

Thus by Remark 13.4.1.10, if $g \in U$ and $j \in \{0, 1, \dots, p-1\}$, then $g_\ell := \pi_\ell(g \circ e_j) \in U$ for $1 \le \ell \le r$. In particular, $g(pn + j) = 0$ if and only if $g_1(n) = g_2(n) = \cdots = g_r(n) = 0$. Given

$$g : \mathbb{N}_0 \to K_0,$$

we let $\chi_g : \mathbb{N}_0 \to \{0, 1\}$ be defined by

$$\chi_g(n) = \begin{cases} 0 & \text{if } g(n) \ne 0; \\ 1 & \text{if } g(n) = 0. \end{cases}$$

Let

(13.4.1.15) $$\mathcal{S} = \{\chi_{g_1} \cdots \chi_{g_t} \ : \ t \ge 0, g_1, \dots, g_t \in U\}.$$

Since $\chi_g^2 = \chi_g$ for all $g \in U$ and U is finite, $\mathcal{S}$ is finite. Moreover, if $g \in U$ and $j \subset \{0, 1, \dots, p-1\}$, then $g_{j,\ell} := g_\ell := \pi_\ell(g \circ e_j) \in U$ for $1 \le \ell \le r$. Then we have

$$(\chi_g \circ e_j)(n) = \prod_{\ell=1}^{r} \chi_{g_\ell}(n),$$

and so we see that if $\chi \in \mathcal{S}$ then $\chi \circ e \in \mathcal{S}$ for all $e \in \Sigma$. Since $\mathcal{S}$ is finite we then have that

$$\chi : \mathbb{N}_0 \to \{0, 1\} \text{ is } p\text{-automatic.}$$

In particular since $f(n) = f_1(n) \in U$, we see that χ_f is p-automatic, and so the set of $n \in \mathbb{N}_0$ such that $f(n) = 0$ is a p-automatic set. This completes the proof of Theorem 13.4.1.3.   □

13.4.2. Derksen's theorem on p-normal sets. In this subsection we give a refinement of Theorem 13.4.0.2, as proven by Derksen [**Der07**] (see Theorem 13.4.2.3). So, if

$$f : \mathbb{N}_0 \to K$$

is a linear recurrence sequence over a field K of characteristic $p > 0$, then the zero set of $f(n)$ is p-automatic. In fact, in this case, the proof shows that if $\chi(n)$ is the characteristic sequence of the zero set of $f(n)$ then for $e \in \Sigma$, $\chi \circ e$ is a finite intersection of characteristic sequences of zero sets of linear recurrences. Moreover, it is easily checked that if $f(n)$ is a simple non-degenerate linear recurrence then $\chi \circ e$ is a finite intersection of characteristic sequences of zero sets of simple non-degenerate linear recurrence sequences.

As stated earlier, the notion of automatic sets and sequences is generally defined using finite-state automata, which, roughly speaking, are machines with a finite set of states, some of which are *accepting states* and the others which are *rejecting states*, and transition rules between the states based on input fed into the machine. Derksen was able to give bounds on the number of states needed in a finite-state automaton that describes the zero set of a linear recurrence sequence in terms of the recurrence and the initial conditions and from this he was able to show that the decidability results mentioned earlier hold.

In fact, by analyzing the types of automata which can occur, Derksen showed that the zero set of a linear recurrence sequence in positive characteristic is much more constrained. To describe Derksen's refinement, we must give the notion of a *p-normal set*.

DEFINITION 13.4.2.1. Let p be a prime number. A subset $S \subseteq \mathbb{N}$ is p-normal if it is a finite union of (possibly finite) arithmetic progressions along with a finite union of sets of the form

$$S_q(c_0; \ldots, c_d) := \{c_0 + c_1 q^{k_1} + \cdots + c_d q^{k_d} : k_1, \ldots, k_d \in \mathbb{N}\} \cap \mathbb{N},$$

where q is a power of p and $c_0, \ldots, c_d$ are nonzero rational numbers satisfying

$$c_0 + \cdots + c_d \in \mathbb{Z} \text{ and } (q-1)c_i \in \mathbb{Z},$$

for $i = 0, \ldots, d$.

REMARK 13.4.2.2. If S and T are p-normal and a and b are natural numbers then $S \cap T$ and $aS + b$ are both p-normal.

Derksen's main result is then the following.

THEOREM 13.4.2.3. [**Der07**, Theorem 1.8] *Let $f(n)$ be a sequence satisfying a recurrence over a field K of characteristic p. Then the set of natural numbers n such that $f(n) = 0$ is a p-normal set.*

To prove Theorem 13.4.2.3, we will again make use of some of the ideas from the theory of finite-state automata. Rather than introduce the formalism of automata, however, we will instead use the more well-known notion of directed graphs.

For the remainder of this subsection, we assume that $g(n)$ is the characteristic sequence of the zero set of a simple non-degenerate linear recurrence. We let

$$g_1(n), \ldots, g_m(n)$$

denote the distinct sequences of the form $g \circ e$ with $e \in \Sigma$, which is a finite set by Theorem 13.4.0.2. By the remarks following the proof of Theorem 13.4.1.3, we have that these are characteristic sequences of intersections of zero sets of simple non-degenerate linear recurrences.

We now construct a directed graph G whose set of vertices is $\{1, 2, \ldots, m\}$ and in which we draw a directed edge with label $\ell \in \{0, 1, \ldots, p-1\}$ from i to j if and only if

$$g_i(pn + \ell) = g_j(n).$$

We then put an equivalence relation on the set of vertices by declaring that $i \sim j$ if there is a directed (possibly empty) path from i to j and a directed path from j to i and we let $[i]$ denote the equivalence class containing the vertex i. We put a partial order on the collection of equivalence classes by declaring that

$$[i] \preceq [j]$$

if and only if there is a directed path (possibly empty) from vertex i to j.

To each path $s = u_1 u_2 \cdots u_n$ on G, where $u_1, \ldots, u_n$ are labeled edges, we can associate a natural number

(13.4.2.4) $$[s]_p := p^{n-1}\ell_1 + p^{n-2}\ell_2 + \cdots + \ell_n$$

where ℓ_i is the label of edge u_i. The following remark can be easily obtained using the formula for geometric series.

REMARK 13.4.2.5. Let $t_1, \ldots, t_r$ be cycles in G and let $s_1, \ldots, s_{r+1}$ be paths on G such that for $i \in \{1, \ldots, r\}$ the terminal vertex of s_i is equal to the initial vertex of t_i and such that the initial vertex of s_i is equal to the terminal vertex of t_{i-1} for $i \in \{2, \dots, r+1\}$. Then

$$\{[s_1 t_1^{m_1} s_2 t_2^{m_2} \cdots s_r t_r^{m_r} s_{r+1}]_p : m_1, \dots, m_r \geq 0\}$$

is p-normal.

To obtain Derksen's refinement from Theorem 13.4.2.3, we first show that the directed paths that begin and end at the same vertex are all generated by a unique cycle.

LEMMA 13.4.2.6. *Let G be the graph described above and let $i \in \{1, \dots, m\}$. If t_1 and t_2 are two paths from i to i then there is a cycle t and natural numbers a and b such that $t_1 = t^a$ and $t_2 = t^b$.*

PROOF. We first show that if t_1 and t_2 have the same length then $t_1 = t_2$. To do this, suppose that we have two distinct paths, t_1 and t_2, from vertex i to itself with the same length. We let r denote the length of t_1 and t_2 and we let

$$\ell_1 = [t_1]_p \text{ and } \ell_2 = [t_2]_p.$$

We next let d be the smallest natural number such that $g_i(n)$ is the characteristic sequence of an intersection of the zero sets of a finite set of simple non-degenerate linear recurrences of length at most d. Then $g_i(n)$ is the characteristic sequence of the zero sets of $h_1, \dots, h_q$, where each h_j is a simple non-degenerate linear recurrence of length at most d. By assumption we have that

$$g_i(n) = 0$$

if and only if

$$h_1(n) = \cdots = h_q(n) = 0.$$

Furthermore, this holds if and only if

$$h_j(p^r n + \ell_1) = h_j(p^r n + \ell_2) \text{ for } j = 1, \dots q.$$

Let $j \in \{1, \dots, q\}$. Then we have

$$h_j(n) = \sum_{i=1}^{e} c_i \alpha_i^n$$

is a simple non-degenerate linear recurrence of length $e \leq d$. Then

$$h_j'(n) := h_j(p^r n + \ell_1) \text{ and } h_j''(n) := h_j(p^r n + \ell_2)$$

are both zero if and only if

$$h_j'(n) := \sum_{i=1}^{e} c_i \alpha_i^{\ell_1} \alpha_i^{p^r n} \text{ and } h_j''(n) := \sum_{i=1}^{e} c_i \alpha_i^{\ell_2} \alpha_i^{p^r n}$$

are both zero. Notice that

$$g_j'(n) := \alpha_1^{\ell_2} h_j'(n) - \alpha_1^{\ell_1} h_j''(n) = \sum_{i=2}^{e} c_i(\alpha_i^{\ell_1} \alpha_1^{\ell_2} - \alpha_i^{\ell_2} \alpha_1^{\ell_1}) \alpha_i^{p^r n}$$

and

$$g_j''(n) := \alpha_e^{\ell_2} h_j'(n) - \alpha_e^{\ell_1} h_j''(n) = \sum_{i=1}^{e-1} c_i(\alpha_i^{\ell_1} \alpha_e^{\ell_2} - \alpha_i^{\ell_2} \alpha_e^{\ell_1}) \alpha_i^{p^r n}$$

are both recurrence sequences of length at most $e - 1 < d$. Since our sequences are non-degenerate, the matrix

$$\begin{pmatrix} \alpha_1^{\ell_1} & -\alpha_1^{\ell_2} \\ \alpha_e^{\ell_1} & -\alpha_e^{\ell_2} \end{pmatrix}$$

is invertible, we have that

$$\{n \colon h_j'(n) = h_j''(n) = 0\} = \{n \colon g_j'(n) = g_j''(n) = 0\}$$

and so $g_i(n)$ is given by

$$\bigcap_{j=1}^{q} \left(Z(h_j') \cap Z(h_j'') \right),$$

which is a finite intersection of zero sets of simple non-degenerate linear recurrences of length less than d, contradicting the minimality of d. It follows that for each positive integer r, there is at most one path of length r from vertex i to itself in G.

We now let t denote the shortest path of positive length from vertex i to itself; if no such path exists, there is nothing to prove. We claim that every path from vertex i to itself is of the form t^n for some n. To see this, suppose that there is a path t' from vertex i to itself that is not of this form. We may choose t' to be the shortest such path with this property. Then we necessarily have that the length of t' is strictly greater than the length of t. We let r' denote the length of t'. Then

$$t^{r'} \text{ and } (t')^r \text{ both have length } rr'$$

and so from what we have just shown we have that

$$t^{r'} = (t')^r.$$

It follows that

$$t' = tu$$

for some path u from i to itself. Moreover, we cannot have that u is a power of t since otherwise t' would also have this property. But now u has length strictly less than that of t', which contradicts the minimality of the length of t'. The result follows.   □

PROOF OF THEOREM 13.4.2.3. We first consider the case when $f(n)$ is a simple non-degenerate linear recurrence sequence and we let $g(n)$ denote the characteristic sequence of the zero set of $f(n)$. We let

$$g_1(n) = g(n), \ldots, g_m(n)$$

denote the distinct sequences of the form $g \circ e$ with $e \in \Sigma$ and we construct a directed graph G whose set of vertices is $\{1, 2, \ldots, m\}$ and in which we draw directed edge with label $\ell \in \{0, 1, \ldots, p-1\}$ from i to j if

$$f_i(pn + \ell) = f_j(n).$$

As previously constructed, we have an equivalence relation on the set of vertices of G and a partial order on the equivalence classes.

By Lemma 13.4.2.6, if there are two distinct paths t_1 and t_2 from edge i to itself then there is some path t and natural numbers m and n such that

$$t_1 = t^m \text{ and } t_2 = t^n.$$

It follows that each equivalence class C has a unique cycle t such that if i and j are in C and q is a path from i to j then q is a subpath of t^n for some n. If C and C' are two equivalence classes with the property that

$$C \prec C'$$

and such that there does not exist an equivalence class C'' such that

$$C \prec C'' \prec C'$$

then for $i \in C$ and $j \in C'$ there are at most finitely many minimal paths from vertex i to j. It follows that for each pair of vertices i, j, the collection of directed paths on G is a finite union of sets of paths of the form

$$\mathcal{P}(s_1, \ldots, s_{r+1}; t_1, \ldots, t_r) := \{s_1 t_1^{m_1} s_2 \cdots s_r t_r^{m_r} s_{r+1} : m_1, \ldots, m_r \geq 0\},$$

where $t_1, \ldots, t_r$ are cycles and s_i are paths with no repeated vertices. We recall that to each path $s = e_1 e_2 \cdots e_n$ on G whose source vertex is 1 we can associate a natural number

$$[s]_p := p^{n-1} \ell_1 + p^{n-2} \ell_2 + \cdots + \ell_n$$

where ℓ_i is the label of edge e_i. Then by construction

$$f([s]_p) = 0 \text{ if and only if } g_1([s]_p) = 1,$$

which occurs if and only if

$$g_j(0) = 1 \text{ where } j \text{ is the terminal vertex of the path } s.$$

We now let T denote the set of j for which $g_j(0) = 1$. Then the collection of paths from vertex 1 to a vertex $j \in T$ is a finite union of sets of the form

$$\mathcal{P}(s_1, \ldots, s_{r+1}; t_1, \ldots, t_r).$$

By Remark 13.4.2.5, the set

$$\{[s_1 t_1^{m_1} s_2 t_2^{m_2} \cdots s_r t_r^{m_r} s_{r+1}]_p : m_1, \ldots m_r \geq 0\}$$

is p-normal. The result now follows.

In general, if $f(n)$ is not simple and non-degenerate, then we have nonzero $\alpha_1, \ldots, \alpha_d \in \overline{K}$ and polynomials $q_1, \ldots, q_d$ such that

$$(13.4.2.7) \qquad f(n) = \sum_{i=1}^{d} q_i(n) \alpha_i^n$$

for all n sufficiently large. We may assume without loss of generality that Equation (13.4.2.7) holds for all $n \in \mathbb{N}_0$. We can find some natural number a such that if α_i / α_j is a root of unity then $\alpha_i^a = \alpha_j^a$. It then follows that for $b \in \{0, 1 \ldots, a-1\}$ then $f(an + b)$ is either identically zero or it is a non-degenerate nonzero linear recurrence sequence. Next notice that for $j \in \{0, 1, \ldots, p-1\}$ we have

$$q_i(pn + j) = q_i(j)$$

and so each $f(pn + j)$ is a simple linear recurrence sequence. It follows that for $j \in \{0, 1, \ldots, ap-1\}$ we either have

$$f(apn + j) = 0 \text{ for all } n,$$

or $f(apn + j)$ is a simple non-degenerate linear recurrence sequence and hence in either case we have that the zero set is p-normal. We now obtain the desired result from Remark 13.4.2.2. $\qquad \square$

CHAPTER 14

Related problems in arithmetic dynamics

In recent years there have been many interesting arithmetic problems studied in the context of dynamics. Some of them had their original motivation in deep conjectures from arithmetic geometry. In the next sections we will survey some of these problems and show their connection to the Dynamical Mordell-Lang Conjecture. In each case, we will first start with the classical problem in arithmetic geometry and then describe the dynamical analogue, which often was formulated in order to generalize the classical case. For a more thorough discussion of the major problems in arithmetic dynamics, we recommend the excellent book of Silverman [**Sil07**] to the interested reader.

14.1. Dynamical Manin-Mumford Conjecture

14.1.1. The classical Manin-Mumford Conjecture. The Manin-Mumford conjecture, proven by Raynaud (see [**Ray83a**] for the case of curves, and [**Ray83b**] for the general case), states that if V is a subvariety of an abelian variety A defined over $\mathbb{C}$, then V contains a Zariski dense set of torsion points if and only if V is a torsion translate of an abelian subvariety of A. The direct implication is immediate since the torsion points are Zariski dense in any abelian subvariety. Over number fields, a stronger theorem, conjectured by Bogomolov and proven by Ullmo [**Ull98**] (in the case of curves embedded in their Jacobian) and Zhang [**Zha98**] (in the case of arbitrary subvarieties of abelian varieties), states that $V \subseteq A$ contains a Zariski dense set of points with Néron-Tate height tending to zero if and only if V is a torsion translate of an abelian subvariety of A (for more details regarding the canonical height on abelian varieties see Section 2.6). The proofs of Ullmo [**Ull98**] and Zhang [**Zha98**] make important use of an equidistribution theorem of Szpiro-Ullmo-Zhang for points of small canonical height on abelian varieties (see [**SUZ97**]). Recently, dynamical analogues of this equidistribution theorem have been proved by various authors [**BR06, CL06, FRL06, Yua08**].

14.1.2. The dynamical formulation. Motivated by the above results, Zhang [**Zha06**] formulated dynamical analogues of both the Manin-Mumford and the Bogomolov conjectures. Essentially, Zhang asked whether a subvariety V of a projective variety X defined over $\overline{\mathbb{Q}}$ must be preperiodic under the action of a *suitable* endomorphism Φ of X once it contains a Zariski dense set of preperiodic points. One has to define carefully the notion of *suitable endomorphism* since otherwise there would be obvious counterexamples to such a statement. Indeed, if

$$X = \mathbb{P}^1 \times \mathbb{P}^1 \text{ and } \Phi = (\varphi_1, \varphi_2)$$

is the endomorphism induced by the action of the polynomials

$$\varphi_1(z) = z^2 \text{ and } \varphi_2(z) = z^3,$$

then clearly the diagonal subvariety of X contains infinitely many Φ-preperiodic points even though it is not preperiodic under the action of Φ. The reason for this is that Φ acts with *different weights* on the two projective lines. This prompted Zhang to ask the Dynamical Manin-Mumford conjecture for *polarizable* endomorphisms Φ, i.e., morphisms for which there exists an ample line bundle $\mathcal{L}$ of the projective variety X such that

$$\Phi^*(\mathcal{L}) \text{ is linearly equivalent to } \mathcal{L}^{\otimes d}$$

for some integer $d \geq 2$. However, there are counterexamples even to this formulation which come from CM endomorphisms of elliptic curves (for more details see [**GTZ11a**] and [**Paz10**]). Again the problem is that for an endomorphism Φ of $E \times E$ (where E is a CM elliptic curve), even if Φ acts with same weights on the elliptic curves (and thus Φ is polarizable), it could be that its action on the two elliptic curves has *different angles* (which can be seen by looking at the induced action on the tangent subspaces). Thus the diagonal subvariety of $E \times E$ might still contain infinitely many Φ-preperiodic points, even though the diagonal is not Φ-preperiodic. Hence the authors of [**GTZ11a**] proposed the following revised version of the Dynamical Manin-Mumford Conjecture:

CONJECTURE 14.1.2.1 ([**GTZ11a**]). *Let X be a projective variety, let*

$$\Phi : X \longrightarrow X$$

be an endomorphism defined over $\mathbb{C}$ with a polarization, and let V be a subvariety of X which has no component included into the singular part of X. Then V is preperiodic under Φ if and only if there exists a Zariski dense subset of smooth points

$$x \in V \cap \mathrm{Prep}_\Phi(X)$$

such that the tangent subspace of V at x is preperiodic under the induced action of Φ on the Grassmannian $\mathrm{Gr}_{\dim(V)}(T_{X,x})$. (Here we denote by $T_{X,x}$ the tangent space of X at the point x.)

In [**GTZ11a**] it was proven that Conjecture 14.1.2.1 holds for algebraic group endomorphisms of abelian varieties, and also for lines V contained in

$$X := \mathbb{P}^1 \times \mathbb{P}^1$$

under the action of

$$(x,y) \mapsto \Phi(x,y) := (f(x), g(y)),$$

where $f, g \in \mathbb{C}(z)$. The proof for abelian varieties is a simple consequence of Raynaud's theorem [**Ray83a**]. On the other hand, the proof for lines contained in $\mathbb{P}^1 \times \mathbb{P}^1$ follows by using the equidistribution results of Baker-DeMarco [**BD11**] and of Yuan-Zhang [**YZ**] for preperiodic points under the action of rational maps. The equidistribution results yield that two rational maps f and g must share the same Julia set (and same equivariant measure on the Julia set—see [**BD11, YZ**]) once f and g share an infinite set of preperiodic points. The equality of the Julia sets is sufficient to infer the desired conclusion in the Dynamical Manin-Mumford Conjecture if neither f nor g is a Lattès map; in the latter case one needs to use the full strength of the hypotheses from Conjecture 14.1.2.1 and a classification of Lattès maps done by Douady and Hubbard [**DH93**].

An outstanding open case of Conjecture 14.1.2.1 is the case when $X = \mathbb{P}^2$. In this case it is expected that one does *not* need the extra assumption from Conjecture 14.1.2.1 regarding the action of Φ on the tangent subspaces, and thus Zhang's original question from [**Zha06**] has a positive answer. In this direction, we note the partial result of Dujardin and Favre [**DF**] towards the Dynamical Manin-Mumford Conjecture for rational self-maps of $\mathbb{P}^2$ which are induced by polynomial automorphisms of $\mathbb{A}^2$.

It is also important to note that Fakhruddin [**Fak14**] proved that Zhang's original conjecture (without the strengthening from Conjecture 14.1.2.1) holds for generic endomorphisms of $\mathbb{P}^N$ since for generic endomorphisms of $\mathbb{P}^N$ there are no proper preperiodic subvarieties, and also no proper subvariety contains infinitely many preperiodic points in this case. We recall that a *generic endomorphism* of $\mathbb{P}^N$ of degree d is given by $N + 1$ homogeneous polynomials of degree d in the variables

$$X_0, \ldots, X_N$$

with the property that the coefficients of all these polynomials are algebraically independent.

14.1.3. Combining the two conjectures. One may formulate a conjecture which combines both the Dynamical Mordell-Lang and the Dynamical Manin-Mumford conjectures as follows.

DEFINITION 14.1.3.1. For an endomorphism Φ of a quasiprojective variety X defined over $\mathbb{C}$, and for a point $\alpha \in X(\mathbb{C})$, we let $\overline{\mathcal{O}}_\Phi(\alpha)$ be the *grand orbit* of α defined as follows:

$$\overline{\mathcal{O}}_\Phi(\alpha) = \{\beta \in X(\mathbb{C}) : \text{there exist } m, n \in \mathbb{N}_0 \text{ such that } \Phi^m(\beta) = \Phi^n(\alpha)\}.$$

CONJECTURE 14.1.3.2. *Let X be a smooth projective variety defined over $\mathbb{C}$, let $V \subseteq X$ be an irreducible subvariety, let $\alpha \in X(\mathbb{C})$, and let Φ be a polarizable endomorphism of X. Then V is preperiodic under Φ if and only if there exists a Zariski dense subset of points $x \in V \cap \overline{\mathcal{O}}_\Phi(\alpha)$ such that the tangent subspace of V at x is preperiodic under the induced action of Φ on the Grassmanian $\mathrm{Gr}_{\dim(V)}(T_{X,x})$.*

If $\overline{\mathcal{O}}_\Phi(\alpha)$ is replaced by $\mathcal{O}_\Phi(\alpha)$ and we drop the condition regarding the action on the tangent space, then the above question reduces to the Dynamical Mordell-Lang conjecture. If α is a preperiodic point, then $\overline{\mathcal{O}}_\Phi(\alpha) \subseteq \mathrm{Prep}_\Phi(X)$ and thus we reduce the above question to the Dynamical Manin-Mumford Conjecture.

Using a strategy similar to the proof of Theorem 11.4.2.1 one can prove (see [**BGT15b**]) the following result regarding *coherent backward orbits* which are subsets of grand orbits (see also Definition 14.1.3.1).

DEFINITION 14.1.3.3. Let X be a quasiprojective variety defined over a field K, let

$$\Phi : X \longrightarrow X$$

be an endomorphism defined over K, and let $\alpha \in X(K)$. A *coherent backward orbit* of x (with respect to Φ) is a sequence $\{x_{-n}\}_{n \geq 0}$ such that

$$x_0 = x$$

and for each $n \geq 0$, we have

$$\Phi(x_{-n-1}) = x_{-n}.$$

Then arguing as in the proof of Proposition 11.4.3.1 (see [**BGT15b**]), one proves that if $V \subseteq X$ is a subvariety, $\{x_{-n}\}_{n \geq 0}$ is a coherent backward orbit of a point in X, and the set

$$S_V := \{n \in \mathbb{N}_0 \colon x_{-n} \in V\}$$

has positive Banch density, then S_V contains an infinite arithmetic progression. Then the following result (see [**BGT15b**]) follows almost identically as the proof of Proposition 11.4.3.1.

THEOREM 14.1.3.4. *Let X be a quasiprojective variety, let $\Phi \colon X \longrightarrow X$ be an endomorphism, let $\{x_{-n}\}_{n \geq 0}$ be a coherent backward orbit of a point $x \in X$, and let $V \subseteq X$ be a subvariety. If the set*

$$S_V := \{n \in \mathbb{N} \colon x_{-n} \in V\}$$

has positive Banach density, then V contains a periodic subvariety.

In [**BGT15b**], Theorem 14.1.3.4 is proven in the more general context of coherent backward orbits under the action of continuous endomorphisms of Noetherian spaces. Actually, similar to the proof of Theorem 11.4.2.1, in [**BGT15b**] the authors prove the more precise result that for any subvariety V, the set S_V is a union of at most finitely many arithmetic progressions along with a set of Banach density 0. Theorem 14.1.3.4 can also be proven (even in the higher generality of coherent backward orbits in Noetherian spaces) by employing measure-theoretic techniques similar to those of [**Gig, Gig14**].

14.2. Unlikely intersections in dynamics

14.2.1. The classical setting. The Dynamical Mordell-Lang Conjecture can be interpreted as saying that a subvariety V of a given quasiprojective variety X is *unlikely* to intersect an orbit $\mathcal{O}_\Phi(\alpha)$ of a point α of X under a self-map Φ in a infinite set of points. More precisely, the intersection can be infinite only if V contains a positive dimensional subvariety which is periodic under Φ (see Conjecture 3.1.3.2), and of course, this is unlikely to happen for a generic subvariety V of X. In this section we describe another problem of *unlikely intersections* in arithmetic geometry which also has a counterpart in algebraic dynamics. For a comprehensive discussion of this exciting direction of research for the unlikely intersections in arithmetic geometry, we refer the reader to the beautiful book of Zannier [**Zan12**]. Roughly speaking, the principle of *unlikely intersections* refers to the fact that if an intersection of two geometric objects has larger dimension, or more generally, is *larger than expected* (if we refer at number of points, for example), then there is a *robust geometric reason* for why this happens. Next we illustrate this principle in the context of algebraic dynamical systems.

14.2.2. The dynamical setting. In [**BD11**], Baker and DeMarco proved the following result.

THEOREM 14.2.2.1 (Baker-DeMarco [**BD11**]). *Let $a, b \in \mathbb{C}$, and let $d \geq 2$ be an integer. Then there exist infinitely many $\lambda \in \mathbb{C}$ such that both a and b are preperiodic for the action of*

$$f_\lambda(x) := x^d + \lambda$$

on $\mathbb{C}$ *if and only if* $a^d = b^d$.

One of the main ingredients in the beautiful proof from [**BD11**] is the theorem on equidistribution for points of small canonical height on Berkovich spaces (see [**BR06, CL06, FRL06**]). In particular, Theorem 14.2.2.1 yields that once there exist infinitely many $\lambda \in \mathbb{C}$ such that both a and b are preperiodic under the action of $x^d + \lambda$, then *for all* $\lambda \in \mathbb{C}$, we have that a is preperiodic if and only if b is preperiodic.

The problem solved in Theorem 14.2.2.1 was originally suggested by Zannier, as a dynamical analogue of a question on families of elliptic curves studied by Masser and Zannier in [**MZ08, MZ10, MZ12**]. In [**MZ08**], Masser and Zannier show that there exist at most finitely many $\lambda \in \mathbb{C}$ such that the points

$$\left(2, \sqrt{2(2-\lambda)}\right) \text{ and } \left(3, \sqrt{6(3-\lambda)}\right)$$

are both torsion on the elliptic curve given by the equation

$$y^2 = x(x-1)(x-\lambda).$$

The techniques used by Masser and Zannier are much different than the ones employed by Baker and DeMarco since they use the method of Pila and Zannier [**PZ08**] based on *o*-minimality (for the definition of *o*-minimality see Definition 10.2.3.2) to count algebraic points on real analytic curves. The problem from [**MZ08, MZ10, MZ12**] was motivated by the Pink-Zilber conjectures in arithmetic geometry regarding unlikely intersections between a subvariety V of a semiabelian variety A and families of algebraic subgroups of A of codimension greater than the dimension of V (see [**BMZ99, Hab09, Pin**]). A thorough treatment of Pink-Zilber conjectures can be found in [**Zan12**].

In [**GHT13, GHT15**] additional instances of the unlikely intersection principle in dynamics are proven which generalize the results from [**BD11**]. In particular, one considers two families of dynamical systems at the same time, and also one allows the starting points also depend on the parameter λ. The following question is implicitly raised in [**GHT13, GHT15**]:

CONJECTURE 14.2.2.2. *Let* X *be a smooth projective curve defined over* $\mathbb{C}$. *Let* $\eta \in X(\mathbb{C})$ *and let* $Y := X \setminus \{\eta\}$. *We let* $\mathbf{A}$ *be the ring of rational functions on* X *defined over* $\mathbb{C}$ *that are regular on* Y. *Suppose that we have rational functions*

$$\mathbf{f}_1 = P_1(x)/Q_1(x) \text{ and } \mathbf{f}_2 = P_2(x)/Q_2(x)$$

such that $P_i, Q_i \in \mathbf{A}[x]$ *and the leading coefficients of* P_i *and of* Q_i *are nonzero constants for* $i = 1, 2$. *For each* $\lambda \in Y(\mathbb{C})$ *and* $i = 1, 2$, *we denote by* $\mathbf{f}_{\lambda,i}$ *the rational function obtained by evaluating each coefficient of* P_i *and of* Q_i *at* λ. *For each* $i = 1, 2$, *let*

$$\mathbf{c}_i = \frac{\mathbf{a}_i}{\mathbf{b}_i}$$

where $\mathbf{a}_i, \mathbf{b}_i \in \mathbf{A}$. *If there exists an infinite family of* $\lambda_n \in Y(\mathbb{C})$ *such that both* $\mathbf{c}_1(\lambda_n)$ *and* $\mathbf{c}_2(\lambda_n)$ *are preperiodic points for the actions of* $\mathbf{f}_{\lambda_n,1}$ *respectively of* $\mathbf{f}_{\lambda_n,2}$, *then at least one of the following conditions holds:*

(a) *for each* $\lambda \in Y(\mathbb{C})$, *we have that* $\mathbf{c}_1(\lambda)$ *is preperiodic for* $\mathbf{f}_{\lambda,1}$;
(b) *for each* $\lambda \in Y(\mathbb{C})$, *we have that* $\mathbf{c}_2(\lambda)$ *is preperiodic for* $\mathbf{f}_{\lambda,2}$;
(c) *for each* $\lambda \in Y(\mathbb{C})$, *we have that* $\mathbf{c}_1(\lambda)$ *is preperiodic for* $\mathbf{f}_{\lambda,1}$ *if and only if* $\mathbf{c}_2(\lambda)$ *is preperiodic for* $\mathbf{f}_{\lambda,2}$.

We note that the results from [**BD11, MZ08, MZ10, MZ12**] can be interpreted as special cases of the above conjecture. Indeed, the point

$$\left(2, \sqrt{2(2-\lambda)}\right) \text{ is torsion for the elliptic curve } y^2 = x(x-1)(x-\lambda)$$

if and only if the point

$$2 \text{ is preperiodic for the Lattès map } \frac{(x^2 - \lambda)^2}{4x(x-1)(x-\lambda)}$$

corresponding to the multiplication-by-2-map on the above elliptic curve. In [**BD11, BWY, MZ08, MZ10, MZ12**] precise relations between the starting points c_1 and c_2 were proven so that—in the instances discussed in those papers—condition (c) is satisfied by c_1 and c_2.

Conjecture 14.2.2.2 was proven in [**GHT13**] when $Y = \mathbb{A}^1$ and each f_i and also c_i is a polynomial. In [**GHT15**] additional cases of Conjecture 14.2.2.2 were proven when X is no longer the projective line, and each f_i is a family of rational maps. Furthermore, a higher dimensional variant of this problem was considered. The key for these extensions lies in the deep equidistribution results of small points on metrized line bundles of Yuan and Zhang [**YZ**] (for more details, see [**GHT15**]).

The problem of finding explicit relations between the starting points c_1 and c_2 so that condition (c) holds is much more delicate. In [**BD11**] it is shown that the explicit relation $a^d = b^d$ holds between the starting points a and b under the assumption that they are simultaneously preperiodic for $x^d + \lambda$ for infinitely many $\lambda \in \mathbb{C}$. This result is obtained using techniques from complex analysis which are specific to families of polynomial mappings. Masser and Zannier [**MZ10, MZ12**] proved that two families of points P_λ and Q_λ on a given family of elliptic curves E_λ are simultaneously torsion points for infinitely many values $\lambda \in \mathbb{C}$ if and only if P_λ and Q_λ are linearly dependent over $\mathbb{Z}$, i.e., on the generic fiber of the elliptic family we have an identity of the form $mP_\lambda + nQ_\lambda = 0$ for some nonzero integers m and n. This special relation was obtained exploiting the group structure of the family of elliptic curves. For a generic family of dynamical systems as in Conjecture 14.2.2.2 one cannot use either of the above tools for finding an explicit relation between the two starting points c_1 and c_2 assuming they satisfy condition (c) from Conjecture 14.2.2.2 (see [**GHT15**] for some partial results in this direction).

The area of unlikeley intersections in a dynamical setting is very rich and grows very fast, as shown by additional new results being proven in [**BD13, GH13, GHT**].

14.3. Zhang's conjecture for Zariski dense orbits

14.3.1. Original setting. Zhang conjectured that if Φ is a polarizable endomorphism of an irreducible projective variety X defined over $\overline{\mathbb{Q}}$, then there exists $\alpha \in X(\overline{\mathbb{Q}})$ such that $\mathcal{O}_\Phi(\alpha)$ is Zariski dense in X (see [**Zha06**, Conj. 4.1.6]). Now, let Y be the union of all proper subvarieties V of X (defined over $\overline{\mathbb{Q}}$) which are Φ-preperiodic. In light of the following lemma, Zhang's conjecture can be reformulated as asserting that $X \neq Y$.

LEMMA 14.3.1.1. *If X, Y, Φ are as above, then Y consists of the points $\alpha \in X$ for which $\mathcal{O}_\Phi(\alpha)$ is not Zariski dense in X.*

PROOF. Pick $\alpha \in Y$, and let $V \subseteq Y$ be a proper Φ-preperiodic subvariety of X such that $\alpha \in V$; moreover, pick $k \geq 0$ and $N \geq 1$ such that $\Phi^{k+N}(V) \subseteq \Phi^k(V)$. Then $\mathcal{O}_{\Phi^N}(\Phi^k(\alpha)) \subseteq \Phi^k(V)$, so

$$\mathcal{O}_\Phi(\alpha) \subseteq \bigcup_{i=0}^{k+N-1} \Phi^i(V).$$

Since $V \neq X$ and X is irreducible, it follows that $\mathcal{O}_\Phi(\alpha)$ is not Zariski dense in X.

Conversely, pick $\alpha \in X \setminus Y$, and let Z be the Zariski closure of $\mathcal{O}_\Phi(\alpha)$. Proposition 3.1.2.14 yields that Z is periodic under Φ (actually, it is fixed by Φ). Therefore, each irreducible component (including the component containing α) is preperiodic under Φ. Since $\alpha \notin Y$, it follows that $Z = X$.   □

On the other hand, a positive answer to the Dynamical Mordell-Lang Conjecture yields that each Zariski dense orbit $\mathcal{O}_\Phi(\alpha)$ intersects any proper subvariety V of the irreducible quasiprojective variety X in at most finitely many points. Indeed, if $\mathcal{O}_\Phi(\alpha) \cap V(\mathbb{C})$ is infinite, then, by the Dynamical Mordell-Lang Conjecture, there exists $k, N \in \mathbb{N}$ such that $\mathcal{O}_{\Phi^N}(\Phi^k(\alpha)) \subseteq V(\mathbb{C})$. Therefore

$$\mathcal{O}_\Phi(\alpha) \subseteq \{\Phi^i(\alpha) : 0 \leq i \leq k-1\} \bigcup \left(\cup_{j=0}^{N-1} \Phi^j(V) \right),$$

and since $\dim(V) < \dim(X)$ it follows that $\mathcal{O}_\Phi(\alpha)$ is not Zariski dense in X.

14.3.2. Medvedev-Scanlon variant. Medvedev and Scanlon [**MS14**] consider a situation inspired by Zhang's conjecture, but which is at one level more general in that they drop the polarizability hypothesis, but in another sense more special in that the map Φ is assumed to be given by a sequence of univariate polynomials. In [**MS14**, Fact 2.25], Medvedev and Scanlon state the characterization of periodic subvarieties of arbitrary dimension in terms of periodic curves (which is proven in Medvedev's PhD thesis [**Med**]).

THEOREM 14.3.2.1 (Medvedev, Scanlon). *Let K be a field of characteristic 0, and let $f_1, \ldots, f_N \in K[z]$ be polynomials of degree $d \geq 2$. Assume that the polynomials*

$$f_1(z), \ldots, f_N(z)$$

are not conjugate to z^d or $T_d(z)$ (the d-th Chebyshev polynomial) for any positive integer d. Let Φ denote the action of $(f_1, \ldots, f_N)$ on $\mathbb{A}^N$ and let V be a Φ-periodic irreducible subvariety of $\mathbb{A}^N$ defined over K. Then V is of the form

$$(14.3.2.2) \qquad\qquad V = \bigcap_{i=1}^{\ell} \pi_{i,j}^{-1}(C_{i,j}),$$

for some $\ell \in \mathbb{N}$, where for each $i, j \in \{1, \ldots, N\}$, the curve $C_{i,j} \subseteq \mathbb{A}^2$ is (f_i, f_j)-periodic, and $\pi_{i,j} : \mathbb{A}^N \longrightarrow \mathbb{A}^2$ is the projection map of $\mathbb{A}^N$ onto the corresponding two coordinates i and j.

The proof of Theorem 14.3.2.1 is based on Medvedev's PhD thesis [**Med**] for classifying group-like σ-degree 1 minimal sets in the model theory of ACFA (the theory of algebraically closed fields with a distinguished automorphism). So, Theorem 14.3.2.1 shows that *all* periodic subvarieties are boolean combinations of pullbacks of periodic plane curves under projection maps from $\mathbb{A}^N$ to various pairs of coordinates of the N-dimensional affine space. Using the above powerful classification result for periodic subvarieties of $\mathbb{A}^N$ under the coordinatewise action of N one-variable polynomials, Medvedev and Scanlon prove the following result.

THEOREM 14.3.2.3 (Medvedev-Scanlon [**MS14**]). *Let K be a field of characteristic 0, let*

$$f_1, \ldots, f_N \in K[z]$$

be polynomials of degree at least equal to 2, and let $\Phi : \mathbb{A}^N \longrightarrow \mathbb{A}^N$ be given by

$$\Phi(x_1, \ldots, x_N) := (f_1(x_1), \ldots, f_N(x_N)).$$

Then there exists $\alpha \in \mathbb{A}^N(K)$ such that $\mathcal{O}_\Phi(\alpha)$ is Zariski dense.

14.3.3. A more general conjecture. Motivated by their result ([**MS14**, Theorem 7.16]) and also by a geometric reformulation of Zhang's conjecture made by Amerik and Campana [**AC08**] (see also [**BGZ**] for a generalization of [**AC08**] to arbitrary semigroups of endomorphisms), Medvedev and Scanlon [**MS14**, Conjecture 7.14] formulated the following strengthening of Zhang's conjecture:

CONJECTURE 14.3.3.1. *Let K be an algebraically closed field of characteristic zero, let X be an irreducible algebraic variety over K, and let*

$$\Phi : X \longrightarrow X$$

be a rational self-map. Suppose there is no positive dimensional algebraic variety Y and dominant rational map $g : X \longrightarrow Y$ for which $g \circ \Phi = g$ generically. Then there is some point $\alpha \in X(K)$ such that $\mathcal{O}_\Phi(\alpha)$ is Zariski dense in X.

Theorem 14.3.2.3 is a special case of Conjecture 14.3.3.1. Amerik, Bogomolov and Rovinsky [**ABR11**] prove other instances of Conjecture 14.3.3.1 using a parametrization of the orbit similar to our p-adic parametrization of orbits under étale maps (see Chapter 4). We will describe in the next subsection how the approach from [**ABR11**] can be used to prove a certain special case of Conjecture 14.3.3.1.

In a somewhat different (but related) direction we also mention the work of Amerik [**Ame11**] who proves the existence of a non-preperiodic algebraic point for a rational self-map of infinite order. Using the p-adic parametrization from [**BGT10**] of the orbit of a smooth point $\alpha \in X$ under a map Φ which is unramified at α, Amerik showed the existence of a non-preperiodic algebraic point which lies sufficiently close p-adically to α. In a (yet, another) related direction, the authors used the same p-adic methods to derive upper bounds for the period of arbitrary dimension periodic subvarieties under the action of étale maps (see [**BGT15a**]); this is discussed in Section 14.4.

14.3.4. Special cases of Conjecture 14.3.3.1. In [**ABR11**] it was proven that if there exists a fixed point $x_0 \in X(\overline{\mathbb{Q}})$ for Φ^N (for some positive integer N) such that the Jacobian of Φ^N at x has multiplicatively independent eigenvalues, then Conjecture 14.3.3.1 has a positive answer. The strategy from [**ABR11**] was to show that choosing a non-preperiodic point x sufficiently close p-adically to x_0 (where p is a suitable prime of good reduction for Φ) yields a Zariski dense orbit $\mathcal{O}_\Phi(x)$ (see [**ABR11**, Corollary 2.7]).

We work now with endomorphisms $\Phi := (f_1, \ldots, f_N)$ of $X := (\mathbb{P}^1)^N$ where each rational function f_i is defined over a number field K and it has degree larger than 1. In addition we assume each f_i is not post-critically finite (PCF), i.e., each f_i has at least one critical point which is not preperiodic. So in order to apply the main result of [**ABR11**] it suffices to find for each f_i some periodic point x_i such that all multipliers λ_i are multiplicatively independent. The following Lemma yields the desired conclusion.

LEMMA 14.3.4.1. *Let K be a number field and let $f \in K(x)$ be a non-PCF rational map of degree larger than 1. Then there exist infinitely many primes $\mathfrak{p}$ of K such that for each such prime there exists some periodic point of f whose multiplier is nonzero and divisible by $\mathfrak{p}$.*

PROOF. Let x_0 be a critical point of $f(x)$ which is not preperiodic. Using [**BGKT12**, Lemma 4.1] (see also Lemma 6.2.2.1) we know there exist infinitely many primes $\mathfrak{p}$ of K (of good reduction for f) such that for some positive integer n we have that $f^n(x_0)$ and x_0 are in the same residue class modulo $\mathfrak{p}$. Therefore there exists some point x_1 in the same residue class as x_0 modulo $\mathfrak{p}$ such that $f^n(x_1) = x_1$. Then

$$(f^n)'(x_1) \equiv (f^n)'(x_0) \equiv 0 \pmod{\mathfrak{p}}.$$

Furthermore, since there exist at most finitely many primes $\mathfrak{p}$ such that x_0 is in the same residue class modulo $\mathfrak{p}$ as a periodic critical point of f, we obtain that there exist infinitely many primes $\mathfrak{p}$ such that for some periodic point $x_{\mathfrak{p}}$ for f the multiplier of $x_{\mathfrak{p}}$ is nonzero and in addition, it is in the maximal ideal $\mathfrak{p}$. $\square$

Lemma 14.3.4.1 shows that indeed we may find periodic points x_i for each f_i with multipliers λ_i multiplicatively independent (since the periodic points x_i can be chosen so that each λ_i is divisible by a different prime than the other λ_j's). Then [**ABR11**, Corollary 2.7] finishes the proof of Conjecture 14.3.3.1 in this case and yields the following result.

THEOREM 14.3.4.2. *Let $f_1, \ldots, f_N \in \overline{\mathbb{Q}}(z)$ be rational maps which are not PCF. Then there exist $\alpha_1, \ldots, \alpha_N \in \mathbb{P}^1(\overline{\mathbb{Q}})$ such that the orbit of $(\alpha_1, \ldots, \alpha_N)$ under the action of $(f_1, \ldots, f_N)$ is Zariski dense in $(\mathbb{P}^1)^N$.*

We expect that Zhang's Conjecture also holds when the maps f_i are PCF, but in this case one needs a different strategy since for any given PCF map f, there exist only finitely many primes $\mathfrak{p}$ such that there exists some f-periodic point x of order n such that $f^n(x) \equiv 0 \pmod{\mathfrak{p}}$. If each f_i is a Lattès map, or each f_i is a monomial, or each f_i is a Chebyshev polynomial, then Zhang's Conjecture follows easily using the fact that each f_i is induced by an endomorphism of an algebraic group, as described in [**MS14**]. However, if each f_i is PCF, but it is neither a Lattès map, nor a Chebyshev polynomial, nor a monomial, then one needs a different approach.

14.4. Uniform boundedness

In [**MS94**], Morton and Silverman conjectured that there is a constant $C(N, d, D)$ such that for any morphism

$$f : \mathbb{P}^N \longrightarrow \mathbb{P}^N$$

of degree d defined over a number field K with $[K : \mathbb{Q}] \leq D$, the number of preperiodic points of f in $\mathbb{P}^N(K)$ is less than or equal to $C(N, d, D)$. This conjecture remains open, but in the case where f has good reduction at a prime $\mathfrak{p}$, a great deal has been proved about bounds depending on $\mathfrak{p}$, N, d, D (see [**Ziea, Pez05, Hut09**]). In [**BGT15a**] we proved a more general uniform boundedness result for the period of arbitrary dimension periodic subvarieties of a quasiprojective variety endowed with an étale endomorphism.

THEOREM 14.4.0.1 ([**BGT15a**]). *Let K be a finite extension of $\mathbb{Q}_p$, let $\mathfrak{o}_v$ be the ring of integers of K, let k_v be its residue field and let e be the ramification index of $K/\mathbb{Q}_p$. Let $\mathcal{X}$ be a smooth $\mathfrak{o}_v$-scheme whose generic fiber X has dimension g, let*

$$\Phi : \mathcal{X} \longrightarrow \mathcal{X}$$

be étale, let $\mathcal{Y}$ be a subvariety of $\mathcal{X}$, and assume there is a point on $\mathcal{Y}(\mathfrak{o}_v)$ which is smooth on the generic fiber of $\mathcal{Y}$. We let $\overline{\mathcal{X}}$ be the special fiber of $\mathcal{X}$, and let r be the smallest nonnegative integer larger than

$$(\log(e) - \log(p - 1))/\log(2).$$

If $\mathcal{Y}$ is preperiodic under the action of Φ, then the length of its orbit is bounded above by

$$p^{1+r} \cdot \# \operatorname{GL}_g(k_v) \cdot \# \overline{\mathcal{X}}(k_v).$$

Once again, the p-adic arc lemma is instrumental in proving this result; for more details, we refer the interested reader to [**BGT15a**].

14.5. Integral points in orbits

According to Siegel's classical theorem (see Theorem 5.5.0.1) we know that a curve of positive genus defined over a number field has at most finitely many points with coordinates in any given finitely generated ring of algebraic integers. Silverman [**Sil93**, Theorem A] later gave a dynamical variant of Siegel's theorem, proving that if

$$f : \mathbb{P}^1 \longrightarrow \mathbb{P}^1$$

is a rational function such that $f \circ f$ is not a polynomial and $\alpha \in \mathbb{P}^1(\overline{\mathbb{Q}})$ is not preperiodic for f, there are only finitely many n such that $f^n(\alpha)$ is integral relative to the point at infinity (see our Definition 14.5.0.1).

DEFINITION 14.5.0.1. Let K be a number field, let $\mathfrak{o}_K$ be the ring of algebraic integers of K, and let S be a finite subset of places of K (containing all archimedean places of K). We say that a point $\beta \in \mathbb{P}^1(K)$ is *S-integral* with respect to a point $\alpha \in \mathbb{P}^1(K)$ if the Zariski closures of β and α in $\mathbb{P}^1(\mathfrak{o}_K)$ do not meet over any primes $v \notin S$.

Similarly, we say that a point $\beta \in \mathbb{P}^1(K)$ is S-integral with respect to a subvariety $V \subseteq \mathbb{P}^1 \times \mathbb{P}^1$ defined over K, if the Zariski closures of β and V in $(\mathbb{P}^1 \times \mathbb{P}^1)(\mathfrak{o}_K)$ do not meet over any primes $v \notin S$.

Corvaja, Sookdeo, Tucker, and Zannier (see [**CSTZ**, Theorem 1.1]) proved the following result which is related to both the Dynamical Mordell-Lang Conjecture (especially the more general Question 3.6.0.1) and also to Silverman's result regarding integral points in orbits.

THEOREM 14.5.0.2. *Let K be a number field, let S a finite set of primes in K, let $f : \mathbb{P}^1 \longrightarrow \mathbb{P}^1$ be a rational function with degree $d \geq 2$ that is not conjugate to a powering map $z^{\pm d}$, and let $\alpha, \beta \in \mathbb{P}^1(K)$ be points that are not preperiodic for f.*

Then the set of $(m,n) \in \mathbb{N}_0^2$ such that $f^m(\alpha)$ is S-integral relative to $f^n(\beta)$ is finite and effectively computable.

Other results regarding integral points in orbits were recently proven (see [**Ghi14, Soo11, BIR08, GI13**]). It is conceivable that the following question has a positive answer.

QUESTION 14.5.0.3. *Let $f : V \longrightarrow V$ be a polarizable self-map of a projective variety, let Z be a codimension 1 subvariety of V, let $\beta \in V(K)$, and let S be a finite set of places of K containing all the archimedean places and all the places of bad reduction for f. Does the set of n such that $f^n(\beta)$ is S-integral relative to β form a finite union of arithmetic progressions?*

This has be proved in the case where V is an semi-abelian variety in [**Tuc14**], using the fact that Conjecture 1.5.0.1 has been proved in that case along with deep results of Faltings [**Fal94**] and Vojta [**Voj96, Voj99**].

Note that a positive answer to Question 14.5.0.3 would imply that there are infinitely many primes p such that $f^n(\alpha)$ meets D modulo p for some n. One obvious possible explanation for this is that it may be that for all sufficiently large n, there is some p such that $f^n(\alpha)$ meets D modulo p but $f^m(\alpha)$ does not meet D modulo p for any $m < n$; in words, after a certain point, one gets a "new" place p for each n.

In dimension 1, some results along these lines have been proven by Bang [**Ban86**], Zsigmondy [**Zsi92**], Schinzel [**Sch74**], and, more recently, Ingram-Silverman [**IS09**] and Faber-Granville [**FG11**]. Understanding this question more generally would give a better understanding of Conjecture 14.6.0.1, since it would say (essentially) that one gets a low density of primes not because one fails to get "new primes" for large n but because one doesn't get that many "new primes" for each given n. In dimension 1, a general result along these lines follows from the *abc* conjecture as explained in [**GNT13**] and [**GNT15**].

The connection with the *abc* and Vojta conjectures (see [**Voj87**], for example) appears to go fairly deep. Recently, Levin [**Lev12**] has given counterexamples to a possible strengthening of Vojta's conjecture, using integral points and dynamics on elliptic surfaces. So we raise the following general question.

QUESTION 14.5.0.4. *What forms of Question 14.5.0.3 are implied by Vojta's conjecture?*

Some work on this question has been done by Levin-Yasufuku [**LY**], Silverman [**Sil13**], and Yasufuku [**Yas11, Yas14, Yas15**], but this connection has yet to be explored in full generality.

14.6. Orbits avoiding points modulo primes

As previously explained in our book, only special instances of Conjecture 1.5.0.1 when the map Φ is ramified are known. In almost all known ramified cases, Φ is given by the coordinatewise action through one-variable rational maps on $(\mathbb{P}^1)^m$, i.e. $\Phi(x_1, \ldots, x_m) = (\varphi_1(x_1), \ldots, \varphi_m(x_m))$ for some rational maps φ_i.

The main difficulty in this case is avoiding the ramification of Φ modulo primes p (assuming all maps are defined over a number field). For example, even the following basic question is at the moment unsolved.

QUESTION 14.6.0.1. *Let $f \in \mathbb{Q}[x]$ be a polynomial of degree ≥ 2 and let $\alpha \in \mathbb{Q}$ be a point such that $\mathcal{O}_f(\alpha)$ does not meet the critical points of f. For what proportion of primes p is there an n such that $f^n(\alpha)$ is congruent to a critical point of f modulo p?*

We believe that the answer is "zero" in most cases. On the other hand, we believe that the answer to Question 14.6.0.1 may often be "one" in higher dimensions (see our heuristics and probabilistic data from Chapter 8). Indeed, it is possible that in some cases, an orbit of a point can fail to pass through a divisor over a number field yet pass through that divisor mod p for every prime p. Finding such an example would demonstrate the limitations of the p-adic parametrization method described in the previous chapters (see Chapters 4 and 7 especially) and indicate the need for a new approach to the Dynamical Mordell-Lang Conjecture. On the other hand, we mention here the following result of Nguyen [**Ngu15**, Theorem 1.1] which provides partial evidence that the p-adic method might work for endomorphisms of $(\mathbb{P}^1)^n$ given by the diagonal action of a 1-variable polynomial (or even a rational map).

THEOREM 14.6.0.2. *Let K be a number field, f be a polynomial of degree at least 2 in $K[X]$, and $\varphi = (f, \ldots, f) : (\mathbb{P}^1_K)^n \longrightarrow (\mathbb{P}^1_K)^n$. Let V be an absolutely irreducible preperiodic curve or hypersurface in $(\mathbb{P}^1_K)^n$, and $P \in (\mathbb{P}^1)^n(K)$ such that the φ-orbit of P does not intersect $V(K)$. Then there are infinitely many primes $\mathfrak{p}$ of K such that the $\mathfrak{p}$-adic closure of the orbit of P does not intersect $V(K_\mathfrak{p})$, where $K_\mathfrak{p}$ is the $\mathfrak{p}$-adic closure of K.*

An alternative way of approaching Conjecture 1.5.0.1 is by inferring its conclusion after studying the completion in the adelic topology of the orbit $\mathcal{O}_\Phi(\alpha)$. More precisely, we expect that assuming X, V, α and Φ are all defined over a number field K, then

$$\overline{V(K) \cap \mathcal{O}_\Phi(\alpha)} = V(\overline{A}_K) \cap \overline{\mathcal{O}_\Phi(\alpha)},$$

where $\overline{A}_K$ are the partial adeles corresponding to the number field K by omitting finitely many places, and for any set $S \subset K \subset \overline{A}_K$, we let $\overline{S}$ be the closure of S in the adelic topology. We expect that a positive answer to our Conjecture 1.5.0.1 may be obtained by using the local information from various primes. This approach may be viewed as complementary to the one described above since in the first instance all we need is one *good* prime p which would allow us to construct a p-adic analytic uniformization of $\mathcal{O}_\Phi(\alpha)$. With the adelic approach, one would use infinitely many primes at once in order to infer the desired result.

For example, assume V does not contain any positive dimensional subvariety which is periodic under the action of Φ; any generic subvariety V would have this property. Then Conjecture 1.5.0.1 predicts that for each $\alpha \in X(K)$ the intersection $V(K) \cap \mathcal{O}_\Phi(\alpha)$ would consist of finitely many points. However, in this case it is conceivable that a stronger, uniform statement would hold.

QUESTION 14.6.0.3. *Let X be a quasiprojective variety defined over a number field K, let $\Phi : X \longrightarrow X$ be a morphism defined over K, and let $V \subseteq X$ be a subvariety defined over K with the property that V contains no positive dimensional subvariety periodic under the action of Φ. Does there exist $m \in \mathbb{N}$ such that $V(\overline{A}_K) \cap \Phi^m(X(\overline{A}_K))$ is finite?*

There are only some partial results for the above Question: either in the case $\dim(V) = 0$ (see [**SV09**]), or X is a semiabelian variety (see [**PV10**] and [**Sun**]), or in the context of Drinfeld modules (see [**GS13**]).

14.7. A Dynamical Mordell-Lang conjecture for value sets

Recently, Jones and Silverman have posed the following question.

QUESTION 14.7.0.1. *Let K be a number field, let Z and X be varieties defined over K, and let $\alpha \in X(K)$. Let $f : X \longrightarrow X$ and $g : Z \longrightarrow Z$ be morphisms of K-varieties. Is $\{n \mid f^n(\alpha) \in g(Z(K))\}$ a finite union of arithmetic progressions?*

As with Question 3.6.0.1, problems arise with automorphism. In particular, the answer to the question is "no" over $\mathbb{Q}$ when

$$f, g : \mathbb{P}^1 \longrightarrow \mathbb{P}^1 \text{ with } f(x) = x + 1 \text{ and } g(x) = x^2.$$

Note however, that this is only a restriction on f not on g, since when g is an automorphism of X, the problem is trivial, and when g is an automorphism (defined over K) of a subvariety of X, the problem is the same as the Dynamical Mordell-Lang conjecture. Moreover, the answer to Question 14.7.0.1 is "yes" when f and g are group endomorphisms of a semiabelian variety X (see [**JCS**]). Hence, in particular, we avoid the sorts of counterexamples that arise in Section 3.6 in the context of Question 3.6.0.1 and in Section 14.1 in the context of the Dynamical Manin-Mumford Conjecture. Treating Question 14.7.0.1 in the case where

$$f, g : \mathbb{P}^1 \longrightarrow \mathbb{P}^1 \text{ and } \deg f > 1$$

is already subtle. Jones, Cahn, and Spears [**JCS**] have shown that when g is a powering map $g(x) = x^m$, the answer to Question 14.7.0.1 is "yes". The set of rational functions f which give infinite progressions of n such that $f^n(\alpha) \in g(\mathbb{P}^1(K))$ (for some given starting point $\alpha \in \mathbb{P}^1(K)$) turns out to be quite complicated, including not only polynomials with roots of multiplicity m but also Lattès maps and other families of rational functions.

One natural restriction one might place on f is that it be "polarizable" in the sense of Subsection 2.2.4. This eliminates the possibility of f being an automorphism of X or acting as an automorphism on any subvariety of X. With this restriction in mind, we present the following strategy for attacking Question 14.7.0.1. We let

$$\Phi : X \times X \longrightarrow X \times Z \text{ given by } \Phi(x, z) = (f(x), z).$$

For each n, we let Y_n be the subvariety of $X \times Z$ given by $f^n(x) = g(z)$ (for $(x, z) \in X \times Z$). Then

$$Y_n = \Phi^{-1}(Y_{n-1}).$$

One might then expect that unless Y is periodic, the components of Y_n to become more geometrically complicated (especially if f is not unramified) as n goes to infinity, and that for sufficiently large n that every component of Y_n will be of general type. Then the Bombieri-Vojta-Lang conjectures (see [**Voj87**], for example) would imply that for sufficiently large Y_n, the set $Y_n(K)$ is contained in a proper (and thus lower-dimensional) Zariski closed subset of Y_n. One might then hope to proceed by induction on the dimension. One problem here is that in general there *are* examples of polarizable morphisms

$$\Phi : V \longrightarrow V$$

of projective varieties V with non-periodic positive dimensional subvarieties W such that for every n, each component of $\Phi^{-n}(W)$ is not of general type. One example is the counterexample to the original formulation of the Dynamical Manin-Mumford conjecture, discussed in Section 14.1.2 (see [**GTZ11a**]). Take an elliptic curve E with complex multiplication, and take ψ_1 and ψ_2 to be two elements of $\mathrm{End}(E)$ with the same odd degree greater than one such that there is no m for which $\psi_1^m = \psi_2^m$ (one might for example take ψ_1 and ψ_2 to be complex conjugates, thinking of $\mathrm{End}(E)$ as a subring of $\mathbb{C}$). Let

$$V = E \times E,$$

and let

$$\Phi : V \longrightarrow V \text{ given by } \Phi(x_1, x_2) = (\psi_1(x_1), \psi_2(x_2)).$$

Let W be the diagonal subvariety consisting of all $(x, x) \in E \times E$. Then W is not periodic but for every n, we have that $\Phi^{-n}(W)$ is a group subvariety of V and thus every component of $\Phi^{-n}(W)$ has genus 1.

On the other hand, the construction above cannot yield any sort of negative answer to Question 14.7.0.1, since Question 14.7.0.1 has a positive answer for any endomorphism of a semiabelian variety. We also believe that the strategy of passing to general type subvarieties in sufficiently high inverse images should work in many cases, in particular when f and g are polynomial morphisms of $\mathbb{P}^1$ of degree greater than one.

CHAPTER 15

Future directions

In this chapter (see Section 15.1) we recall the known results towards the Dynamical Mordell-Lang Conjecture and also speculate about the future of the conjecture in two directions. One direction is to present from our perspective what would be the first still unknown cases of the conjecture but which are both most interesting and also best suited for the known methods (see Section 15.2). The second direction is presenting possible generalizations of the conjecture (see Section 15.4).

15.1. What is known?

As stated in Conjecture 1.5.0.1, the Dynamical Mordell-Lang conjecture is a statement regarding an endomorphism Φ of a quasiprojective variety X, and a subvariety V of X. We know the Dynamical Mordell-Lang Conjecture holds in the following instances:

(1) For unramified endomorphisms Φ of any quasiprojective variety (see [**BGT10**], and also our Chapter 4). In particular, the Dynamical Mordell-Lang conjecture holds for all endomorphisms of a semiabelian variety (see [**GT09**], and also our Chapter 9).

(2) For all endomorphisms $\Phi := (f_1, \ldots, f_N)$ of $\mathbb{A}^N$, where each f_i is a one-variable polynomial, and the subvariety V is a line (see [**GTZ08, GTZ12**], and also our Chapter 5).

(3) For certain subvarieties V of $X = (\mathbb{P}^1)^N$ and for endomorphisms $\Phi := (f_1, \ldots, f_N)$ of X, where each f_i is a rational function satisfying certain hypotheses as in [**BGKT12, BGHKST13**] (see also our Chapter 7).

(4) For all endomorphisms of $\mathbb{A}^2$ (see [**Xieb**] and Section 10.3 for the case of polynomial birational endomorphisms of $\mathbb{A}^2$).

All of these results hold for an arbitrary starting point $\alpha \in X$. However, there are other known cases of the Dynamical Mordell-Lang Conjecture which hold when α is in a certain basin of attraction for the map Φ (as explained in Section 10.1; see [**GT09, Sca**]). Also, there are a few known extensions of the Dynamical Mordell-Lang principle to the case of endomorphisms of Riemann surfaces (see [**NW13**]) and to real-analytic maps under certain hypotheses (see [**Sca11**]).

15.2. What is next?

It would be very interesting to solve the following instances of the Dynamical Mordell-Lang Conjecture; we list the next special cases in what we consider to be increasing order of their difficulty:

(a) $X = (\mathbb{P}^1)^N$ and $\Phi(x_1, \ldots, x_N) := (f_1(x_1), \ldots, f_N(x_N))$ where each f_i is a one-variable rational map.

(b) $X = \mathbb{P}^2$, V is a plane curve, and Φ is any endomorphism.

Case (a) might be proven using a refinement of the argument from [**BGHKST13**] since it is expected (see the heuristics we introduced in Chapter 8) that for such endomorphisms Φ there always exists a prime number p such that each starting point α_i falls into a quasiperiodic domain for f_i, where $\alpha := (\alpha_1, \ldots, \alpha_N)$.

Case (b) is harder, even though Xie [**Xieb**] proved the special case when Φ reduces to an endomorphism of $\mathbb{A}^2$. Now, if one attempts to use the p-adic methods described in this book, heuristically we would still expect that there exists a prime number p such that the orbit of α under Φ avoids modulo p the ramification divisor of Φ (see the heuristics from Chapter 8), however proving that such a prime number exists would be more difficult. Also note that (as explained in Chapter 8) this same approach is not expected to work for endomorphisms of $\mathbb{P}^N$ for $N \geq 5$. On the other hand, we mention again Fakhruddin's result [**Fak14**], which shows that for generic endomorphisms of $\mathbb{P}^N$, Conjecture 1.5.0.1 does indeed hold.

15.3. Varieties with many rational points

The study of the Dynamical Mordell-Lang Conjecture for endomorphisms Φ of an arbitrary variety X (defined over a number field K) leads to deep questions in Diophantine geometry. Indeed, if $V \subset \mathbb{P}^N$ is a subvariety which contains a Zariski dense set of points from a given orbit $\mathcal{O}_\Phi(\alpha)$ of a point $\alpha \in \mathbb{P}^N$ under an endomorphism Φ of $\mathbb{P}^N$, then V contains (in particular) a Zariski dense set of points which are all *rational*. So, if both Φ and α are defined over a number field K, then $V(K)$ is Zariski dense in V. This imposes strict geometric restrictions on V; for example, V is *not* of general type (for more details, see [**Lan83**]). In the special case when V is a curve, then V must be of genus 0 or 1 (by Faltings classical theorem [**Fal83**]). So, in the case of curves, generically the intersection $V(K) \cap \mathcal{O}_\Phi(\alpha)$ is indeed finite and thus the Dynamical Mordell-Lang Conjecture holds, but it is *very hard* to show that in the cases when V is already of small genus (most notably, when V is a rational curve), then the intersection is indeed infinite *only* if V is periodic under Φ.

One may also ask whether the Dynamical Mordell-Lang Conjecture holds for *rational* self-maps Φ (see Question 3.2.0.1); it is conceivable that the same statement would hold, but we do not yet have examples where this is true and the map Φ is not already regular.

15.4. A higher dimensional Dynamical Mordell-Lang Conjecture

One can formulate a more general Dynamical Mordell-Lang conjecture by replacing the orbit of a point by the orbit of a subvariety.

CONJECTURE 15.4.0.1. *Let X be a quasiprojective variety defined over $\mathbb{C}$, let $\Phi : X \longrightarrow X$ be an endomorphism, and let Y and V be irreducible subvarieties of X. Is the set of $n \in \mathbb{N}_0$ such that $\Phi^n(Y) \subseteq V$ a finite union of arithmetic progressions?*

A special case of Conjecture 15.4.0.1 was asked by Bell and Lagarias [**BL15**] for automorphisms Φ of affine varieties X. Also in [**BL15**] the above problem is placed in a purely algebraic context by discussing inclusions of ideals; see the following result from [**BL15**].

THEOREM 15.4.0.2 (Bell-Lagarias [**BL15**]). *Let K be any field of characteristic zero and let A be a finitely generated commutative K-algebra. If $\sigma : A \longrightarrow A$ is a*

K-algebra automorphism and I and J are ideals of A, then $\{n \in \mathbb{Z} : J \subseteq \sigma^n(I)\}$ is a finite union of complete doubly infinite arithmetic progressions augmented by a finite set.

We explain below how one can deduce Conjecture 15.4.0.1 from the Dynamical Mordell-Lang Conjecture. To do this, we require a basic result about varieties over an uncountable algebraically closed field.

LEMMA 15.4.0.3. *Let k be a field and let K be an extension of k such that the dimension of K as a k-vector space is strictly less than $\#k$. Then K is algebraic over k.*

PROOF. Let $x \in K \setminus k$. Then $\{1/(x - \lambda) : \lambda \in k\}$ is a set of cardinality $\#k$ and hence must be linearly dependent. Let

$$\sum_{i=1}^{m} c_i/(x - \lambda_i) = 0$$

be a non-trivial dependence. Clearing denominators, we see that there is a nonzero polynomial $P(t) \in k[t]$ such that $P(x) = 0$. It follows that K is algebraic over k. $\square$

In the following arguments we identify a subvariety with the set of its points over an uncountable algebraically closed field.

PROPOSITION 15.4.0.4. *Let k be an uncountable algebraically closed field and let X be an irreducible quasiprojective variety over k. Then X cannot be written as a union of countably many proper closed subsets.*

PROOF. We may immediately reduce to the case where X is affine. We let A denote the ring of regular functions of X. Suppose that there exist closed subsets $Y_1, Y_2, \ldots$ of X such that $X = \bigcup_{i=1}^{\infty} Y_i$. For each $i \in \mathbb{N}$, we let $f_i \in I(Y_i) \subseteq A$. Then $X = \bigcup_{i=1}^{\infty} V(f_i)$. Let S be the multiplicatively closed subset of A generated by the f_i and let $B = S^{-1}A$. Then since A is finitely generated and S is countable, we see that B is at most countably infinite dimensional as a k-vector space. Let P be a maximal ideal of B. Then B/P is a field extension of k and is countable-dimensional as a k-vector space. It follows that $B/P \cong k$ by Lemma 15.4.0.3 and since $A/(P \cap A)$ embeds in B/P, we see that $P \cap A$ is a maximal ideal of A (more precisely, $A/(P \cap A) \overset{\sim}{\to} k$). It follows that there is some point $x \in X$ that is not in the zero set of any of the f_i, a contradiction. $\square$

CONJECTURE 1.5.0.1 YIELDS CONJECTURE 15.4.0.1. Assume that $\dim(Y) \geq 1$. For each $n \in \mathbb{N}_0$ we let

$$Y_n = Y \cap (\Phi^n)^{-1}(V).$$

If $\Phi^n(Y)$ is not contained in V, then Y_n is a proper subset of Y, and since Y is irreducible (and Y_n is closed by construction), we conclude that $\dim(Y_n) < \dim(Y)$. Now let

$$S := \{n \in \mathbb{N}_0 : \Phi^n(Y) \subseteq V\}$$

and let

$$U := \bigcup_{n \notin S} Y_n(\mathbb{C}).$$

As explained above, we know that for each $n \notin S$ we have $\dim(Y_n) < \dim(Y)$, and thus U is a proper subset of $Y(\mathbb{C})$ by Proposition 15.4.0.4. So, let $\alpha \in Y(\mathbb{C}) \setminus$

U. Then $\Phi^n(\alpha) \in V(\mathbb{C})$ precisely when $n \in S$. So, assuming the Dynamical Mordell-Lang Conjecture holds, we conclude that S is a finite union of arithmetic progressions, as desired. $\qquad\square$

The above proof coupled with Theorem 11.1.0.7 yields the following result.

THEOREM 15.4.0.5. *Let K be an uncountable algebraically closed field, let X be a quasiperojective variety defined over K, let $\Phi : X \longrightarrow X$ be an endomorphism defined over K, let $V \subseteq X$ and $Y \subseteq X$ be subvarieties defined over K. Then the set $\{n \in \mathbb{N}_0 : \Phi^n(Y) \subseteq V\}$ is a union of at most finitely many arithmetic progressions along with a set of Banach density 0.*

Bibliography

[AS92]   M. Abramowitz and I. A. Stegun, *Handbook of mathematical functions with formulas, graphs, and mathematical tables*, Dover Publications Inc., New York, 1992, Reprint of the 1972 edition.

[AB12]   B. Adamczewski and J. P. Bell, *On vanishing coefficients of algebraic power series over fields of positive characteristic.* Invent. Math. **187** (2012), 343–393.

[AG09]   A. Akbary and D. Ghioca, *Periods of orbits modulo primes*, J. Number Theory **129** (2009), 2831–2842.

[AS03]   J.-P. Allouche and J. Shallit, *Automatic sequences. Theory, applications, generalizations.* Cambridge University Press, Cambridge, 2003.

[AK70]   A. Altman and S. Kleiman, *Introduction to Grothendieck duality theory*, Lecture Notes in Mathematics, Vol. 146, Springer-Verlag, Berlin, 1970.

[Ame11]   E. Amerik, *Existence of non-preperiodic algebraic points for a rational self-map of infinite order*, Math. Res. Lett. **18** (2011), 251–256.

[ABR11]   E. Amerik, F. Bogomolov, and M. Rovinsky, *Remarks on endomorphisms and rational points*, Compositio Math. **147** (2011), 1819–1842.

[AC08]   E. Amerik and F. Campana, *Fibrations méromorphes sur certaines variétés à fibré canonique trivial*, Pure Appl. Math. Q. **4** (2008), 509–545.

[AM69]   M. F. Atiyah and I. G. Macdonald, *Introduction to commutative algebra*, Addison-Wesley Publishing Co., Reading, Mass.-London-Don Mills, Ont., 1969.

[Aut01]   P. Autissier, *Points entiers sur les surfaces arithmétiques*, J. reine. angew. Math **531** (2001), 201–235.

[Bac91]   E. Bach, *Toward a theory of Pollard's rho method*, Inform. and Comput. **90** (1991), 139–155.

[Bak09]   M. Baker, *A finiteness theorem for canonical heights attached to rational maps over function fields*, J. Reine Angew. Math. **626** (2009), 205–233.

[BD11]   M. Baker and L. DeMarco, *Preperiodic points and unlikely intersections*, Duke Math. J. **159** (2011), 1–29.

[BD13]   _____, *Special curves and post-critically-finite polynomials*, Forum Math. π **1** (2013), e3, 35 pp.

[BH05]   M. Baker and L.-C. Hsia, *Canonical heights, transfinite diameters, and polynomial dynamics*, J. reine angew. Math. **585** (2005), 61–92.

[BIR08]   M. Baker, S.-I. Ih, and R. Rumely, *A finiteness property of torsion points*, Algebra Number Theory **2** (2008), no. 2, 217–248.

[BR06]   M. Baker and R. Rumely, *Equidistribution of small points, rational dynamics, and potential theory*, Ann. Inst. Fourier (Grenoble) **56** (2006), 625–688.

[BR10]   _____, *Potential theory and dynamics on the Berkovich projective line*, Mathematics Surveys and Monographs **159**, American Mathematical Society, Providence, RI, 2010.

[Ban86]   A. S. Bang, *Taltheoretiske Undersøgelse*, Tidsskrift Mat. **4** (1886), 70–80, 130–137.

[Bea91]   A. Beardon, *Iteration of rational functions*, Complex analytic dynamical systems. Graduate Texts in Mathematics **132**. Springer-Verlag, New York, 1991.

[Bel06]   J. P. Bell, *A generalized Skolem-Mahler-Lech theorem for affine varieties*, J. London Math. Soc. (2) **73** (2006), 367–379.

[BBY12]   J. P. Bell, S. N. Burris, K. Yeats, *On the set of zero coefficients of a function satisfying a linear differential equation*, Math. Proc. Cambridge Philos. Soc. **153** (2012), 235–247.

[BGT10]   J. P. Bell, D. Ghioca, and T. J. Tucker, *The dynamical Mordell-Lang problem for étale maps*, Amer. J. Math. **132** (2010), 1655–1675.

[BGT15a]   ———, *Applications of p-adic analysis for bounding periods of subvarieties under étale maps*, Int. Math. Res. Not. IMRN, vol. 2015, 3576–3597.

[BGT15b]   ———, *The Dynamical Mordell-Lang problem for Noetherian spaces*, Funct. Approx. Comm. Math. **53** (2015), no. 2, 313–328.

[BGZ]   J. P. Bell, D. Ghioca, and Z. Reichstein, *On a dynamical variant of a theorem of Rosenlicht*, Ann. Scuola Norm. Sup. Pisa Cl. Sci. (3) to appear.

[BL15]   J. P. Bell and J. Lagarias, *A Skolem-Mahler-Lech theorem for iterated automorphisms of K-algebras*, Canad. J. Math. **67** (2015), no. 2, 286–314.

[Ben05]   R. L. Benedetto, *Heights and preperiodic points for polynomials over function fields*, Int. Math. Res. Not. **62** (2005), 3855–3866.

[BGKT10]   R. L. Benedetto, D. Ghioca, P. Kurlberg, and T. J. Tucker, *A gap principle for dynamics*, Compositio Math. (2010), **146** 1056–1072.

[BGKT12]   ———, *A case of the dynamical Mordell-Lang conjecture* (with an Appendix by U. Zannier), Math. Ann. **352** (2012), 1–26.

[BGHKST13]   R. L. Benedetto, D. Ghioca, B. A. Hutz, P. Kurlberg, T. Scanlon, and T. J. Tucker, *Periods of rational maps modulo primes*, Math. Ann. **355** (2013), 637–660.

[BBM13]   M. Bennett, Y. Bugeaud, and M. Mignotte, *Perfect powers with few binary digits and related Diophantine problems*, Ann. Sc. Norm. Super. Pisa Cl. Sci. (5) **12** (2013), no. 4, 941–953.

[Ber90]   V. Berkovich, *Spectral theory and analytic geometry over non-Archimedean fields*, Mathematical Surveys and Monographs **33**, American Mathematical Society, Providence, RI, 1990.

[BS96]   F. Beukers and H. P. Schlickewei, *The equation $x + y = 1$ in finitely generated groups*, Acta Arith. **78** (1996), 189–199.

[Bez89]   J.-P. Bézivin, *Une généralisation du théorème de Skolem-Mahler-Lech*, Quart. J. Math. Oxford Ser. (2) **40** (1989), 133–138.

[BT00]   Y. F. Bilu and R. F. Tichy, *The Diophantine equation $f(x) = g(y)$*, Acta Arith. **95** (2000), 261–288.

[BG06]   E. Bombieri and W. Gubler, *Heights in Diophantine geometry*, New Mathematical Monographs, vol. 4, Cambridge University Press, Cambridge, 2006.

[BMZ99]   E. Bombieri, D. Masser, and U. Zannier, *Intersecting a curve with algebraic subgroups of multiplicative groups*, IMRN **20** (1999), 1119–1140.

[BS66]   A. I. Borevich and I. R. Shafarevich, *Number theory*, Translated from the Russian by Newcomb Greenleaf. Pure and Applied Mathematics, Vol. 20, Academic Press, New York, 1966.

[Bos99]   V. Bosser, *Minorations de formes linéaires de logarithmes pour les modules de Drinfeld*, J. Number Theory **75** (1999), no. 2, 279–323.

[Bou98]   N. Bourbaki, *Lie groups and Lie algebras. Chapters 1–3*, Elements of Mathematics (Berlin), Springer-Verlag, Berlin, 1998, Translated from the French, Reprint of the 1989 English translation.

[Bou06]   ———, *Éléments de mathématique. Algèbre commutative. Chapitres 8 et 9*, Springer, Berlin, 2006, Reprint of the 1983 original.

[BH88]   B. Branner and J. H. Hubbard, *The iteration of cubic polynomials. I. The global topology of parameter space.*, Acta Math. **160(3-4)** (1988), 143-206.

[BEK12]   X. Buff, A. L. Epstein, and S. Koch, *Böttcher coordinates*, Indiana Univ. Math. J. **61** (2012), no. 5, 1765–1799.

[CS93]   G. S. Call and J. H. Silverman, *Canonical heights on varieties with morphisms*, Compositio Math. **89** (1993), 163–205.

[CG93]   L. Carleson and T. W. Gamelin, *Complex dynamics*, Springer-Verlag, New York, 1993.

[Cha41]   C. Chabauty, *Sur les points rationnels des courbes algébriques de genre supérieur à l'unité*, C. R. Acad. Sci. Paris **212** (1941), 882–885.

[CL06]   A. Chambert-Loir, *Mesures et équidistribution sur les espaces de Berkovich*, J. Reine Angew. Math. **595** (2006), 215–235.

[CH08a]   Z. Chatzidakis and E. Hrushovski, *Difference fields and descent in algebraic dynamics. I*, J. Inst. Math. Jussieu **7** (2008), 635–686.

[CH08b]   _____, *Difference fields and descent in algebraic dynamics. II*, J. Inst. Math. Jussieu **7** (2008), 687–704.

[Col85]   R. F. Coleman, *Effective Chabauty*, Duke Math. J. **52** (1985), no. 3, 765–770.

[CZ00]   P. Corvaja and U. Zannier, *On the Diophantine equation $f(a^m, y) = b^n$*, Acta Arith. **94** (2000), 25–40.

[CZ13]   P. Corvaja and U. Zannier, *Finiteness of odd perfect powers with four nonzero binary digits*, Ann. Inst. Fourier (Grenoble) **63** (2013), no. 2, 715–731.

[CSTZ]   P. Corvaja, V. Sookdeo, T. J. Tucker, and U. Zannier, *Integral points in two-parameter orbits*, J. Reine Angew. Math., to appear.

[CS93]   S. D. Cutkosky and V. Srinivas, *On a problem of Zariski on dimensions of linear systems*, Ann. of Math. (2) **137** (1993), no. 3, 531–559.

[DR55]   H. Davenport and K. F. Roth, *Rational approximations to algebraic numbers*, Mathematika **2** (1955), 160–167.

[Del74]   P. Deligne, *La Conjecture de Weil. I*, Inst. Hautes Études Sci. Publ. Math. **43** (1974), 273–307.

[Dem]   L. Demangos, *Some examples toward a Manin-Mumford conjecture for abelian uniformizable T-modules*, preprint.

[BWY]   L. DeMarco, X. Wang, and H. Ye, *Torsion points and the Lattes family*, to appear in Amer. J. Math.

[vdDG06]   L. van den Dries and A. Günaydin, *The fields of real and complex numbers with a small multiplicative group*, Proc. London Math. Soc. **93** (2006), 43–81.

[vdDMM94]   L. van den Dries, A. Macintyre, and D. Marker, *the elementary theory of restricted analytic fields with exponentiation*, Ann. Math. **140** (1994), 183–205.

[vdDM94]   L. van den Dries and C. Miller, *On the real exponential field with restricted analytic functions* Israel J. Math. **85** (1994), 19–56.

[vdDS84]   L. van den Dries and K. Schmidt, *Bounds in the theory of polynomial rings over fields. A nonstandard approach*, Invent. Math. **76** (1984), no. 1, 77–91.

[Den92a]   Laurent Denis , *Diophantine geometry on Drinfeld modules*, The arithmetic of function fields (Columbus, OH, 1991), 285–302, Ohio State Univ. Math. Res. Inst. Publ., 2, de Gruyter, Berlin, 1992.

[Den92b]   _____, *Hauters canoniques et modules de Drinfeld*, Math. Ann. **294** (1992), 213–223.

[Den94]   _____, *Géométrie et suites récurrentes*, Bull. Soc. Math. France **122** (1994), 13–27.

[Der07]   H. Derksen, *A Skolem-Mahler-Lech theorem in positive characteristic and finite automata*, Invent. Math. **168** (2007), 175–224.

[DM12]   H. Derksen and D. Masser, *Linear equations over multiplicative groups, recurrences, and mixing I*, Proc. London Math. Soc. **104** (2012), 1045–1083.

[DF01]   J. Diller and C. Favre, *Dynamics of bimeromorphic maps of surfaces*, Amer. J. Math. **123** (2001), 1135–1169.

[DH93]   A. Douady and J. H. Hubbard, *A proof of Thurston's topological characterization of rational functions*, Acta Math. **171** (1993), 263–297.

[DF]   R. Dujardin and C. Favre, *The dynamical Manin-Mumford problem for plane polynomial automorphisms*, preprint.

[Eve95]   J.-H. Evertse, *The number of solutions of decomposable form equations* Invent. Math. **122** (1995), 559–601.

[ESS02]   J.-H. Evertse, H. P. Schlickewei, and W. M. Schmidt, *Linear equations in variables which lie in a multiplicative group*, Ann. of Math. (2) **155** (2002), 807–836.

[FG11]   X. Faber and A. Granville, *Prime factors of dynamical sequences*, J. Reine Angew. Math. **661** (2011), 189–214.

[Fak14]   N. Fakhruddin, *The algebraic dynamics of generic endomorphisms of $\mathbb{P}^n$*, Algebra & Number Theory **8** (2014), no. 3, 587–608.

[Fal83]   G. Faltings, *Endlichkeitssätze für abelsche Varietäten über Zahlkörpern*, Ivent. Math. **73** (1983), 349–366.

[Fal91]   _____, *Diophantine approximation on abelian varieties*, Ann. of Math. (2) **133** (1991), no. 3, 549–576.

[Fal94]   ———, *The general case of S. Lang's conjecture*, Barsotti Symposium in Algebraic Geometry (Abano Terme, 1991), 175–182, Perspect. Math. **15**, Academic Press, San Diego, CA, 1994.

[Fav]   C. Favre, *Dynamique des applications rationnelles*, PhD thesis, Université Paris-Sud XI - Orsay, 2000.

[FJ11]   C. Favre and M. Jonsson, *Dynamical compactification of* $\mathbb{C}^2$, Ann. Math. (2) **173** (2011), 211–248.

[FRL06]   C. Favre and J. Rivera-Letelier, *Équidistribution quantitative des points de petite hauteur sur la droite projective*, Math. Ann. **335** (2006), 311–361.

[Fur81]   H. Furstenberg, *Poincaré recurrence and number theory*, Bull. Amer. Math. Soc. **5** (1981), 211–234.

[Ghi05]   D. Ghioca, *The Mordell-Lang theorem for Drinfeld modules*, Int. Math. Res. Not. IMRN **53** (2005), 3273–3307.

[Ghi06]   ———, *Equidistribution for torsion points of Drinfeld modules*, Math. Ann. **336** (2006), 841–865.

[Ghi08a]   ———, *Mordell exceptional locus for subvarieties of the additive group*, Math. Res. Lett. **15** (2008), 43–50.

[Ghi08b]   ———, *The isotrivial case in the Mordell-Lang Theorem*, Tran. Amer. Math. Soc. **360** (2008), 3839–3856.

[Ghi10]   ———, *Towards the full Mordell-Lang conjecture for Drinfeld modules*, Canad. Math. Bull. **53** (2010), 95–101.

[Ghi14]   ———, *Integral points for Drinfeld modules*, J. Number Theory **140** (2014), 93–121.

[GH13]   D. Ghioca and L.-C. Hsia, *Torsion points in families of Drinfeld modules*, Acta Arith. (2013), **161** 219–240.

[GHT13]   D. Ghioca, L.-C. Hsia, and T. J. Tucker, *Preperiodic points for families of polynomials*, Algebra & Number Theory (2013), **7** 701–732.

[GHT15]   ———, *Preperiodic points for families of rational maps*, Proc. London Math. Soc. (3) **110** (2015), no. 2, 395–427.

[GHT]   ———, *Unlikely intersection for two-parameter families of polynomials*, Int. Math. Res. Not. IMRN to appear.

[GM06]   D. Ghioca and R. Moosa, *Division points on subvarieties of semiabelian varieties*, IMRN **19** (2006), 1–23.

[GNT15]   D. Ghioca, K. Nguyen, and T. J. Tucker, *Portraits of preperiodic points under rational maps*, Math. Proc. Cambridge Philos. Soc. **159** (2015), no. 1, 165–186.

[GS13]   D. Ghioca and T. Scanlon, *Algebraic equations on the adélic closure of a Drinfeld module*, Israel J. Math. **194** (2013), 461–483.

[GS]   ———, *Density of orbits of endomorphisms of abelian varieties*, Tran. Amer. Math. Soc. to appear.

[GT08a]   D. Ghioca and T. J. Tucker, *Equidistribution and integral points for Drinfeld modules*, Trans. Amer. Math. Soc. **360** (2008), 4863–4887.

[GT08b]   ———, *A dynamical version of the Mordell-Lang conjecture for the additive group*, Compositio Math. **164** (2008), 304–316.

[GT09]   ———, *Periodic points, linearizing maps, and the dynamical Mordell-Lang problem*, J. Number Theory **129** (2009), 1392–1403.

[GT10]   ———, *Proof of a dynamical Bogomolov conjecture for lines under polynomial actions*, Proc. Amer. Math. Soc. **138** (2010), 937–942.

[GTZ11a]   D. Ghioca, T. J. Tucker and S. Zhang, *Towards a dynamical Manin-Mumford Conjecture*, Int. Math. Res. Not. IMRN 2011, no. 22, 5109–5122.

[GTZ08]   D. Ghioca, T. J. Tucker, and M. E. Zieve, *Intersections of polynomial orbits, and a dynamical Mordell-Lang conjecture*, Invent. Math. **171** (2008), 463–483.

[GTZ11b]   ———, *The Mordell-Lang question for endomorphisms of semiabelian varieties*, J. Theor. Nombres Bordeaux **23** (2011), 645–666.

[GTZ12]   ———, *Linear relations between polynomial orbits*, Duke Math. J. (2012), **161** 1379–1410.

[Gig]   W. Gignac, *Equidistribution of preimages in nonarchimedean dynamics*, PhD thesis, University of Michigan, 2013.

[Gig14]   ———, *Measures and dynamics on Noetherian spaces*, J. Geom. Anal. **24** (2014), no. 4, 1770–1793.

[Gos96]   D. Goss, *Basic Structure of Function field Arithmetic*, Springer-Verlag, Berlin, 1996.

[GW10]   U. Görtz and T. Wedhorn, *Algebraic geometry I. Schemes with examples and exercises*, Advanced Lectures in Mathematics. Vieweg + Teubner, Wiesbaden, 2010. viii+615 pp.

[GI13]   D. Grant and S.-I. Ih, *Integral division points on curves*, Compos. Math. **149** (2013), no. 12, 2011–2035.

[GNT13]   C. Gratton, K. Nguyen, and T. J. Tucker, *ABC implies primitive prime divisors in arithmetic dynamics*, Bull. London Math. Soc. **45** (2013), no. 6, 1194–1208.

[Hab]   S. Häberli, *Kummer theory of Drinfeld modules*, Master's Thesis, ETH Zürich, May 2011.

[Hab09]   P. Habegger, *Intersecting subvarieties of abelian varieties with algebraic subgroups of complementary dimension*, Invent. Math. **176** (2009), 405–447.

[Hal57]   G. af Hällström, *Über halbvertauschbare Polynome*, Acta Acad. Abo. **21** (1957), no. 2, 20 pp.

[HJM15]   S. Hamblen, R. Jones, and K. Madhu, *The density of primes in orbits of $z^d + c$*, Int. Math. Res. Not., 2015(7):1924-1958.

[Har77]   R. Hartshorne, *Algebraic geometry*, Springer-Verlag, New York, 1977.

[Hei83]   J. Heintz, *Definability and fast quantifier elimination in algebraically closed fields*, Theoret. Comput. Sci. **24** (1983), no. 3, 239–277.

[HY83]   M. Herman and J.-C. Yoccoz, *Generalizations of some theorems of small divisors to non-archimedean fields*, Geometric dynamics (Rio de Janeiro, 1981), Lecture Notes in Math., no. 1007, Springer-Verlag, Berlin, 1983, pp. 408–447.

[Hin88]   M. Hindry, *Autour d'une conjecture de Serge Lang*, Invent. Math. **94** (1988), 575–603.

[HS00]   M. Hindry and J. H. Silverman, *Diophantine geometry. An introduction*, Graduate Texts in Mathematics **201**. Springer-Verlag, New York, 2000. xiv+558 pp

[Hru96]   E. Hrushovski, *The Mordell-Lang conjecture for function fields*, J. Amer. Math. Soc. **9** (1996), 667–690.

[Hru99]   ———, *Proof of Manin's theorem by reduction to positive characteristic*, 197–205. *Model Theory and Algebraic Geometry*, Lecture Notes in Mathematics **1696**, Elisabeth Bouscaren (Ed.), 1999.

[Hut09]   B. Hutz, *Good reduction of periodic points on projective varieties*, Illinois J. Math. **53** (2009), 1109–1126.

[Iit76]   S. Iitaka, *Logarithmic forms of algebraic varieties*, J. Fac. Sci. Univ. Tokyo Sect. IA Math. **23** (1976), no. 3, 525–544.

[IS09]   P. Ingram and J. H. Silverman, *Primitive divisors in arithmetic dynamics*, Math. Proc. Cambridge Philos. Soc. **146** (2009), 289–302.

[Jos02]   J. Jost, *Riemannian geometry and geometric analysis*, third ed., Universitext, Springer-Verlag, Berlin, 2002.

[KS14]   S. Kawaguchi and J. H. Silverman, *Examples of dynamical degree equals arithmetic degree*, Michigan Math. J. **63** (2014), no. 1, 41–63.

[KRS05]   D. S. Keeler, D. Rogalski, and J.T. Stafford, *Naïve noncommutative blowing up*, Duke Math. J. **126** (2005), 491–546.

[Jon08]   R. Jones, *The density of prime divisors in the arithmetic dynamics of quadratic polynomials*, J. Lond. Math. Soc. (2) **78** (2008), no. 2, 523–544.

[Jon13]   R. Jones, *Galois representations from pre-image trees: an arboreal survey*, Pub. Math. Besançon. (2013), 107–136

[Jon15]   ———, *Fixed-point-free elements of iterated monodromy groups*, Trans. Amer. Math. Soc., 367(3):2023-2049, 2015.

[JCS]   R. Jones, J. Cahn, and J. Spear, *Powers in orbits of rational functions: cases of an arithmetic dynamical Mordell-Lang conjecture*, in preparation.

[JKMT]   J. Juul, P. Kurlberg, K. Madhu, and T. J. Tucker, *Wreath products and proportions of periodic points*, to appear in Int. Math. Res. Not. IMRN.

[Lan60]   S. Lang, *Integral points on curves*, Publ. Math. IHES **6** (1960), 27–43.

[Lan83]   ———, *Fundamentals of Diophantine geometry*, Springer-Verlag, New York, 1983.

[Lan02]   ———, *Algebra*, third ed., Springer-Verlag, New York, 2002, xvi+914 pp.

[Lau84]   M. Laurent, *Equations diophantiennes exponentielles*, Invent. Math. **78** (1984), 299–327.

[Lec53]   C. Lech, *A note on recurring series*, Ark. Mat. **2** (1953), 417–421.

[Lev12]   A. Levin, *The exceptional set in Vojta's conjecture for algebraic points of bounded degree*, Proc. Amer. Math. Soc. **140** (2012), no. 7, 2267–2277.

[LY]   A. Levin and Y. Yasufuku *Integral points and orbits of endomorphisms on the projective plane*, preprint.

[Mah35]   K. Mahler, *Eine arithmetische Eigenshaft der Taylor-Koeffizienten rationaler Funktionen*, Proc. Kon. Ned. Akad. Wetensch. **38** (1935), 50–60.

[Mah58]   ———, *An interpolation series for continuous functions of a p-adic variable*, J. Reine Angew. Math. **199** (1958), 23–34.

[Mah61]   ———, *A correction to the paper "an interpolation series for continuous functions of a p-adic variable"*, J. Reine Angew. Math. **208** (1961), 70–72.

[MZ08]   D. Masser and U. Zannier, *Torsion anomalous points and families of elliptic curves*, C. R. Math. Acad. Sci. Paris **346** (2008), 491–494.

[MZ10]   ——— *Torsion anomalous points and families of elliptic curves*, Amer. J. Math. **132** (2010), 1677–1691.

[MZ12]   ——— *Torsion points on families of squares of elliptic curves*, Math. Ann. **352** (2012), 453–484.

[Mat86]   H. Matsumura, *Commutative ring theory*, Cambridge Studies in Advanced Mathematics, vol. 8, Cambridge University Press, Cambridge, 1986, Translated from the Japanese by M. Reid.

[Med]   A. Medvedev, *Minimal sets in $ACFA$*, PhD thesis, UC Berkeley, 2007.

[MS14]   A. Medvedev and T. Scanlon, *Invariant varieties for polynomial dynamical systems*, Ann. Math. (2) **179** (2014), no. 1, 81–177.

[Met00]   C. Methfessel, *On the zeros of recurrence sequences with non-constant coefficients*, Arch. Math. (Basel) **74** (2000), 201–206.

[MS03]   R. Moosa and T. Scanlon, *The Mordell-Lang Conjecture in positive characteristic revisited*, Model Theory and Applications (eds. L. Bélair, P. D'Aquino, D. Marker, M. Otero, F. Point, & A. Wilkie), 2003, 273–296.

[MS04]   ———*F-structures and integral points on semiabelian varieties over finite fields*, Amer. J. Math. **126** (2004), 473–522.

[MS94]   P. Morton and J. H. Silverman, *Rational periodic points of rational functions*, Internat. Math. Res. Notices (1994), no. 2, 97–110.

[MS95]   ———, *Periodic points, multiplicities, and dynamical units*, J. Reine Angew. Math. **461** (1995), 81–122.

[Mum65]   D. Mumford, *A remark on Mordell's conjecture*, Amer. J. Math. **87** (1965), 1007–1016.

[MZ]   P. Müller and M. E. Zieve, *On Ritt's polynomial decomposition theorems*, submitted for publication.

[Nek05]   V. Nekrashevych, *Self-similar groups*, volume 117 of Mathematical Surveys and Monographs. American Mathematical Society, Providence, RI, 2005.

[NW13]   T. W. Ng and M.-X. Wang, *Ritt's theory on the unit disk*, Forum Math. **25** (2013), 821–851.

[Ngu15]   K. Nguyen, *Some arithmetic dynamics of diagonally split polynomial maps*, Int. Math. Res. Not. IMRN, vol. 2015, 1159–1199.

[NZa]   K. Nguyen and M. E. Zieve, *Algebraic dynamics of split morphisms associated to rational maps having generic ramification*, preprint.

[NZb]   ———*Functional equations and the curves $f(X) = g(Y)$ for generic f and arbitrary g*, preprint.

[Odo85]   R. W. K. Odoni, *The Galois theory of iterates and composites of polynomials*, Proc. London Math. Soc. (3) **51** (1985), no. 3, 385–414.

[OS]   A. Ostafe and M. Sha, *On the quantitative Dynamical Mordell-Lang Conjecture*, preprint.

[Paz10]   F. Pazuki, *Zhang's conjecture and squares of abelian surfaces*, C. R. Math. Acad. Sci. Paris (2010), **348** 483–486.

[Pez05]   T. Pezda, *Cycles of polynomial mappings in several variables over rings of integers in finite extensions of the rationals. II*, Monatsh. Math. **145** (2005), no. 4, 321–331.

[Pet]   C. Petsche, *On the distribution of orbits in affine varieties*, preprint, arXiv:1.1401.3425.

[PZ08]   J. Pila and U. Zannier, *Rational points in periodic analytic sets and the Manin-Mumford conjecture*, Atti Accad. Naz. Lincei Cl. Sci. Fis. Mat. Natur. Rend. Lincei (9) Mat. Appl. **19** (2008), 149–162.

[Pil98]   A. Pillay, *The model-theoretic content of Lang's conjecture*, Model Theory and Algebraic Geometry, Lecture Notes in Math. **1696** (E. Bouscaren, ed.), Springer-Verlag, Berlin, 1998, pp. 101–106.

[STP04]   J. Piñeiro, L. Szpiro, and T. Tucker, *Mahler measure for dynamical systems on $\mathbb{P}^1$ and intersection theory on a singular arithmetic surface*, Geometric methods in algebra and number theory (F. Bogomolov and Y. Tschinkel, eds.), Progress in Mathematics 235, Birkhäuser, 2004, pp. 219–250.

[Pin]   R. Pink, *A common generalization of the conjectures of André-Oort, Manin-Mumford, and Mordell-Lang*, unpublished manuscript, 2005.

[Pol75]   J. M. Pollard, *A Monte Carlo method for factorization*, Nordisk Tidskr. Informationsbehandling (BIT) **15** (1975), no. 3, 331–334.

[Poo95]   B. Poonen, *Local height functions and the Mordell-Weil theorem for Drinfeld modules*, Compositio Math. **97** (1995), 349–368.

[Poo14]   ———, *p-adic interpolation of iterates*, Bull. London Math. Soc. **46** (2014), no. 3, 525–527.

[PV10]   B. Poonen and J. F. Voloch, *The Brauer-Manin obstruction for subvarieties of abelian varieties over function fields*. Ann. Math. **171** (2010), 511–532.

[Ray83a]   M. Raynaud, *Courbes sur une variété abélienne et points de torsion*, Invent. Math. **71** (1983), 207–233.

[Ray83b]   ———, *Sous-variétés d'une variété abélienne et points de torsion*, Arithmetic and geometry, vol. I, Progr. Math., vol. 35, Birkhäuser, Boston, MA, 1983, pp. 327–352.

[Rit20]   J. F. Ritt, *On the iteration of rational functions*, Trans. Amer. Math. Soc. **21** (1920), 348–356.

[RL03]   J. Rivera-Letelier, *Dynamique des fonctions rationnelles sur des corps locaux*, Astérisque (2003), no. 287, 147–230, Geometric methods in dynamics. II.

[Rob00]   A. M. Robert, *A course in p-adic analysis*, Graduate Texts in Mathematics, vol. 198, Springer-Verlag, New York, 2000.

[Rub83]   L. A. Rubel, *Some research problems about algebraic differential equations*, Trans. Amer. Math. Soc. **280** (1983), 43–52.

[Sca02]   T. Scanlon, *Diophantine geometry of the torsion of a Drinfeld module*, J. Number Theory **97** (2002), no. 1, 10–25.

[Sca11]   ———, *A Euclidean Skolem-Mahler-Lech-Chabauty method*, Math. Res. Lett. **18** (2011), 835–842.

[Sca]   ———, *Analytic relations on a dynamical orbit*, preprint.

[SY14]   T. Scanlon and Y. Yasufuku, *Exponential-polynomial equations and dynamical return sets*, Int. Math. Res. Not. IMRN 2014, no. 16, 4357–4367.

[Sha74]   I. R. Shafarevich, *Basic algebraic geometry*, study ed., Springer-Verlag, Berlin, 1977, Translated from the Russian by K. A. Hirsch, Revised printing of Grundlehren der mathematischen Wissenschaften, Vol. 213, 1974.

[Sch74]   A. Schinzel, *Primitive divisors of the expression $a^n - b^n$ in algebraic number fields*, J. Reine Angew. Math. **268/269** (1974), 27–33.

[Sch99]   W. M. Schmidt, *The zero multiplicity of linear recurrence sequences*, Acta Math. **182** (1999), 243–282.

[Sch00]   ———, *Zeros of linear recurrence sequences*, Publ. Math. Debrecen **56** (2000), 609–630.

[SW88]   H. Sharif and C. F. Woodcock, *Algebraic functions over a field of positive characteristic and Hadamard products*, J. London Math. Soc. (2) **37** (1988), no. 3, 395–403.

[Sie29]   C. L. Siegel, *Über einige Anwendungen Diophantischer Approximationen*, Abh. Preuss. Akad. Wiss. Phys. Math. Kl. (1929), 41–69. (Reprinted as pp. 209–266 of his Gesammelte Abhandlungen I, Springer, Berlin, 1966.)

[Sil83]　　　　J. H. Silverman, *Heights and the specialization map for families of abelian varieties*, J. Reine Angew. Math. **342** (1983), 197–211.

[Sil93]　　　　————, *Integer points, Diophantine approximation, and iteration of rational maps*, Duke Math. J. **71** (1993), 793–829.

[Sil07]　　　　————, *The arithmetic of dynamical systems*, Graduate Texts in Mathematics **241**, Springer, New York, 2007, x+511 pp.

[Sil08]　　　　————, *Variation of periods modulo p in arithmetic dynamics*, New York J. Math. **14** (2008), 601–616.

[Sil13]　　　　————, *Primitive divisors, dynamical Zsigmondy sets, and Vojta's conjecture*, J. Number Theory **133** (2013), no. 9, 2948–2963.

[SV13]　　　　J. H. Silverman and B. Viray, *On a uniform bound for the number of exceptional linear subvarieties in the Mordell-Lang conjecture*, Math. Res. Lett. **20** (2013), no. 3, 547–566.

[SV09]　　　　J. H. Silverman and J. F. Voloch, *A Local-Global Criterion for Dynamics on $\mathbb{P}^1$*, Acta Arith. **137** (2009), 285–294.

[Sko34]　　　　T. Skolem, *Ein Verfahren zur Behandlung gewisser exponentialer Gleichungen und diophantischer Gleichungen*, Comptes Rendus Congr. Math. Scand. (Stockholm, 1934) 163–188.

[Soo11]　　　　V. Sookdeo, *Integer points in backward orbits*, J. Number Theory **131** (2011), 1229–1239.

[Sri91]　　　　V. Srinivas, *On the embedding dimension of an affine variety*, Math. Ann. (1991), **289** 125–132.

[SL96]　　　　P. Stevenhagen and H. W. Lenstra Jr., *Chebotarev and his density theorem*, Math. Intelligencer **18** (1996), 26–37.

[Sto92]　　　　M. Stoll, *Galois groups over $\mathbf{Q}$ of some iterated polynomials*, Arch. Math. (Basel) **59** (1992), no. 3, 239–244.

[Sun]　　　　　C.-L. Sun, *Adélic points of subvarieties of isotrivial semi-abelian varieties over a global field of positive characteristic*, preprint, arXiv:1.005.4998

[Sze75]　　　　E. Szemerédi, *On sets of integers containing no k elements in arithmetic progression*, Acta Arithm. **27** (1975), 199–245.

[SUZ97]　　　　L. Szpiro, E. Ullmo, and S. Zhang, *Équirépartition des petits points* (French) [Uniform distribution of small points], Invent. Math. **127** (1997), 337–347.

[Ull98]　　　　E. Ullmo, *Positivité et discrétion des points algébriques des courbes*, Ann. of Math. (2) **147** (1998), 167–179.

[Tuc14]　　　　T. J. Tucker, *Integer points in arithmetic sequences*, Bull. Inst. Math. Acad. Sin. (N.S.) **9** (2014), no. 4, 633–639.

[Voj87]　　　　P. Vojta, *Diophantine approximations and value distribution theory*, Lecture Notes in Mathematics, vol. 1239, Springer-Verlag, Berlin, 1987.

[Voj89]　　　　———— *Mordell's conjecture over function fields*, Invent. Math. **98** (1989), 115–138.

[Voj96]　　　　———— *Integral points on subvarieties of semiabelian varieties. I*, Invent. Math. **126** (1996), 133–181.

[Voj99]　　　　————, *Integral points on subvarieties of semiabelian varieties. II*, Amer. J. Math. **121** (1999), no. 2, 283–313.

[Wan]　　　　　M. Wang, *Endomorphisms of uniformization spaces*, preprint.

[Wil96]　　　　A. J. Wilkie, *Model completeness results for expansions of the ordered field of real numbers by restricted Pfaffian functions and the exponential function* J. Amer. Math. Soc. **9** (1996), 1051–1094.

[Xie14]　　　　J. Xie, *Dynamical Mordell-Lang Conjecture for birational polynomial morphisms on $\mathbb{A}^2$*, Math. Ann. **360** (2014), no. 1-2, 457–480.

[Xiea]　　　　　————, *Intersection of valuation rings in $k[x, y]$*, preprint available on arXiv:1403.6052.

[Xieb]　　　　　————, *The Dynamical Mordell-Lang Conjecture for polynomial endomorphisms of the affine plane*, preprint available on arXiv:1503.00773.

[Yas11]　　　　Y Yasufuku, *Vojta's conjecture and dynamics*, Algebraic number theory and related topics 2009, RIMS Kôkyûroku Bessatsu, B25, Res. Inst. Math. Sci. (RIMS), Kyoto, 2011, pp. 75–86.

[Yas14]　　　　————, *Deviations from S-integrality in orbits on $\mathbb{P}^N$*, Bull. Inst. Math. Acad. Sin. (N.S.) **9** (2014), no. 4, 603–631.

[Yas15]   _____, *Integral points and relative sizes of coordinates of orbits in* $\mathbb{P}^N$, Math. Z. **279** (2015), no. 3-4, 1121–1141.

[Yu90]   K. Yu, *Linear forms in p-adic logarithms. II*, Compositio Math. **74** (1990), 15–113.

[Yua08]   X. Yuan, *Big line bundles over arithmetic varieties*, Invent. Math. **173** (2008), 603–649.

[YZ]   X. Yuan and S.-W. Zhang, *Calabi theorem and algebraic dynamics*, preprint, 23 pages.

[Zan09]   U. Zannier, *Lecture notes on Diophantine analysis*, Appunti. Scuola Normale Superiore di Pisa (Nuova Serie) [Lecture Notes. Scuola Normale Superiore di Pisa (New Series)], vol. 8, Edizioni della Normale, Pisa, 2009, With an appendix by Francesco Amoroso.

[Zan12]   _____, *Some problems of unlikely intersections in arithmetic and geometry*, Annals of Mathematics Studies, vol. 181, Princeton University Press, Princeton, NJ, 2012, With appendixes by David Masser.

[Zha95]   S. Zhang, *Small points and adelic metrics*, J. Algebraic Geometry **4** (1995), 281–300.

[Zha98]   _____, *Equidistribution of small points on abelian varieties*, Ann. of Math. (2) **147** (1998), 159–165.

[Zha06]   _____, *Distributions in algebraic dynamics*, Survey in Differential Geometry, vol. 10, International Press, 2006, pp. 381–430.

[Ziea]   M. E. Zieve, *Cycles of polynomial mappings*, PhD thesis, University of California at Berkeley, 1996

[Zieb]   _____, *Decompositions of Laurent polynomials*, preprint.

[Zsi92]   K. Zsigmondy, *Zur Theorie der Potenzreste*, Monatsh. Math. Phys. **3** (1892) 265–284.

Index

Selected Published Titles in This Series

For a complete list of titles in this series, visit the
AMS Bookstore at **www.ams.org/bookstore/survseries/**.